Mechatronic Systems and Process Automation

Model-Driven Approach and Practical Design Guidelines

Mechatronic Systems and Process Automation

Model-Driven Approach and Practical Design Guidelines

Patrick Kaltjob

CRC Press
Taylor & Francis Group
Boca Raton London New York

CRC Press is an imprint of the
Taylor & Francis Group, an **informa** business

CRC Press
Taylor & Francis Group
6000 Broken Sound Parkway NW, Suite 300
Boca Raton, FL 33487-2742

First issued in paperback 2020

© 2018 by Patrick Kaltjob
CRC Press is an imprint of Taylor & Francis Group, an Informa business

No claim to original U.S. Government works

ISBN 13: 978-0-367-73502-9 (pbk)
ISBN 13: 978-0-8153-7079-6 (hbk)

Visit the Taylor & Francis Web site at
http://www.taylorandfrancis.com

and the CRC Press Web site at
http://www.crcpress.com

To the Holy Trinity and Saint Mary

Special thanks to Stella, Emmanuelle, Naomi, Lukà, and David

To Aaron[t], Thomas, Olive, and Anne

Contents

Preface

The high level of efficiency requirements in the design of automated systems has led to the development of new applications, including functionally integrated systems such as mechatronics and operationally integrated processes. Such applications have proven to significantly enhance product consistency, quality, and operating safety, as well as the productivity and safety of industrial processes. Due to their behavioristic similarities, mechatronic systems and automated processes are usually designed by combining their time-driven and event-driven characteristics within a hybrid control scheme integrating sequential logic control with feedback control strategies and eventually with data acquisition and remote operation monitoring capabilities.

However, the current engineering literature on the design of process automation and mechatronic systems usually presents discrete-event and discrete-time systems separately, rather than in a practical integrated approach. Challenges in the development of this unique approach are (1) the model-driven design of automation systems and processes, (2) the integration of automation field devices and equipment through a data acquisition and communication network that is compliant with performance constraints, and (3) the incorporation of low-level regulatory control with high-level supervisory control functions.

This book is intended to revisit the design concept of process automation and mechatronic systems. By reviewing the theoretical and practical knowledge on time-driven and event-driven systems, it offers an integrated approach and a general design methodology for the modeling, analysis, automation, control, networking, monitoring, and sensing of various systems and processes, from single electrical-driven machines to large-scale industrial processes. Furthermore, it covers design applications for several engineering disciplines (mechanical, manufacturing, chemical, electrical, computer, and biomedical) through real-life mechatronics design problems (e.g., hybrid vehicle, driverless car, newborn incubator, and elevator motion) and industrial process automation case studies (e.g., power grid, wind generator, crude oil distillation, brewery bottle filling, and beer fermentation).

Through this book, the reader should acquire methods for (1) model formulation, analysis, and auditing of single electrical-driven machine and multivariable process operations; (2) model-driven design of software and hardware required for machine control and process automation; (3) selection and configuration of automation system field buses and network protocols; (4) sizing and selection of electrical-driven actuating systems (including electric motors), along with their commonly used electrotransmission elements and binary actuators; (5) selection and calibration of devices for process variable measurement and detection, as well as a compliant data acquisition and computer interface; and (6) design of a process operation monitoring and fault management system. Hence, the book is organized into six chapters.

Chapter 1 gives a brief conceptual definition and classification of electrical-driven systems and technical processes. A functional decomposition of the automation system architecture is presented, along with some examples to illustrate automation system components for sensing, actuating, computing, signal conditioning, and communicating. Furthermore, the generic structure of electromechanical systems with embedded automation function (i.e., mechatronics) is described, along with the interconnection structure of synchronized electromechanical systems (i.e., process automation). Generic design requirements for process automation are outlined, and major steps of design projects in industrial automation are covered.

The integration of electric actuating systems (alternating current, direct current, and stepper motors; control valves; heaters; solenoids; etc.) with mechanical (e.g., belts and screw wheels), fluidic (pumps), or thermal transmission elements enables us to drive actions related to the dynamics of solid, liquid, and gas material or chemical reactions. In Chapter 2, classical dynamics models for electrical-driven actuating systems are presented, along with their associated transmission element dynamics. For each type of motor, the starting and operating conditions are derived, along with the technical specifications, suitable speed control strategies, and computer interface requirements. Then, the resulting dynamical models for all the subcomponents are combined to derive the entire process model. Similarly, binary actuators, such as electroactive polymers, piezoactuators, shape alloys, solenoids, and even nanodevices, are technically described and modeled. Eventually, a motor selection and sizing procedure based on various characteristics (e.g., motion profile, load torque, and operational conditions) is presented.

In Chapter 3, models of discrete-event process operations are derived through Boolean functions, by using sequential or combinatorial logic-based techniques to capture the relationship between the state outputs of process operations and the state inputs of their transition conditions. Hence, a logic controller design methodology for process operations (discrete-event systems) is described, from the process description and its functional analysis to formal modeling techniques, such as truth tables and K-maps, sequence table analysis and switching theory, state diagrams (Mealy and Moore), or even state function charts. In addition, logic controller implementation strategies using sequential logic circuit design methods, such as hardwired relay network (even solid-state logic devices), μController, programmable logic controller, digital signal processing, and even field-programmable gate array, are presented. Furthermore, some programming languages of industrial logic controllers, such as ladder diagrams, structured text, state functional charts (SFCs), instruction lists, and function blocks, are outlined. The sketching of typical wiring diagrams related to the documentation of an automation project is discussed, along with some design strategies, such as fail-safe design and interlocks. Eventually, illustrative examples covering key logic controller design steps are presented, from process schematics and involved input/output (I/O) equipment listings to state transition tables, I/O Boolean functions, and timing diagrams. Examples of industrial process automation include breweries, traffic management, fruit packaging, cement portico scratchers, wood band saws, sea port gantry cranes, elevators, and car parking. Overall, the chapter covers (1) the modeling methodology of discrete-event processes with single and multiple concurrent or mutually exclusive operating cycles, and (2) the logic controller design methodology to derive I/O Boolean functions based on truth tables and Karnaugh maps, switching theory or state diagrams, logic programming languages, wiring and electrical diagrams, and piping and instrumentation and process flow diagrams.

Chapter 4 presents a generic design and implementation methodology for process monitoring and fault management combining control strategies (logic and feedback) with supervisory algorithms to ensure safe operations within discrete-event industrial processes. First, functional and operational process requirements are used to define process control and supervision systems with respect to process data gathering, as well as process data analysis and reporting. Subsequently, some components of the supervision and control architecture are presented, such as (1) database structure, (2) data acquisition, and (3) monitoring and decision support units. Then, the formal modeling of fault management strategies is outlined in terms of fault classification and diagnosis, as well as integration of control strategies. This design methodology enables the characterization of the monitoring system through the formulation of process tasks from their operational boundaries, data-type definition, and visual encoding and interaction techniques. Some industrial cases illustrate the design of process monitoring and fault management systems (e.g., cement pouzzolona drying process, brewery bottle washing process, or induction motors). Eventually, based on cooperative requirements, a distributed control design approach to synchronize the process operations of interconnected subprocess components is described. It is illustrated through a variety of industrial applications, such as voltage control electrical power grids or distributed manufacturing control. Overall, chapter topics include process data collection, supervisory control and data acquisition (SCADA), process data reporting, distributed control system design, failure detection and prevention, process FAST and SADT decomposition methods, process start and stop operating mode graphical analysis, SFCs, and fault-tolerant processes.

Chapter 5 describes a spectrum of digital and analog sensing and detecting methods, as well as the technical characterization of devices commonly encountered in industrial automation processes. Among sensors presented, there are motion sensors (position, distance, velocity and acceleration, and flow), force sensors, pressure or torque sensors, noncontact and contact temperature sensors and detectors, proximity sensors, light sensors, and smart sensors. The following sensing methods are covered: resistive, optoelectric, Hall effect, variable reluctance, piezoresistive, capacitive, piezoelectric, strain gauge, photoresistor, photodiode, ultrasonic, phototransistor, triangulation, measuring wheel, radar, echelon, capacitive and inductive proximity, thermistor (NPC and PTC), infrared radiation, thermodiode, and thermocouple. Binary detectors, such as noncontact and contact detectors (e.g., capacitive proximity, pressure switches and vacuum switches, radiofrequency identification (RFID)–based tracking and detection, or electromechanical contact [switch]), are described. In addition, some smart sensors based on electrostatic, piezoresistive, piezoelectric, and electromagnetic sensing principles are presented. Furthermore, interfaces for logic-level I/O devices are described, including multiplexers, filters, and converters (digital-to-analog and analog-to-digital), using various conversion techniques (i.e., successive approximation, dual-slope analog to digital converter (ADC), delta-encoded ADC, etc.). The sizing and selection procedures of measurement and detecting devices are accordingly presented. Then, data acquisition unit operations are described, with an emphasis on data gathering, logging, and processing from the bus structure toward the computing unit. A selection methodology of the sample period is also outlined. Overall, chapter topics include some measurement devices of process variables (e.g., force, speed, position, pH, temperature, pressure, and gas and liquid chemical content), RFID detection, signal conversion, sensor characteristics (resolution, accuracy, range, etc.), computer control interface, signal conditioning, data logging and processing, converters, and smart sensors.

Chapter 6 presents a review of industrial data communication networks applied to process automation with their technical characteristics, such as bus architectures and topologies, communication techniques and protocols, media, address decoding, transmission rate, and interrupt interfacing. Among the industrial communication network protocols covered are Profibus, WordFIP, Hart, 4-20mA, and Modbus. Furthermore, the audit and benchmarking of industrial networks through performance modeling are presented, especially in terms of (1) network transmission latency, (2) congestion rate, (3) throughput, (4) network utilization, and (5) data losses and efficiency. Hence, a generic procedure is developed for the sizing and selection, and performance evaluation of communicating protocols and cabling equipment is proposed. Overall, chapter topics include quality of service for industrial networks, transmission component systems, transmission delay evaluation modeling, real-time information of industrial processes via the SCADA system, and actuating system remote control.

This textbook emphasizes the analysis of the real-life industrial environment and the integration of process automation system components through the suitable sizing, selection, and tuning of actuating, sensing, transmitting, and computing or controlling units. It allows self-study via comprehensive and straightforward step-by-step modular procedures. As such, the reader is expected, at the conclusion of the textbook, to have fully mastered (1) the design requirements and design methodology for process automation systems, (2) the sizing and selection of an industrial data transmission network and protocols, (3) the sizing and selection of the devices involved in industrial process data acquisition, and (4) the sizing and selection of actuating equipment for industrial processes. Examples and industrial case studies are used to illustrate formal modeling, hybrid controller design, data acquisition devices selection, and transmission network architecture sizing for various types of processes.

This book was conceived to develop the reader's skills for engineering-based problem solving, engineering system design, critical analysis, and automation system implementation. Suggested teaching plans are as follows: (1) Chapters 1, 3, and 5 for mechatronic students with computer hardware and software programming experience; (2) all chapters, specifically Section 4.4 of Chapter 4, for advanced automation and process control students with control theory background; (3) Chapters 1, 2, and 6 for electric machines and instrumentation students with computer hardware and software programming experience; and (4) Chapters 1, 2, 4 (specifically Section 4.4), 5, and 6 for automation field engineers aiming to design or migrate process automation systems.

Author

Patrick Kaltjob has been an associate professor of electrical engineering and telecommunications at Ecole Polytechnique, UY1 since 2005. After pursuing a degree in undergraduate engineering programs at Laval and McGill (BScEng, 1996), he received graduate degrees from the University of Wisconsin–Madison (MSc, 1999, PhD, 2003). He was a project investigator at Werkzeugmachinenlabor (WZL) RWTH-Aachen University, Germany (2004). His interests cover distributed control systems, smart grid, and biomedical systems. Since 2003, he has been a senior engineering consultant for industrial processes and automation solution integrators and for global corporations operating in various industrial sectors, including cement production, electric power networks, and oil refineries. He is an author of numerous journal articles and conference proceedings.

1

Introduction to Mechatronic Systems and Process Automation

1.1 Introduction

Historically, automation applications into electrical-driven systems and processes (from material transformation to information processing) have enhanced their computing capability and functionality while significantly reducing their operating cost and complexity. Furthermore, embedding such applications within products using miniaturized sensing devices (e.g., microelectromechanical systems [MEMS] or nanosensors) for the assessment of their operation status allows us to design smart products capable of carrying diagnostic and error correction advisories within process operations (e.g., a collision avoidance system in a driverless car).

Recent technology advancements offer a plethora of automated systems and process solutions, each with design-specific requirements and compliance constraints. Hence, it is suitable to lay out a generic design procedure for mechatronic systems and process automation, especially in (1) controlling and monitoring electrical-driven systems; (2) sizing and selecting devices related to information processing and computing, electrical-driven actuation, sensing and detection, and data acquisition and communication; (3) integrating those devices with respect to compliance constraints; and (4) integrating multifunctional operating supervisory and control applications.

1.2 Definitions and Classifications

Electrical-driven systems are machines transforming current, voltage, or other electrical power into mechanical, fluidic, pneumatic, hydraulic, thermal, or chemical power. Hence, those systems can be classified according to their functional objectives as either specialized machines performing specific operations or multipurpose and adjustable machines. *Automation systems* are a set of technologies enabling computing algorithms or signal processing devices to use the signal emitted from analog or digital detecting, sensing, and communicating devices, to perform automatic operations of systems or process actuation. Such systems are expected to operate routine tasks, independent of human intervention, with performance superior to that of manual operation. Thus, automation systems aim to provide the necessary input signals to achieve the desired patterns of variation of specific process variables or sequences. Therefore, the functions of automation systems embedded into electromechanical systems (*machine or product automation*) are called *mechatronic systems*. Figure 1.1 is a schematic of a typical electric cooking machine with automation system for speed and temperature control, combined with monitoring indicators for cooking time and cooking stage, as well as the control panel allowing the selection of spin speed and cooking program.

Such a design combination allows us to enhance the product or machine functionality and reduce the operating and maintenance costs. This is done by integrating the data processing and computing taking place within a field device or machine (e.g., cooking or washing machines and navigation systems). Among commonly encountered automated machines or products are those with (1) embedded automation functions, (2) dedicated

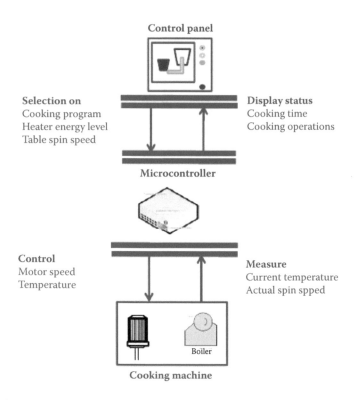

FIGURE 1.1
Example of a washing machine and its associated input and output variables.

automation functions, or (3) automation functions limited to a couple of sensors and actuators.

A *technical process* is the entirety of all interacting machines within a process that either transforms or transports and stores material, energy, or information. Such technical processes can be classified according to their operational objectives as follows:

- Transportation-related process, such as material handling processes, energy flow processes, and information transmission processes
- Transformation-related processes, such as chemical processes, manufacturing processes, and power generation and storage processes

Technical processes can be characterized according to their functional objectives, such as

- Processes characterized by a continuous flow of material or energy (e.g., cement drying process, electric power distribution, and paper production). Here, the process variables are physical state variables with a continuous range of values, such as temperatures in a heating system. The process parameters are related to the physical properties of the process (e.g., power transmission network impedance and liquefied gas density). Process control consists of maintaining the process state on a determined level or trajectory. In this case, the process dynamics models can be obtained through differential equations.
- Processes characterized by discrete-event operations representing different process states, such as devices activation or deactivation during the start-up or shutdown of a turbine. Here, process variables are binary signals indicating the discrete status of devices or machines involved in process operations and changes in logic devices (e.g., activating events from ON/OFF switches). Their process discrete-event models can be obtained through Boolean functions or logic flowcharts.
- Processes characterized by identifiable objects that are transformed, transported, or stored, such as silicon-based wafer production, data processing, and storage operations. Here, process variables indicate the state changes of objects and can have a continuous range of values (i.e., the temperature of a slab in a clogging mill or the size of a part in a store) or binary variables. Those variables can also be nonphysical variables (i.e., type, design, application, or deposit number) assigned to the objects. Figure 1.2, depicting the bottle washing substation, illustrates an example of noncontinuous processes (discrete event or object related).

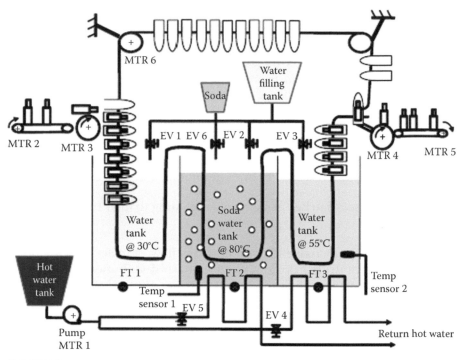

FIGURE 1.2
Brewery bottle washing process schematic. MTR, motor; EV, electrovalve; FT, Manual empty tank valve.

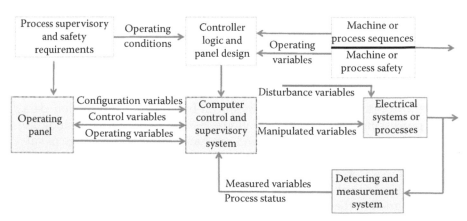

FIGURE 1.3
Generic computer-based industrial supervisory and control block diagram.

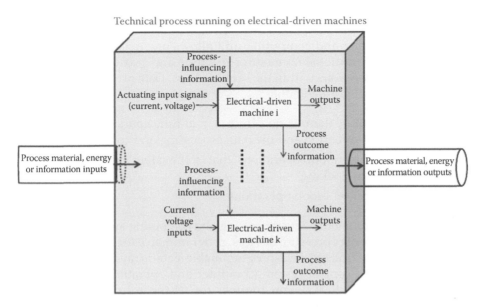

FIGURE 1.4
Relationship between technical processes and automation systems.

Process automation consists of the coordinating operations of several automated systems (electromechanical machines) with the aim to achieve a specific objective (e.g., purification and water plants, energy distribution, gas supply, and city traffic light management). They can be remotely monitored, configured, or controlled (activated or deactivated) from a single operator station. Thus, it is usually required to ensure the integration of a large amount of automation devices (from data processing and computing units to measuring and detecting units). Figure 1.3 illustrates the generic components in the design of a process automation system with control and supervisory functions. An example of the relationship between automated systems and processes is illustrated in Figure 1.4.

1.3 Generic Automation System Architecture and Components

Through the functional system decomposition, the automation system architecture can be divided into the following units: data processing and computing, electrical-driven actuating, measuring and detecting, data acquisition and transmitting, and signal conditioning. All those units are presented in the subsequent paragraphs.

1.3.1 Data Processing and Computing Unit

The data processing and computing unit allows one to (1) control and regulate machine operations, (2) monitor machine and process operations, or
(3) coordinate operations within the same process. Data processing and computing could be performed either

- Off-line, that is, there is no direct or real-time connection between
  the process execution and the data processing and computing unit
- On-line for open-loop operations, that is, the safety of process operations and interlocking
- On-line for closed-loop operations

Commonly encountered data processing and computing automation devices
are (1) digital signal processing devices, programmable logic controllers and
devices, microcontrollers, field-programmable gate arrays, and so on, and
(2) the distributed control system (computer and operation history server).
Those devices execute program routines for (1) the acquisition of process
variables, (2) process condition monitoring and exception handling (i.e., executing process safety operations), (3) the control of machine operations (i.e.,
activation or deactivation of the motor, and tracking of motor speed), and
(4) the archiving and sharing of process data with other automation devices
through the communication network.

1.3.2 Data Acquisition and Transmission Unit

Data acquisition and transmission units are used to (1) interface with various control devices (e.g., operator panel, and detecting and measuring field
devices), (2) transport process data between network nodes, (3) integrate process data from different sources on a single platform, and (4) integrate control functions (e.g., machine control and process control). Those units operate
through data transfer platforms and their data distribution service protocols.
They can be designed based on the Open Systems Interconnection model,
which could be summarized as follows:

- A physical layer, being either a wire or wireless connection, such as
  twisted-pair wiring, a fiber-optic cable, or a radio link, and the commutation unit connecting the network to the devices (e.g., field buses for
  data transfer between primary controllers and control field devices).
- The network, transmission, and transport layers, performing functions such as data routing over the network, data flow control, packet
  segmentation and desegmentation, error control, and clock synchronization. In addition, those layers provide a mechanism for packet
  tracking and the retransmission of failing ones.
- The session and presentation layers, mainly used for data formatting.

1.3.3 Electrical-Driven Actuating Unit

Electrical-driven actuating units convert voltage or current signals from the computing unit into appropriate input forms (mechanical, electrical, thermal, fluidic, etc.) for the execution of machine and process operations. Then those converted signals produce variations in the machine physical variables (e.g., torque, heat, or flow) or amplify the energy level of the signal, causing changes in the process operation dynamics. Some examples of actuating units are relays, magnets, and servomotors.

1.3.4 Measuring and Detecting Unit

Measuring and detecting units consist of low-power devices, such as sensors and switch-based detectors interfacing with involved electrical-driven machines in the process operations. As such, they convert physical-related output signals from the actuating unit into voltage or binary signals readily to be used within the data processing and computing unit. Some key functions of those devices are (1) data acquisition related to the change of machine variables and (2) conversion of the machine-gathered signals into electrical or optical signals. Depending on the nature of the process signal generated, a signal conditioner can be added.

1.3.5 Signal-Conditioning Unit

Signal-conditioning units convert the form of the signal generated by the sensing device into another suitable form (mostly electrical). The signal-conditioning unit can also be embedded within the sensing devices. An example of such a unit is a resistance temperature detector (RTD). Here, a change in the temperature of its environment is converted into a voltage signal reflecting its resistance change through a Wheatstone's bridge. Here, the bridge is a signal-conditioning module.

Figure 1.5 depicts all major automation system components.

1.4 Examples of Product and Industrial Process Automation

Process automation aims to achieve various objectives: synchronize, control, and sequence process operations or detect and monitor process status. Table 1.1 presents some typical process automation objectives and their corresponding automation functions, along with some illustrative examples. In addition, some examples of typical devices used in process or machine automation are presented in Table 1.2.

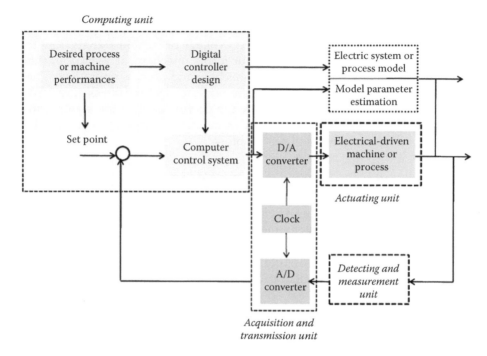

FIGURE 1.5
Generic automation system block diagram. D/A, digital to analog; A/D, analog to digital.

TABLE 1.1

Some Industrial Processes and Automation Functions

Process Automation Objectives	Automation Functions	Examples of Industrial Process Automation
Process detection and diagnostics	Malfunction handling, performance monitoring, failure detection, control loop detuning, actuator saturation	Condition monitoring of petrochemical process variables and parameters (flow rate, temperature, etc.)
Process monitoring and assessing	Recoding and forcing process variables through sensors and actuators, level measuring, changing process parameters	Remote measurement, along with communication, configuration, and control (SCADA) of a large-scale food process data
Process security and emergency	Interlocking, process safety operations, emergency operations in case of detected failure modes	Integrated anti-collision system for a railways traffic management
Process operation control	Real-time measuring, controlling, or regulating process variables	Distributed fermentation temperature measurement and control for a brewery process

TABLE 1.2

Generic Automation System Design Steps, Related Documentation, and Expected Results

Project Step	Design Tasks	Documents Provided	Resulting Documents
1. Process preliminary studies and analysis (design steps 1–3)	• Process functional analysis	• Process schematics and diagrams (PFD, PID, etc.) • Process performance benchmarking • Process operational and failure modes	• Automation system design requirements • FAST-based process decomposition • Power, utilities, and wiring diagrams • Listing of I/O process and operating conditions
2. Process operating and dynamics modeling (design step 3)	• Automation system sizing and selection (processing and computing, sensing, and other units) • Hybrid process modeling (discrete time and discrete event)	• Operating conditions and set points • Process data acquisition, as well as sensing and detecting selection	• Process model (time-based process analysis, including estimation of process dead time and time delays) • Process operating sequences (including failure modes and interlocks) • Process equipment sizing and selection
3. Process automation system design (design step 4)	• Automation solution design • Automation solution and data integration	• Process I/O variables • Process operating sequences	• Industrial automation algorithm • I/O variables' definition and mnemonics
4. Process performance analysis and controller tuning (design step 5)	• Process performance benchmarking • Automation process objectives assessment	• Industrial standards (with similar optimal process data) • Industrial-specific safety policy	• Comparative I/O process data per standard • Classification of performance (energy consumption, productivity, cost reduction)
5. Process documentation and operator training (design step 6)	• Failure or disaster recovery procedure • Operating procedure • Process automation commissioning documents	• Escalation policy and mitigation technique (remedy)	• All documents collected above • Operations and process control tests • Validation and personnel training

Example 1.1

A crude oil distillation process using a boiler-based temperature control within a hybrid system is illustrated in Figures 1.6 and 1.7, along with the resulting block diagram. There are two process variables to be measured: a continuous variable, the temperature, and a binary variable, the flame presence in the combustion chamber or boiler.

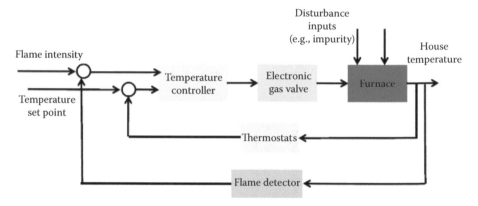

FIGURE 1.6
Block diagram of boiler temperature regulation within a furnace.

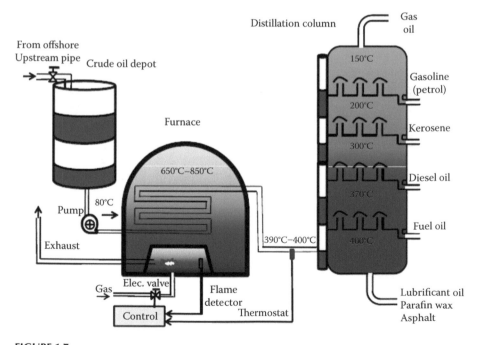

FIGURE 1.7
Schematic of a crude oil distillation process with its boiler temperature control system. Elec, electric-driven.

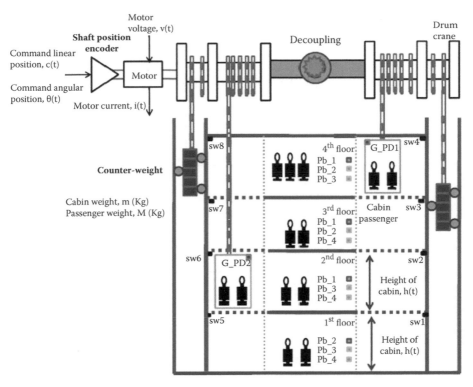

FIGURE 1.8
Crane-based vertical motion control system schematic.

Example 1.2

An automation system for a crane-based vertical motion process is illustrated in Figure 1.8, while its feedback block diagram and the logic control connections are shown in Figure 1.9.

Example 1.3

A milk beverage processing factory is illustrated in Figure 1.10. In such a large-scale process, a supervisory control and data acquisition (SCADA) system is used to collect geographically plant-wide data through an industrial network, to archive and define process sequences to be executed. The block diagram depicting the relevant components of such a SCADA-oriented process automation system is presented in Figure 1.11.

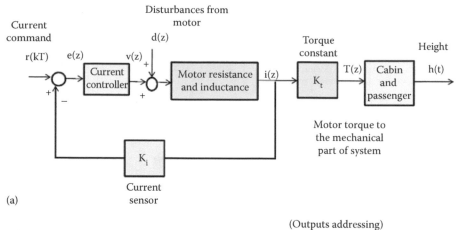

(a)

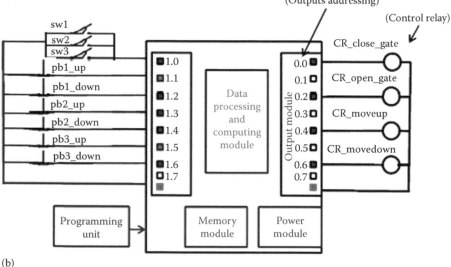

(b)

FIGURE 1.9
(a) Block diagram of the crane motion feedback control system. (b) Connection diagram of the crane motion logic control system.

1.5 Generic Automation System Objectives and Design Methodology

A project related to automation system design or migration has either

- Time-based objectives, such as improving process productivity or service through an extension of automation functions

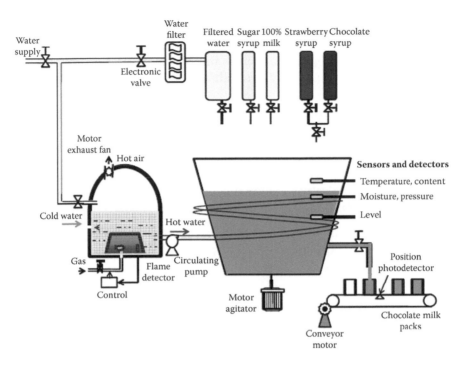

FIGURE 1.10
Milk-based beverage processing factory schematic.

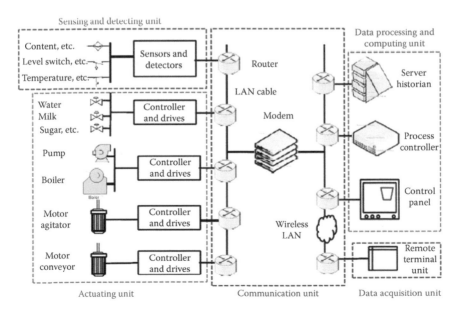

FIGURE 1.11
Block diagram with SCADA components for a milk-based beverage processing system.

- Cost-based objectives, such as reducing the cost of process production or service by improving the performance of the automated process
- Quality-based objectives, such as improving process operation efficiency or service quality through an extension of process control functions
- Control system compound design objectives, such as delivering new service or product, or replacing or extending it
- Monitoring and analysis objectives related to process control system design

To fulfill these objectives, the management steps of mechatronics and a process automation design or migration project are usually as follows:

1. Process preliminary studies based on process schematics (piping and instrumentation diagrams [P&IDs], process flow diagrams [PFDs], etc.)
2. Process performance audit
   a. Process operations performance assessment
   b. Process performance objectives and controller specifications
3. Process functional analysis and modeling
   a. Functional analysis system technique (FAST) method
   b. Graphical analysis of start and stop operating modes (GEMMA in French)
4. Process controller design
   a. Evaluation of storage requirements (e.g., memory size), control system throughput, and human–machine interface requirements
   b. Process controller software design
   c. Controller hardware design (particularly panel and instrumentation design)
   d. Process control design (distribution and control architecture integration)
5. Process controller solution execution
   a. Actuation feedback tuning
   b. Operation monitoring
6. Process control commissioning
   a. Operation and process control test
   b. Validation and personnel training
   c. Exploitation

The execution of an engineering project covers activities related to control system design or migration, as well as installation within industrial processes. As such, major steps could be summarized, as shown in Figure 1.12.

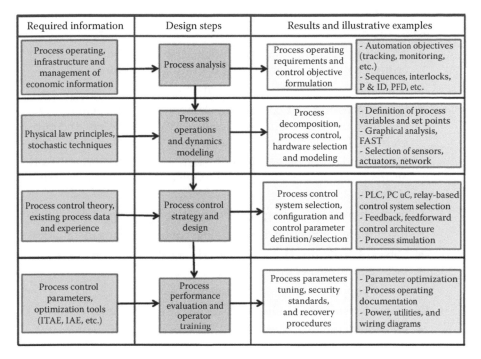

Required information	Design steps	Results and illustrative examples	
Process operating, infrastructure and management of economic information	Process analysis	Process operating requirements and control objective formulation	- Automation objectives (tracking, monitoring, etc.) - Sequences, interlocks, P & ID, PFD, etc.
Physical law principles, stochastic techniques	Process operations and dynamics modeling	Process decomposition, process control, hardware selection and modeling	- Definition of process variables and set points - Graphical analysis, FAST - Selection of sensors, actuators, network
Process control theory, existing process data and experience	Process control strategy and design	Process control system selection, configuration and control parameter definition/selection	- PLC, PC uC, relay-based control system selection - Feedback, feedforward control architecture - Process simulation
Process control parameters, optimization tools (ITAE, IAE, etc.)	Process performance evaluation and operator training	Process parameters tuning, security standards, and recovery procedures	- Parameter optimization - Process operating documentation - Power, utilities, and wiring diagrams

FIGURE 1.12
Generic automation design steps and dependencies. ITAE, integrated of absolute value of error; IAE, integral of absolute error; PC, personal computer; uC= Microcontroller .

Automation project activities require the production of specific process control documentation, such as

- Process schematics for functional analysis containing information on
  - Modes of operation and sequence of operations
  - Failure modes, emergency shutdown, and alarm
  - Operator interactions, and signals and sensors
  - Recovery procedures
- Process diagrams, especially
  - Logic diagrams: symbol and interpretation
  - P&I drawings from flow diagrams and process description
  - Electrical schematics, wiring and connection diagram, panel layout, and operator diagram
  - Automation field instrument lists with technical specifications for various instruments, such as flowmeters, thermistors, transmitters, converters, isolators, multipliers, and control valves

- Documents for automation process equipment sizing and selection, especially on
  - Actuating elements, such as motor drives, control valves, and orifices with estimation of their operating conditions
  - Sensing elements, such as flow sensors
  - Data processing and computing elements, such as memory size, and scan cycle of distributed control systems (DCSs) and programmable logic controller (PLC) systems
  - Power and utilities equipment, such as UPS batteries based on load calculations of various devices from the automation system
- Automation process cabling and wiring diagrams, especially
  - Instrument connection drawings and commissioning installation
  - Cables dispatching diagram, including cable junction boxes and their technical specifications
  - Power supply and protection schematics
- Diagrams for safety compliance and power requirements
  - Process interlock device interconnection drawings
  - Panel and cabinet localization and their rack arrangement
  - Compliance evaluation of vendor drawings and system design as per standards and norms
  - Power interconnection and distribution diagrams and electrical device placement, such as the number of repeaters between nodes
- Documents on sizing and selection of data communication network components
- Documents on selection criteria of data storage, acquisition, and transmission network devices and protocols, including their type and physical media, such as
  - Controller unit input/output (I/O) module sizing and selection
  - I/O addressing
  - Data logging and process data archiving
  - Database formulation
- Documents on automation software structure and operations
  - Control feedback loops (e.g., proportional-integral-derivative [PID]) and control logics design using programming languages such as LD, FBD, SFC, STL, or IL
  - Process automation system configuration and instrument calibration

- Automation system maintenance support procedures and documents, including
  - I/O process interface for discrete and analog field devices, as well as memory organization
  - System power, such as automation system power and field power
  - Diagnostics and detection of external automation field device failure, I/O module replacement procedure, I/O process wiring check, operational testing (factory acceptance testing), definition of emergency stops, and safety control relays
  - Power supply and PLC memory card replacement procedures
  - Troubleshooting procedures and listing of automation software errors

Table 1.2 summarizes process automation major steps and documents required as well as those that are expected to be provided, while Figure 1.13 presents automation project management activities.

Exercises and Conceptual Problems

1.1. Describe each of the following automated electrical-driven systems or processes based on their possible control objectives (specify feedback or logic control strategies):

a. Power generation from nuclear power plant

b. Robot land mine detection device

c. Sport-based wearable smart textiles with sensing functions

d. Coating of a pharmaceutical drug process

e. Egg incubator

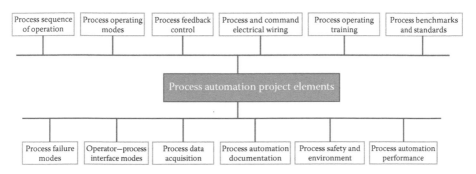

FIGURE 1.13
Overview of activities related to process automation project.

    f. Fruit picker robot harvesting machine

    g. Autoguided camera for detection and tracking of moving objects

    h. Intelligent water treatment and distribution process

    i. Power dispatcher over a network grid

    j. Oil distillation process

1.2. For each of the following automated systems or processes, outline their possible control objectives, as well as the automation devices and electric sensing components involved:

    a. Power generation from solar power plant

    b. Three-dimensional (3-D) printing robot for customized prosthetic heart valves

    c. High-speed automatic baggage handling and dispatching system network

    d. Flight traffic (arrival and departure) control along runways

    e. Remote intravascular microsurgery robot

1.3. The human body is composed of many innate feedback and feedforward control loops. For example, insulin is a pancreas-generated hormone that regulates the blood glucose concentration. A diabetic patient has the inability to produce a significant blood level of insulin. He has to rely on an artificial pancreas to regulate his blood glucose level by providing automatic insulin injections several times per day. Based on the patient's diet or physical and psychological activities, the amount of insulin necessary to maintain a normal blood glucose level must be determined.

    a. Derive the control objectives for the regulation of his blood glucose level and list all input and output process variables.

    b. Sketch a closed-loop control block diagram of the blood glucose level using a real-time blood glucose measurement device and a continuous insulin infusion pump.

    c. Draw a schematic of the instrumentation to be used (actuating, sensing and signal conditioning, etc.) and corresponding to the control block diagram in (b).

1.4. Heart surgery requires regulation of the patient's blood pressure around a desired value. This can be achieved by adjusting the infusion rate of vasoactive drugs into the patient's vessels. In addition to the effect of vasoactive drugs, blood pressure is also affected by the level of anesthetic given to the patient.

    a. Using feedback control strategies, sketch a detailed block diagram for a control system that enables the measurement of the blood pressure and regulates the infusion rate of a vasoactive drug accordingly.

    b. From the resulting block diagram, identify automation devices that could be used in the measuring or detecting unit, as well as in the actuating unit.

1.5. Similarly to blood glucose level or blood pressure regulation, the regulation of the human body blood temperature is performed through the evaporation of water from the skin over the capillary network.

    a. Formulate a block diagram for the human body blood temperature control problem (including internal feedback and feedforward control loops).

    b. Integrate the above block diagram with the one for blood pressure regulation (Exercise 1.4) and the one for blood glucose level control (Exercise 1.3) as an overall blood regulation (pressure and temperature) schematic.

    c. From the precedent block diagram, identify devices that could be used in the measuring or detecting unit, as well as in the actuating unit, of this multilevel automated blood regulation process.

    d. List at least 10 I/O process variables.

1.6. An autoguided missile is required to follow a predetermined flight trajectory by regulating the angle between its axis of motion and its velocity vector. The adjustment of the missile angle ϕ is achieved through the thrust angle β, which is the angle between the thrust direction and the axis of the missile, as illustrated in Figure 1.14. Draw a block diagram of an automated positioning system for the missile angle using a gyroscope to measure its angle and a motor to adjust the thrust angle.

1.7. An inkjet printer consists of two actuating devices: a motor for printer head positioning and a piezoelectric transducer for ink injection by the printer head. The head position is measured by a photosensor-based position encoder, and the estimated position error is transmitted to each actuator controller. Draw the block diagram for the automated two-dimensional (2-D) printing process of this inkjet printer. Deduce an equivalent block diagram in the case of a 3-D printer.

1.8. A control project aims to upgrade the automated airport passenger bag handling system depicted in Figure 1.15, from the ticket center to the right aircraft, in accordance with transport security agency requirements. Among those requirements, there should be (1) no single point of failure or (2) a baggage dimension checkpoint. All passenger bags should pass through an integrated security system consisting of a series of security scanners, while being directed automatically to their final destination by barcode scanning points along the long-distance conveyor system.

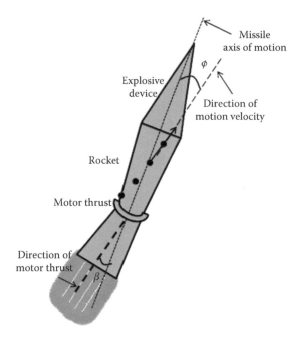

FIGURE 1.14
Missile trajectory.

- a. List all input and output handling system variables involved.
- b. Build the corresponding block diagram and a schematic showing the automation devices used for the measuring or detecting unit, as well as the actuating unit.

1.9. It is desired to control and monitor a power solar plant using a SCADA system. A polar solar plant consists of a photovoltaic power generation station, power inverters and trackers, a meteorological station, a power metering station, and a soiling monitoring station. The SCADA system should meet the following requirements:

- a. It should be able to (i) interact with field devices (relays, breakers, etc.), (ii) remotely troubleshoot plant operations, and (iii) track and control the power flow.
- b. The operator control panel should be user-friendly, with all necessary light indicators and an alarm of the management system that signals key process operation failures, as well as push buttons for plant activation and deactivation.

List all input and output variables involved in the solar plant operating process. Build the corresponding block diagram, including the monitoring and control automation devices.

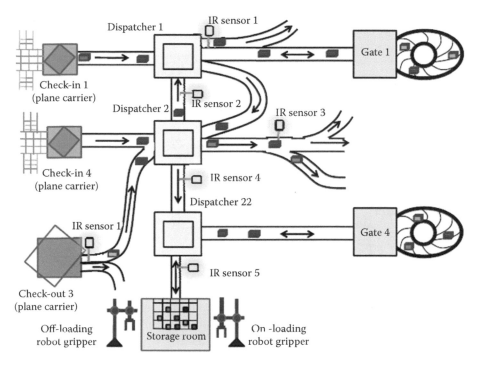

FIGURE 1.15
Automated airport baggage handling system. IR, infrared detector.

1.10. Consider the dialysis blood process illustrated in Figure 1.16. In order to design an automated dialysis blood processing system like that shown in the figure, list all input and output dialysis blood process variables involved and draw the corresponding control block diagram.

1.11. The following are among the embedded machine automation (mechatronic systems) found in modern vehicles:

- Antilock brake system (ABS)
- Traction control system (TCS)
- Vehicle dynamics control system (VDC)
- Electric ignition system for fuel–air combustion
- Engine management and transmission control
- Airbag activation
- Air-conditioning system
- Seat belt control system
- Mirror control system
- Instrument cluster

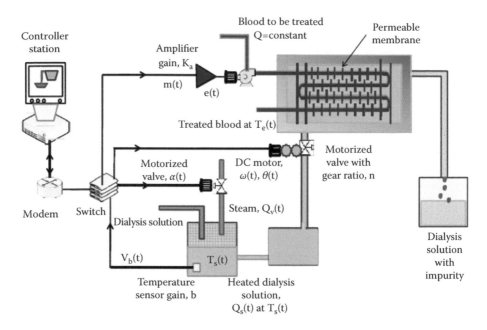

FIGURE 1.16
Dialysis blood processing system.

- Climate and vehicle front light activation control
- Park distance control system
- Self parallel parking device
- Averaging of the fuel tank level
- Alcohol test and engine activation
- Window lift system
  a. Based on either comfort, safety, emission reduction, or autonomous and intelligent cruise control automation objectives, classify the mechatronic systems listed above.
  b. For five mechatronic systems out of the above list, define the possible input and output variables involved. (Hint: Each variable is associated with an automation field device.)

1.12. Consider a batch process of the conveyor oven illustrated in Figure 1.17. List all input and output process variables involved (binary and continuous), and draw the block diagram for the automation of the conveyor oven temperature and the connection schematic of the programmable logic control interface with process field devices.

1.13. In the gold mining extraction operation, a SCADA system allows one to monitor and remotely control: primary and secondary crushing

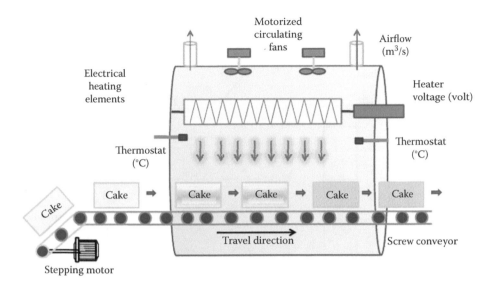

FIGURE 1.17
Cake conveyor oven.

motors: raw material conveying motors: ball milling; operations related to chemical pumps; underground water distribution during extraction operations; and operations related to leading and tailing stacking. List all input and output variables involved in the monitoring and control of such gold mining extraction processes. Build the corresponding block diagram, including the monitoring and control automation devices.

1.14. List the input and output process variables for the following telematic systems:

a. Millimeter-wave radar for the detection of a surrounding object's distance in an unmanned car collision avoidance management system.

b. Autoguided vehicle connected to a GPS-based traffic monitoring device

c. 3-D patient anatomy monitor in a robot-aided surgery process

d. Artificial liver implant system

e. Automatic dishwasher with a friendly user interface

f. Automatic vacuum cleaner with a friendly user interface

g. Automatic microwave with a friendly user interface

1.15. An unmanned electric vehicle has an embedded automation system for speed and direction control, as well as traffic light monitoring and motion synchronization with other users. Considering the

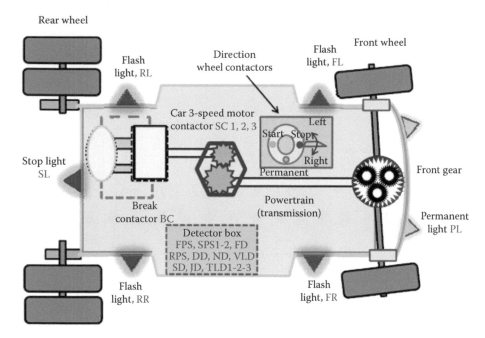

FIGURE 1.18

Chassis of a driverless vehicle. DD, day detector; FD, front car presence detector; FPS, front position sensor; JD, junction detector; ND, night detector; RPS, relative position sensor; SD, side car presence detector; SPS1-2, side position sensors 1 and 2; TLD 1-2-3, traffic light detector (red, yellow, green); VLD, visibility length detector.

devices for such a driverless vehicle shown in Figure 1.18, list all possible input and output variables involved. Draw the corresponding control block diagram.

Bibliography

Åström K.J., B. Wittenmark, *Computer-Controlled Systems: Theory and Design*, Courier Dover Publications, Mineola, NY, 2011.

Bailey D., E. Wright, *Practical SCADA for Industry*, Newnes (copyrighted Elsevier), Boston, 2003.

Erickson K., J. Hedrick, *Plant Wide Process Control*, Wiley, Hoboken, NJ, 1999.

Franklin G.F., M.L. Workman, D. Powell, *Digital Control of Dynamic Systems*, 3rd ed., Addison-Wesley Longman Publishing, Boston, 1997.

Golnaraghi F., B.C. Kuo, J.A. Adams, *Automatic Control*, 9th ed., Wiley, Hoboken, NJ, 2009.

Goodwin G.C., S.F. Graebe, M.E. Salgado, *Feedback Control of Dynamic Systems*, Prentice Hall, Upper Saddle River, NJ, 2001.

Groover M.P., *Automation, Production Systems and Computer-Integrated Manufacturing*, 3rd ed., Prentice-Hall, Upper Saddle River, NJ, 2007.

Kuo B., *Digital Control Systems*, Oxford University Press, Oxford, 1995.

Luyben, W.L., M.L. Luyben, *Essentials of Process Control*, McGraw-Hill, New York, 1997.

Marlin T.E., *Process Control: Design Processes and Control Systems for Dynamic Performance*, 2nd ed., McGraw-Hill, New York, 2000.

Ogata K., *Modern Control Engineering*, 4th ed., Prentice Hall, Upper Saddle River, NJ, 2004.

Powell F., A. Emami-Naeini, *Control System Design*, 4th ed., Prentice Hall, Upper Saddle River, NJ, 2002.

Seborg D., T.F. Edgar, D. Mellichamp, J. Doyle, *Process Dynamics and Control*, 3rd ed., Wiley, Hoboken, NJ, 2011.

Siouris G.M., *Missile Guidance and Control Systems*, Springer, Berlin, 2003.

Smith C.A., A.B. Corripio, *Principles and Practice of Automatic Process Control*, 2nd ed., Wiley, New York, 1997.

2

Electrical-Driven Actuating Elements: Modeling and Selection

2.1 Introduction

Electrical-driven actuating systems convert the electric power energy supplied to their actuator into mechanical energy or work. Among those actuating systems there are devices such as motors (e.g., direct current [DC] motors, alternating current [AC] motors, and stepper motors), control valves, heaters, solenoids, switches (e.g., relays, solid-state devices such as thyristors, and metal oxide semiconductor field-effect transistors [MOSFETs]), and electromagnets. Usually, those actuating systems are associated with mechanical, fluidic, or thermal transmission elements, in order to drive actions related to the dynamics of solid, liquid, and gas material or to chemical reactions. In this chapter, classical models for electrical-driven actuating systems, along with the dynamics of their associated transmission elements, are presented. Then, a methodology for the sizing and selection of those actuating systems is developed, based on their operational characteristics, such as torque–speed curves, load profile, and operational conditions. Some process models involving various transmission elements are covered to illustrate industrial applications.

2.2 Electrical-Driven Actuating Systems

Electric actuators can generate either binary output signal (i.e., contact relay) or continuous signal (stepper motor). They can be classified based on their electrostatic, electromagnetic, or electrothermal design principles. Their coupled transmission elements can produce force to motion, pressure to flow, heat, and so forth. Notice that recent advances in nanotechnologies have led to the development of low-scale actuators (nano or micro). Table 2.1 summarizes commonly encountered electrical-driven actuating systems.

TABLE 2.1

Typical Electrical-Driven Actuating System

Actuator Class	Actuator Types	Description
Electrical		
Binary actuator (switching circuits)	Thyristor, bipolar, solid-state relay, diode, MOFSET, transistor	Electronic switching devices offering high-frequency response
Electromagnetic (electromechanical)	DC motor	Wound field (separately excited) where speed varies with armature voltage or field current
		Wound field (shunt) with constant-speed application and low starting torque
		Wound field (series) with high starting torque and speed varies with low load
		Wound field (compound) with low starting torque
		Permanent magnet (variable speed but fixed magnetic field)
		Brushless (electronic commutation) with fast response
	AC motor	Induction motor
		Synchronous motor
		Universal motor (AC/DC power supply)
	Stepper motor	Hybrid (change pulse signal into rotational motion)
		Variable reluctance (precise motion but switching devices required)
Electromagnetic	Relay contact, electromagnet, solenoid	Applications requiring large force
Electrohydraulic	Electromagnetic cylinder	Linear motion cylinder
	Electrovalves	Directional valves, pressure valves
	Electropump, fan, blower	Hydraulic motor (gear, vane, piston, rotary, reciprocating)
Electrostatic actuator	Piezoelectric motor	Voltage proportional to applied stress (e.g., laser printer)
Electrothermal	Heater	
	Furnace	
Low-scale material devices (micro/ nano)	Shape memory alloys	Temperature sensitive
	Dielectric elastomer actuators	Spring sensitive
	Ionic polymers	
	Electrostrictive	
	Piezoelectric material	Micropumps micromotor, microvalve, microrobot (micromanipulation, cave exploration)
	Electroactive polymers	
	Electromagnetics	Binary electromagnetic force (i.e. door locks, HVAC control)

2.2.1 Electromechanical Actuator

Electromechanical actuators usually operate based on the electrostatic principle (electrostatic actuators) or on the electromagnetic principle (electromagnetic actuators). Electrostatic actuators use charged particle motion within the induced field to produce the adequate force, vibration, pressure, and temperature variation necessary for the displacement of a membrane or beam. Among those actuators there are piezoelectric actuators, which deliver strain-based actuation from piezomaterials generating a voltage proportional to the applied mechanical deformation. Such piezomaterials use either ceramics (i.e., lead-zirconate-titanate) or polymers (i.e., polyvinylidene fluoride). Surrounded by electrodes or attached to electrodes, they allow strain-based actuation up to a few kilovolts per millimeter. Among the microdevice applications there are (1) position systems, such as micromanipulation systems (microrobot with piezolegs), vibration oscillation suppression systems, valves, micropumps in artificial hearts, and ultrasonic motors, and (2) handling systems, such as microgrippers and laser printers using charged particle movement based on the electrophotographic principle.

Electromagnetic actuators are based on either electromagnetic force (Lorentz) or electromagnetic induction (Faraday) principles. They consist of induced magnetic fields around a ferrous stator causing an object to move within the air gap separating them. Hence, those electromagnetic actuators convert electrical energy into rotational or linear mechanical energy. Among those actuators there are linear solenoids consisting of an energized solenoid (coil) that generates a magnetic field along which a ferromagnetic object can move, causing an increase in flux linkage. Their typical applications include actuation of valves, switches, and electromechanical relays. There are also rotary solenoids, where a ball moves along inclined bearing raceways. Other electromagnetic actuators include electromechanical motors, such as DC motors, AC motors, and stepper motors. As summarized in Table 2.1, there are various types of electric motors, including series-excited DC motors, DC shunt motors, and DC separately excited motors; AC asynchronous motors, AC synchronous motors, and brushless AC motors; stepper motors; and linear motion motors. Those motors have the following components: a stator, a rotor, a field coil, a commutating device (for controlling DC motor current flow), an armature rotor winding, and the brush or slip rings. Due to their encountered applications, the modeling and selection methodology of various types of motors is presented in the following sections.

2.2.2 DC Motor Dynamics Modeling

DC motors have two distinct circuits: an outside set of coils, called the field stator circuit, and an inside set, called the armature rotor circuit. From Faraday's laws of electromagnetic induction, when a DC voltage, $V_f(t)$, supplies an excitation, the resulting force within the coil causes the rotation of the rotor armature (i.e., due to generating torque). If there is a separate field winding circuit, applying Kirchhoff's voltage law into the field and armature circuits respectively yields

$$V_f(t) = R_f i_f(t) + L_f \frac{di_f(t)}{dt} \tag{2.1}$$

and

$$V_a(t) = R_a i_a(t) + L_a \frac{di_a(t)}{dt} + E_b(t) \tag{2.2}$$

where R_f and L_f are the resistance and inductance of the field winding, respectively. The current, $i_f(t)$, generated in the winding produces a magnetic field inducing the rotation of the rotor in the armature circuit, such that the armature voltage is $V_a(t)$ and the armature circuit current is $i_a(t)$. R_a and L_a are the resistance and inductance of the armature winding, and E_b is the induced armature voltage. Considering one coil of wire (conductor), the ampere laws of the force are such that

$$F(t) = i l \times \mathbf{B} = Bli(t) \tag{2.3}$$

where l is the axial length of wire within the field, $\mathbf{B}$ is the magnetic field flux density produced, and $i(t)$ is the current flowing into the wire. Thus, the torque generated from the effect of two forces is given by

$$T(t) = 2F(t)r \tag{2.4}$$

where r is the distance between the center line of the shaft and the conductor. Then, applying the same effect within the motor armature winding, the torque for one armature conductor is given by

$$T(t) = B_v L i_a(t)r \tag{2.5}$$

where r is the radius of the armature winding circular structure, $i_a(t)$ is the current passing through the armature winding, L is the axial length of the armature winding, and B_v is the magnetic flux density at a pole, such as

$$B_v = \frac{\phi}{A} \tag{2.6}$$

Assuming that the coil has N turns, the motor torque is given by

$$T_m(t) = 2NB_v l i_a(t)r \tag{2.7}$$

where A is the area of the coil and ϕ is the flux or pole in weber. The resulting motor torque equation is equivalent:

$$T_m(t) = K\phi\, i_a(t) = K_t i_a(t) \tag{2.8}$$

where the torque constant K_t depends on the coil geometry (N, B_v, l, and r) for any particular motor in N.m.A^{-1}. The direction of the current flow is the rotation direction of the motor shaft. Thus, using Faraday's law, the generated electromotive force (emf) voltage in the motor by several coils wound on the rotor is determined by

$$E_b(t) = 2NB_v lr\big(\omega_m(t)\big) = K_t \omega_m(t) \tag{2.9}$$

Hence, during the transient period, the applied armature voltage, $V_a(t)$, varies with the current (producing a torque proportional to emf, $E_b(t)$) until it equals the applied voltage at steady state such that

$$E_b(t) = V_a(t) - R_a i_a(t) \tag{2.10}$$

Here, if any current flows in the coil, there is no further acceleration (i.e., the rotor turns at a constant speed). Hence, the electrical power delivered is equivalent to the mechanical power, such as

$$E_b(t) i_a(t) = \omega(t) T_m(t) \tag{2.11}$$

The electrical design configurations of the armature and field windings are used to classify DC motors. As such, the field windings are either self-excited (i.e., winding terminals are interconnected to the same source voltage) or separately excited (i.e., different voltage sources). Furthermore, in the case of self-excited motors, the field windings are either in series or in parallel.

2.2.2.1 Self-Excited Series DC Motor

Here, the voltage source and field and armature windings are in series configuration, as shown in Figure 2.1. When the load current increases, the armature and field resistance drops, inducing a decrease of voltage, $E_b(t)$. From this circuit series configuration, the magnetic flux varies proportionally to the current passing through the armature winding. Those motors are called series-wound DC motors or universal motors, as they operate from an AC or DC voltage source.

From the circuit configuration schematic shown in Figure 2.1a, the voltage equation results in

$$V_a(t) = E_b(t) + (R_a + R_f) i_a(t) + (L_a + L_f)\frac{di_a(t)}{dt} \tag{2.12}$$

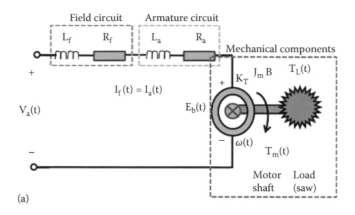

(a)

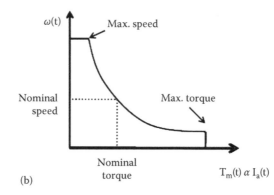

(b)

FIGURE 2.1
(a) Self-series-excited DC motor equivalent circuit configuration. (b) Torque–speed curve of self-series-excited DC motor.

During normal operation (no saturation), the rotor torque yields

$$T_m(t) = \left(K_f i_a(t) \right) K i_a(t) = K_f K i_a(t)^2 \qquad (2.13)$$

where K_f is a constant depending on the number of field winding turns and its physical construction. At steady-state operating conditions, the armature current results in

$$i_a(t) = \frac{V_a(t)}{R_a + R_f + KK_f \omega(t)} \qquad (2.14)$$

while the rotor torque developed yields

$$T_m(t) = K_f K \frac{V_a(t)^2}{\left(R_a + R_f + KK_f \omega(t) \right)^2} \qquad (2.15)$$

From constant voltage, $V_a(t)$, there is an inverse proportionality relationship between the angular velocity of the motor shaft and the torque square root, as illustrated in Figure 2.1b by torque–speed characteristics.

2.2.2.2 Self-Excited Shunt DC Motor

Here, the field and armature windings are in parallel configuration with the applied voltage supply, causing the current through the field winding to be independent of the armature current, as illustrated in Figure 2.2. When the armature current increases, the speed increases due to the armature reaction. Consequently, the motor torque rises linearly until the armature reaction affects the field. Hence, the current resulting from the supply is given by (for $R_s = 0$)

$$i_L(t) = i_f(t) + i_a(t) \tag{2.16}$$

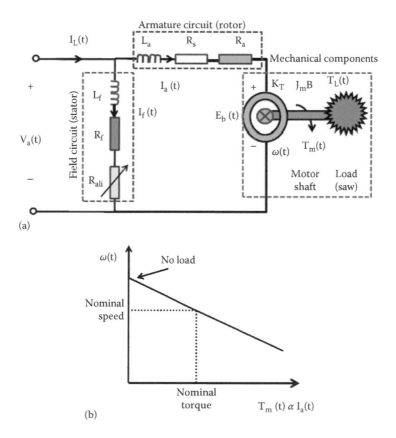

FIGURE 2.2
(a) Shunt DC motor equivalent circuit configuration. (b) Shunt DC motor torque–speed curve.

From the circuit analysis, the induced voltage, current, and mechanical power equations result in

$$V_a(t) = R_{eq} i_L(t) + L_{eq} \frac{di_L(t)}{dt} + E_b(t) \tag{2.17}$$

with

$$R_{eq} = \frac{R_a R_f}{R_a + R_f}$$

$$L_{eq} = \frac{L_a L_f}{L_a + L_f}$$

Hence, at steady state, the torque–speed relationship is given by

$$\omega(t) = \frac{V_a(t)}{K\phi} - \frac{R_{eq}}{(K\phi)^2} T_m(t) \tag{2.18}$$

2.2.2.3 Separately Excited DC Motor

With separately excited DC motors, the armature windings and field windings are individually energized by a separate DC source, as illustrated in Figure 2.3. Applying Kirchhoff's voltage law to the field and the armature circuits respectively yields

$$\begin{cases} V_a(t) = E_b(t) + R_a i_a(t) + L_a \dfrac{di_a(t)}{dt} \\[2mm] V_f(t) = R_f i_f(t) + L_f \dfrac{di_f(t)}{dt} \end{cases} \tag{2.19}$$

Then, at steady state, the generated rotor torque is given by

$$T_m(t) = \left(\frac{K\phi}{R_a} \right) \left(V_a(t) - K\phi\omega(t) \right) \tag{2.20}$$

The relationship between the torque and the speed for constant $V_a(t)$ and flux ϕ is a straight drooping line, as with the shunt DC motor.

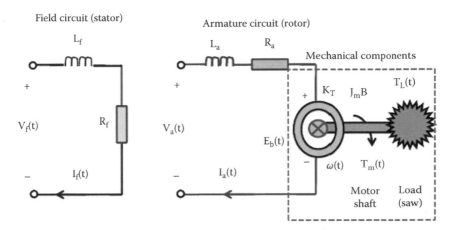

FIGURE 2.3
Separately excited DC motor equivalent circuit configuration.

2.2.2.4 Compound-Wound DC Motor

With compound-wound DC motors, two stator windings are connected, one in series and one in parallel (or shunt) with the armature windings, as shown in Figure 2.4. Two types of arrangements are possible in compound DC motors: (1) cumulative compounding, when the magnetic fluxes generated by both series and shunt field windings are in the same direction (i.e., additive), and (2) differential compounding, when the two fluxes are in opposition.

From the equivalent circuit, the induced voltage is given by

$$V_a(t) = E_b(t) + R_{eq}i_L(t) + L_{eq}\frac{di_L(t)}{dt} \qquad (2.21)$$

with

$$R_{eq} = \frac{(R_a + R_{f2})R_{f1}}{(R_a + R_{f2} + R_{f1})}$$

$$L_{eq} = \frac{L_{f1}L_{f2}}{L_{f1} + L_{f2}}$$

where the rotor torque developed here yields

$$T_m(t) = K(\phi_{series} + \phi_{shunt})i_a(t) \qquad (2.22)$$

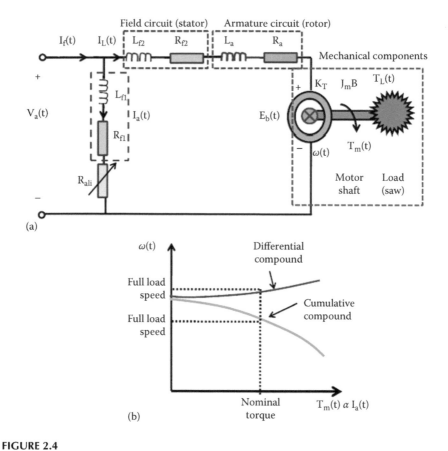

FIGURE 2.4
(a) Compound-wound DC motor equivalent circuit configuration. (b) Compound-wound DC motor torque–speed characteristics.

At steady state,

$$\begin{cases} V_a(t) = K\phi\omega(t) + R_{eq}\left(i_f(t) + i_a(t)\right) \\ i_a(t) = \dfrac{T_m(t)}{K\phi I_f(t)} \\ i_f(t) = \dfrac{V_a(t)}{R_f} \end{cases} \tag{2.23}$$

This is equivalent to

$$V_a(t) = K\phi\omega(t) + R_{eq}\left(\frac{V_a(t)}{R_f} + \frac{T_m(t)}{K\phi}\frac{R_f}{V_a(t)}\right) \tag{2.24}$$

As illustrated in Figure 2.4, the torque–speed relationship is given by

$$\omega(t) = \frac{1}{K\phi}\left[V_a(t)\left(1 - \frac{R_{eq}}{R_f} - \frac{R_{eq}T_m(t)}{K\phi}\frac{R_f}{V_a(t)^2} \right)\right] \tag{2.25}$$

2.2.2.5 Permanent Magnet DC Motor

In a DC permanent magnet motor, stator magnetic fields are permanent magnets and the current is supplied into the armature windings through commutators (brushes), as illustrated in Figure 2.5. The equivalent electrical circuit of the armature coil has an inductance, L_a, in series with a resistance, R_a, and an induced voltage, $E_b(t)$. With a constant magnetic flux, the torque–speed relationship is given by

$$\omega(t) = \frac{V_a(t)}{K\phi} - \frac{R_a(t)}{(K\phi)^2}T_m(t) \tag{2.26}$$

Here, the torque increases proportionally to the current and independently of the shaft velocity. Such a motor configuration can deliver high torque at low speed. The assumptions used in the development of DC motor models with a linear torque–speed relationship are

1. The torque provided by the bearings of the motor has negligible frictional losses.
2. The inductance of the motor is zero at steady state.

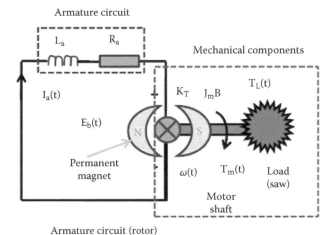

FIGURE 2.5
Permanent magnet DC motor equivalent circuit configuration.

3. The motor armature resistance is independent of speed and temperature changes (unless the motor is operating at very high temperatures).
4. The motor constant, K, and the resistance, R, are known from the manufacturer.

2.2.2.6 Speed Control Methods of DC Motors

Process requirements for a wide range of speed variation of electrical motors can be fulfilled with DC motors using variable speed drives. The speed drives have the ability to match the speed and torque with the process requirements independently of the load. For DC motors with negligible armature inductance and without load, the induced voltage, $E_b(t)$, is given by

$$E_b(t) = -R_a i_a(t) + V_a(t) = K\phi\omega(t) \tag{2.27}$$

This could be rewritten at steady state or rated speed as

$$\omega(t) = \frac{-R_a i_a(t) + V_a(t)}{K\phi} \tag{2.28}$$

Based on Equation 2.28, subsequently at steady state, the speed, $\omega(t)$, can be controlled by one of these methods: (1) armature voltage control (through $V_a(t)$ variation), (2) field control (through ϕ variation), and (3) armature resistance control (through R_a variation). Hence, solid-state circuits are used to modify DC motor torque–speed characteristics by adjusting either the armature voltage, $V_a(t)$ (through thyristor-based circuits in an AC-to-DC converter for an AC voltage supply or by using chopper circuits for a DC voltage supply), or the field current, $i_f(t)$ (through thyristor-based circuits to supply additional field winding), or both.

2.2.2.6.1 Armature Voltage Control, $V_a(t)$

This method of speed control is suitable for separately excited DC motors. Here, R_a and ϕ are kept constant and the DC drives aim to keep the rated motor torque constant at any speed, as long as it is lower than the rated motor speed. At steady state, the armature voltage decreases such that

$$E_b(t) = V_a(t) \approx K\phi\omega(t) \tag{2.29}$$

For constant flux ϕ, the speed, $\omega(t)$, changes linearly with $V_a(t)$, as shown in Figure 2.6. Such variable speed drives can be built by either (1) using thyristors through the variation of the firing phase angle from the bridge configuration (e.g., full-wave 12-pulse bridge, full-wave 6-pulse bridge, or half-wave

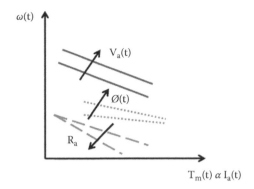

FIGURE 2.6
Torque–speed curves for varying R_a, ϕ, and $V_a(t)$.

three-pulse bridge), relative to the AC supply voltage, as illustrated in Figure 2.7, or (2) using the variation of the ON and OFF time ratio (duty cycle) of the DC voltage supply from chopper circuits or switch mode converters (such as Jones, Morgan, and oscillation). Those choppers transform a constant magnitude of DC input signal into a variable magnitude of DC signal by inverting it using a DC-to-AC converter, passing it through an AC transformer, and then rectifying it through an AC-to-DC converter circuitry. Among other design alternatives there are (1) rectifiers (a combination of bistable devices like diodes and thyristors for phase control) to convert single- or three-phase AC signal (either voltage, current, or frequency) into constant and smooth DC voltage signal, and to pass it through a chopper circuit, in order to obtain an adjustable mean DC voltage, and (2) Ward Leonard drives (separately excited

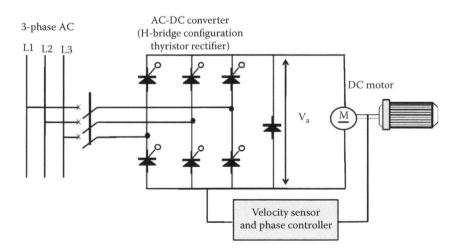

FIGURE 2.7
Typical thyristor-controlled DC drive.

DC generators), where a three-phase AC voltage is supplied to a fixed-speed induction motor. Those devices are suitable for shunt-wound DC motors.

2.2.2.6.2 Field Control φ

Field control is a speed control method where R_a and $V_a(t)$ are kept constant. Here, the variation of the motor flux is derived by connecting a variable resistance, R_s, in series with the field winding for shunt-wound and compound-wound DC motors. This is suitable for a series DC motor. Hence, recalling that R_a is proportional to $1/\phi$, and assuming that magnetic linearity ϕ is proportional to the field current, $i_f(t)$, which is inversely proportional to $\omega(t)$, Equation 2.28 leads to

$$\omega(t) \propto \frac{V_a(t) - i_a(t)R_a}{i_f(t)} \tag{2.30}$$

An adjustable resistance allows us to vary this field current. The resulting torque–speed curves are depicted in Figure 2.6. With field control, the applied armature voltage, V_a, is maintained constant. Hence, reduction of the flux could increases the speed above about three or four times the rated speed while decreasing the torque proportionally.

2.2.2.6.3 Armature Resistance Control, R_a

Here, the armature resistance is kept constant such that the applied voltage can be varied by connecting in series a variable resistance, R_s, with the armature circuit. From torque–speed characteristics, it is known that

$$T_m(t) \propto \frac{K\phi}{R_a}\left(V_a(t) - K\phi\omega(t)\right) \tag{2.31}$$

The effect of varying the armature resistance value on the torque–speed curve is illustrated in Figure 2.8. This control method allows a continuous speed control while $i_f(t)$ is maintained constant.

The three methods indicated above can be applied to DC shunt motors. Except the commonly used armature resistance control method, the other two speed control methods can be applied for DC series motors.

2.2.3 AC Motor Technical Specifications and Dynamics Modeling

Asynchronous and synchronous electric motors are the two main categories of AC motors. AC motors are either single phase or polyphase, usually with two or three phases energizing poles from three same magnitude AC voltage sources with the same magnitude. The frequency of the exciting voltages (in number of cycles per second) determines the speed at which the

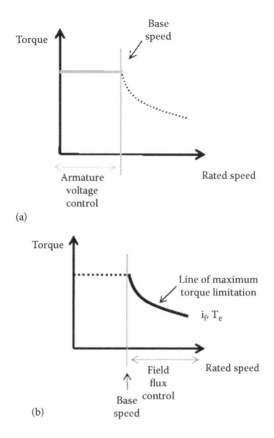

(a)

(b)

FIGURE 2.8
(a) Variation of speed with external armature resistance. (b) Variation of speed with motor flux (using variable resistance, Rs).

resulting magnetic field rotates (rev/s). Hence, AC motors are constant-speed electric motors operating in absolute synchronism either with line frequency or without. As an asynchronous motor, the induction motor consists of an AC transformer (alternator) connected to a secondary configuration (wound rotor type or squirrel cage type) that produces a rotating magnetic flux. This flux generates a field voltage proportional to the angular speed, and causes the torque from the electromagnetic induction between the stator and the rotor. The gap between the rotor speed (actual speed), N, and the stator speed (synchronous speed), N_s, is called the slip, s, and it is given as a percentage of the synchronous speed, such as

$$s = \frac{N_s - N}{N_s} \tag{2.32}$$

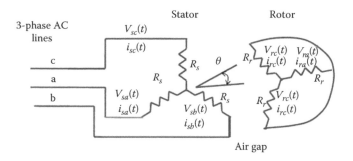

FIGURE 2.9
Schematic of a two-pole induction motor.

With an induction motor, the number of poles per phase, p, is rounded up as

$$p = \frac{120\,f}{N_s} \qquad (2.33)$$

where f is the frequency of current sources in cycles per second, and N_s is the rated (synchronous) speed or the field speed in revolutions per minute (rpm). At full load speed, the torque increases to T_{max}, and the slip decreases from the maximum torque when the electromechanical torque is matched by the mechanical torque.

For example, a schematic of a two-pole induction motor is shown in Figure 2.9 with the stator consisting of a 120° place three-phase winding a, b, and c. When the three AC currents are flowing through those windings, an air gap flux density is induced and rotates at an angular frequency ω_s of the stator currents. The induced rotor currents crossing over the air gap flux produce a motor torque on the rotor. Hence, the power of an AC motor is given by

$$\text{Power (kW)} = \frac{\text{Torque (N.m)} \times \text{Speed (RPM)}}{9550} \qquad (2.34)$$

and the energy is given by

$$\text{Energy (kWh)} = \frac{\text{Torque (Nm)} \times \text{Speed (RPM)} \times \text{Time (Hours)}}{9550} \qquad (2.35)$$

Usually, AC motors do not operate at a speed less than about one-third of the base speed due to thermal constraints. Single-phase induction motors are found in low-power-requirement devices, such as dishwasher motors or small-precision equipment. A commonly encountered universal motor is a single-phase induction motor similar to a series-wound DC motor, which can be converted to either an AC or DC voltage source. Synchronous motors consist of a rotor (permanent magnet or winding coil) connected to a slip ring for commutation, such that its rotating rotor speed, N_s, is synchronized with the frequency of the AC voltage source (rotating magnetic field related to the frequency of the AC source currents, f). A *polyphase synchronous AC motor* has several stator windings for smoother operations, such as the three-phase AC induction motor encountered in seaport hoist drives, offshore oil pump drives, and high-speed train locomotive drives.

There are various starting techniques for induction motors, such as the (1) start-delta starting technique (reducing the starting current by about a third), (2) autotransformer starting technique (reducing the supply voltage by half and reducing the starting current and torque up to a quarter), and (3) rotor resistance technique (reducing the starting current and increasing the torque). Among braking techniques of induction motors there are (1) plugging, which consists of reversing the magnetic field rotation, and (2) dynamic braking, which consists of switching the voltage supply from AC to DC, resulting in a unidirectional field for the alignment of the rotor.

2.2.3.1 Induction Motor Dynamics Modeling

Commonly encountered induction motor models have three-phase motor variables (voltages, currents, and magnetic flux), which are expressed in state-space vectors. The resulting state-space models are valid for any instantaneous variation of voltage and current, capturing their steady-state and transient dynamical characteristics. Because an induction motor displays time-variant electrical and mechanical system dynamics, a commonly used modeling strategy is to approximate it by a DC-like motor model after successive transformations. The resulting induction motor model returns a constant input voltage and frequency. Hence, using Faraday's laws for the three stator phases, a global expression in the matrix form is

$$\begin{bmatrix} V_{sa}(t) \\ V_{sb}(t) \\ V_{sc}(t) \end{bmatrix} = R_s \begin{bmatrix} i_{sa}(t) \\ i_{sb}(t) \\ i_{sc}(t) \end{bmatrix} + \frac{d}{dt} \begin{bmatrix} \psi_{sa}(t) \\ \psi_{sb}(t) \\ \psi_{sc}(t) \end{bmatrix} \tag{2.36}$$

Similarly, the three rotor phases can be written as

$$\begin{bmatrix} V_{ra}(t) \\ V_{rb}(t) \\ V_{rc}(t) \end{bmatrix} = R_r \begin{bmatrix} i_{ra}(t) \\ i_{rb}(t) \\ i_{rc}(t) \end{bmatrix} + \frac{d}{dt} \begin{bmatrix} \psi_{ra}(t) \\ \psi_{rb}(t) \\ \psi_{rc}(t) \end{bmatrix} \tag{2.37}$$

where R_s, R_r are stator and rotor winding equivalent resistances, while $V_{sa,b,c}(t)$, $V_{ra,b,c}(t)$, $\psi_{sa,b,c}(t)$, $\psi_{ra,b,c}(t)$, $i_{sa,b,c}(t)$, and $i_{ra,b,c}(t)$ are stator and rotor three-phase motor a, b, and c voltages, fluxes, and currents, respectively. In fact, each flux is the result of an interaction between all phases' currents such that for all fluxes from the stator and rotor, it can be expressed by

$$
\begin{bmatrix}
\psi_{sa}(t) \\
\psi_{sb}(t) \\
\psi_{sc}(t) \\
\psi_{ra}(t) \\
\psi_{rb}(t) \\
\psi_{rc}(t)
\end{bmatrix}
=
\begin{bmatrix}
l_s & m_s & m_s m_1 & m_3 & m_2 \\
m_s & l_s & m_s m_2 & m_1 & m_3 \\
m_s & m_s & l_s m_3 & m_2 & m_1 \\
m_1 & m_2 & m_3 l_r & m_r & m_r \\
m_3 & m_1 & m_2 m_r & l_r & m_r \\
m_2 & m_3 & m_1 m_r & m_r & l_r
\end{bmatrix}
\begin{bmatrix}
i_{sa}(t) \\
i_{sb}(t) \\
i_{sc}(t) \\
i_{ra}(t) \\
i_{rb}(t) \\
i_{rc}(t)
\end{bmatrix}
\tag{2.38}
$$

where l_s is the phase self-inductance of the stator, l_r is the phase self-inductance of the rotor, m_s is the inductance between two stators' phases, and m_r is the inductance between two rotors' phases. In addition,

$$
m_1 = m_{sr}\cos(\theta) \tag{2.39}
$$

$$
m_2 = m_{sr}\cos\left(\theta - \frac{2*\pi}{3}\right) \tag{2.40}
$$

$$
m_3 = m_{sr}\cos\left(\theta + \frac{2*\pi}{3}\right) \tag{2.41}
$$

where m_{sr} is the maximum inductance between one stator phase and one rotor phase, and θ is the angle between the same number of stator and rotor phases. To ease the calculation, the number of current components can be reduced through a transformation converting a three-phase induction motor model a, b, and c into a stationary two-orthogonal phase system α and β. Figure 2.10 shows the graphical representations of the space vectors and their projection over the quadrature-phase axes α and β.

Recalling that the total root mean square (RMS) current flowing across a variable speed drive is derived as the square root of the sum of the harmonic currents' squares, and considering that the a and α axes are aligned, the relationships between the quadrature-phase stator currents $i_{s\alpha}$ and $i_{s\beta}$ and the three-phase stator currents are given by

$$
i_{s\alpha} = k\left(i_{sa} - \frac{1}{2}i_{sb} - \frac{1}{2}i_{sc}\right) \tag{2.42}
$$

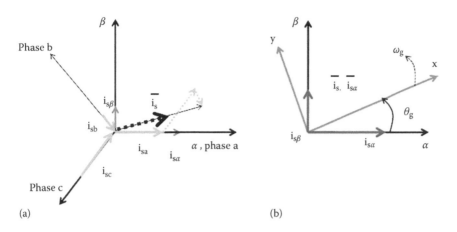

FIGURE 2.10
(a) Stator current space vector. (b) General reference frame space.

$$i_{s\beta} = k\frac{\sqrt{3}}{2}(i_{sb} - i_{sc})$$ (2.43)

$i_{s\alpha}$ and $i_{s\beta}$ are the currents in the stator reference frame α and β. Hence, it is possible to express all variables in the rotor reference frame for vector control purposes through a transformation of stator current a, b, c components into $s\alpha$ and $s\beta$ components:

$$i_{s\alpha}(t) = \sqrt{\frac{2}{3}}i_{sa}(t) - \frac{1}{\sqrt{6}}i_{sb}(t) - \frac{1}{\sqrt{6}}i_{sc}(t)$$ (2.44)

$$i_{s\beta}(t) = \frac{1}{\sqrt{2}}i_{sb}(t) - \frac{1}{\sqrt{2}}i_{sc}(t)$$ (2.45)

where this transformation is given by $T_{23} = \sqrt{\dfrac{2}{3}}\begin{bmatrix} 1 & -\dfrac{1}{2} & -\dfrac{1}{2} \\ 0 & \dfrac{\sqrt{3}}{2} & -\dfrac{\sqrt{3}}{2} \end{bmatrix}$. Thus,

using Equation 2.36, the dynamics equations become

$$T_{23}[V_{abcs}(t)] = T_{23}\left\{R_s[i_{abcs}(t)] + \frac{d[\psi_{abcs}]}{dt}\right\}$$ (2.46)

$$[V_{\alpha\beta s}(t)] = R_s T_{23}[i_{abcs}(t)] + \frac{d}{dt} T_{23}[\psi_{abcs}] \tag{2.47}$$

leading to

$$[V_{\alpha\beta s}(t)] = R_s[i_{\alpha\beta s}(t)] + \frac{d}{dt}[\psi_{\alpha\beta s}] \tag{2.48}$$

Similarly, the rotor voltage dynamics equations result in

$$\begin{cases} V_{r\alpha}(t) = R_r i_{r\alpha}(t) + \dfrac{d\psi_{dr}(t)}{dt} - \omega\, \psi_{r\beta}(t) \\[2mm] V_{r\beta}(t) = R_r i_{r\beta}(t) + \dfrac{d\psi_{r\beta}(t)}{dt} + \omega\, \psi_{r\alpha}(t) \end{cases} \tag{2.49}$$

This is equivalent to

$$\begin{bmatrix} \psi_{\alpha\beta s} \\ \psi_{\alpha\beta r} \end{bmatrix} = \begin{bmatrix} \psi_{\alpha s} \\ \psi_{\beta s} \\ \psi_{\alpha r} \\ \psi_{\beta r} \end{bmatrix} = \begin{bmatrix} L_s & 0 & & M.P(\theta) \\ 0 & L_s & & \\ & M.P(-\theta) & L_r & 0 \\ & & 0 & L_r \end{bmatrix} \begin{bmatrix} i_{\alpha s} \\ i_{\beta s} \\ i_{\alpha r} \\ i_{\alpha s} \end{bmatrix}$$

$$= \begin{bmatrix} L_s & 0 & & M.P(\theta) \\ 0 & L_s & & \\ & M.P(-\theta) & L_r & 0 \\ & & 0 & L_r \end{bmatrix} \begin{bmatrix} i_{\alpha\beta s} \\ i_{\alpha\beta s} \end{bmatrix} \tag{2.50}$$

with $P(\theta) = \begin{bmatrix} \cos\theta & -\sin\theta \\ \sin\theta & \cos\theta \end{bmatrix}$ and $L_s = l_s - m_s$, $L_r = l_r - m_r$, and $M = \dfrac{3}{2}m_{sr}$.

It is desired to transform induction motor variables behaving like a time-invariant DC motor at steady state. These $i_{s\alpha}$ and $i_{s\beta}$ variables need to be converted from the reference frame (α, β) into the d-q reference frame. Because the latter reference is moving at the same angular frequency of $i_{s\alpha}$ and $i_{s\beta}$, the resulting i_{sd} and i_{sq} variables are time and angular velocity independent. Considering that the d-axis is in the same direction as the rotor flux, the transformation is depicted in Figure 2.11, where θ_{field} is the relative flux position of the rotor.

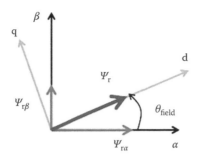

FIGURE 2.11
Park transformation.

Consider a squirrel cage induction motor, with the representation of stator and rotor winding depicted in the complex *d-q* equivalent circuit of an induction motor (neglecting rotor leakage inductance). Here, the voltage feeds a squirrel cage induction motor model in a *d-q* synchronously rotating frame, and it ensures that i_{qs} delivers the desired electromagnetic torque such that

$$\theta_s = \theta + \theta_r \qquad (2.51)$$

Then, the stator and rotor components become

$$[x_{\alpha\beta s}] = P(\theta_s)[x_{dqs}] \qquad (2.52)$$

$$[x_{\alpha\beta r}] = P(\theta_s)[x_{dqr}] \qquad (2.53)$$

After transformations of the motor quantities from one reference frame to the general reference frame, which is the reference frame attached to the rotor flux linkage space vector with the direct axis (*d*) and quadrature axis (*q*) (*d-q* coordinates), the motor model is as follows:

$$\begin{cases} V_{ds}(t) = R_s i_{ds}(t) + \dfrac{\psi_{ds}(t)}{dt} - \omega_s \psi_{qs}(t) \\[3mm] V_{qs}(t) = R_s i_{qs}(t) + \dfrac{d\psi_{qs}(t)}{dt} - \omega_s \psi_{ds}(t) \end{cases} \qquad (2.54)$$

$$\begin{cases} V_{dr}(t) = 0 = R_r i_{dr}(t) + \dfrac{d\psi_{dr}(t)}{dt} - (\omega_s - \omega)\psi_{qr}(t) \\[3mm] V_{qr}(t) = 0 = R_r i_{qr}(t) + \dfrac{d\psi_{qr}(t)}{dt} + (\omega_s - \omega)\psi_{dr}(t) \end{cases} \qquad (2.55)$$

Similarly, the space vectors of the rotor voltages and rotor flux linkages in the general reference frame can be expressed as

$$
\begin{cases}
\psi_{ds}(t) = L_s i_{ds}(t) + M i_{dr}(t) \\
\psi_{qs}(t) = L_s i_{qs}(t) + M i_{qr}(t) \\
\psi_{dr}(t) = L_r i_{dr}(t) + M i_{ds}(t) \\
\psi_{qr}(t) = L_r i_{qr}(t) + M i_{qs}(t)
\end{cases}
\tag{2.56}
$$

with α, β stator orthogonal components being the system coordinates, $V_{s\alpha,\beta}$ the stator voltages, $i_{s\alpha,\beta}$ the stator currents, $V_{r\alpha,\beta}$ the rotor voltages, $i_{r\alpha,\beta}$ the rotor currents, $\psi_{s\alpha,\beta}$ the stator magnetic fluxes (in V·s), and $\psi_{r\alpha,\beta}$ the rotor magnetic fluxes (in V·s). Hence, the different voltages can be expressed in the *d-q* coordinates such that

$$
V_{ds}(t) = R_s i_{ds}(t) - \frac{d\theta_s(t)}{dt}\psi_{qs} + \frac{d\psi_{ds}(t)}{dt}
\tag{2.57}
$$

$$
V_{qs}(t) = R_s i_{qs}(t) + \frac{d\theta_s(t)}{dt}\psi_{ds}(t) + \frac{d\psi_{qs}(t)}{dt}
\tag{2.58}
$$

$$
V_{dr}(t) = R_r i_{dr}(t) - \frac{d\theta_s(t)}{dt}\psi_{qr}(t) + \frac{d\psi_{dr}(t)}{dt}
\tag{2.59}
$$

$$
V_{qr}(t) = R_r i_{qr}(t) + \frac{d\theta_s(t)}{dt}\psi_{dr}(t) + \frac{d\psi_{qr}(t)}{dt}
\tag{2.60}
$$

Thus, the electromechanical model of a three-phase induction motor consists of five ordinary differential equations:

$$
\frac{di_{ds}(t)}{dt} = \frac{\left(R_s L_r i_{ds}(t) - M^2 i_{qs}(t)\frac{d\theta(t)}{dt} - R_r M i_{dr}(t) - L_r M i_{dr}(t)\frac{d\theta(t)}{dt} - L_r V_{ds}(t) \right)}{M - L_r L_S}
\tag{2.61}
$$

$$
\frac{di_{qs}(t)}{dt} = \frac{\left(R_s L_r i_{qs}(t) + M^2 i_{ds}(t)\frac{d\theta(t)}{dt} - R_r M i_{dr}(t) + L_r M i_{dr}(t)\frac{d\theta(t)}{dt} - L_r V_{qs}(t) \right)}{M - L_r L_S}
\tag{2.62}
$$

$$\frac{di_{dr}(t)}{dt} = \frac{-\left(R_s M i_{ds} - M L_s i_{qs}(t)\frac{d\theta(t)}{dt} - R_r L_s i_{dr} - L_r L_s i_{qr}(t)\frac{d\theta(t)}{dt} - M V_{ds}(t) \right)}{M - L_r L_S}$$

(2.63)

$$\frac{di_{qr}(t)}{dt} = \frac{-\left(R_s M i_{qs}(t) + M L_s i_{ds}(t)\frac{d\theta(t)}{dt} - R_r L_s i_{qr}(t) + L_r L_s i_{dr}(t)\frac{d\theta(t)}{dt} - M V_{qs}(t) \right)}{M - L_r L_S}$$

(2.64)

$$T_m(t) = M\left(i_{qs}(t)i_{dr}(t) + i_{ds}(t)i_{qr}(t) \right)$$

(2.65)

where $i_{ds}(t)$, $i_{qs}(t)$ are the transformed stator currents in the stator fixed frame; $i_{dr}(t)$, $i_{qr}(t)$ are the transformed rotor currents in the stator fixed frame, $V_d(t)$; $V_q(t)$ are the transformed stator voltages in the stator fixed reference frame; $\frac{d^2\theta(t)}{dt^2}$ is the rotor angular acceleration in rad/s²; and $\frac{d\theta(t)}{dt}$ is the rotor angular velocity in rad/s. R_s, R_r, L_s, L_r, M, J, B are, respectively, the stator phase resistance (in Ω), the rotor phase resistance (in Ω), the stator phase inductance (in H), the rotor phase inductance (in H), the mutual (stator-to-rotor) inductance (in H), the moment of inertia of the rotor and the load (kg − m²), and the damping coefficient of the bearing (kg − m²/s). The electromagnetic torque can be expressed by using space vector quantities as

$$T_m(t) = \frac{3}{2}P_p\left(\psi_{s\alpha}(t)\, i_{s\beta}(t) - \psi_{s\beta}(t)i_{s\alpha}(t) \right)$$

(2.66)

Thus, the mechanical model of the induction motor is given by

$$J\frac{d\omega(t)}{dt} + B\omega(t) = \frac{n_p}{2}\left(T_m(t) - T_L(t) \right)$$

(2.67)

n_p is the number of motor pole pairs (in H)m and $T_m(t)$ is the electromagnetic torque (in Nm). This motor rotates at a general speed of $\omega_g(t) = d\theta_g(t)/dt$ (i.e., with direct and quadrature axes $[x,y]$), where $\theta_g(t)$ is the angle between the direct axis of the stationary reference frame (α) attached to the stator and the real x-axis of the general reference frame.

2.2.3.2 Speed Control of AC Motors

The speed of an AC motor depends on three variables: (1) the fixed number of stator winding sets (known as poles) defining its base speed, (2) the frequency of the AC voltage or current supply, and (3) the load torque defining the slip. Induction motor speed can be varied through

1. A variable-frequency drive enabling the conversion of a fixed-frequency AC source into a variable-frequency AC signal. This is implemented using a AC-to-DC converter, such as a variable-voltage inverter, pulse width modulation (PWM) inverter, current-controlled inverter, or even cycloconverter.

2. A variation of the rotor circuit resistance by inserting an AC chopper with a high frequency and an ON/OFF time that can vary.

3. A variation of the stator voltage by using AC regulators.

4. An injection of slip frequency emfs to the rotor circuit.

5. An AC motor design modification by changing the number of stator coil poles.

Based on the converter type and AC motor type, AC motor drives can be classified into four categories: (1) thyristor-based voltage drives, capable of bistate force switching due to gate signal conditions and suitable for induction motor control; (2) transistor-based volts per hertz (V/f) and vector pulse width modulator drives (PWM/cycloconverter), used to control an induction motor; (3) transistor-based drives with natural commutating state (ON/OFF) capability, such as the bridge-commutated drives suitable to control synchronous motors; and (4) (PWM) voltage vector drives, used to control permanent magnet AC motors. In the frequency control strategy using the PWM, the supply DC voltage and current are smoothed and fed into a PWM inverter in order to produce a controlled variable-magnitude and -frequency AC voltage signal. The cycloconverter (AC/AC) requires a large number of thyristors and is appropriate for induction motors and synchronous motors.

In summary, AC motor control methods are either (1) scalar control through regulation of the command input voltage magnitude or (2) vector control through regulation of both command input magnitude and phase. The AC motor speed is controlled by using the inverter that can alter the AC power frequency based on the PWM. Typical industrial applications of the AC motor include ball mills, rotary cement kilns, large crushers, mine winders, and mine hoists. Table 2.2 summarizes some motor drives used in electric motor control.

2.2.4 Stepper Motors

A stepper motor is an incremental electric actuator consisting of several wound fields and a permanent magnet. Each pulse of current input signal received by one of the stator windings generates a rotational motion by a specific number of degrees (fixed angular step). Such incremental motors can be either unipolar with n rotor poles for $360/n$ steps or bipolar. Their

TABLE 2.2

Some Motor Drives and Their Corresponding Electric Motors

Motor Drive Types	Drives	Motor Types
Thyristor based	Six-pulse bridge current	Synchronous motor
	Cycloconverter	Induction motor
		Synchronous motor
	Thyristor voltage controller	Induction motor
Transistor based	Matrix converter	Induction motor
	Current PWM	Induction motor
	Voltage PWM	Induction motor
		Permanent magnet motor

rotors could have as many as 200 poles producing up to 1.8° per step; that is, a stepper motor can move per pulse by 90°, 45°, 18°, or even a fraction of a degree.

Typical stepper motors are permanent magnet, hybrid motor, and variable reluctance motors. A variable reluctance magnet stepper motor has a toothed rotor and a stator winding. Hence, the rotation is induced by the reduction of the magnetic reluctance between the stator poles and rotor poles. The motor step angle is determined by the number of teeth on the rotor, while the excitation type determines the stepping rates. A hybrid stepper motor has two multitoothed armatures covering a cylindrical permanent magnet and defining the motor step angle between 0.9° and 5°. In the case of the permanent magnet stepper motor, the alignment of its stator-generated magnetic field with its permanent magnet ensures its rotation by a step angle between 45° and 120°.

For stepper motors, positioning drives are used to generate pulse signals from the power source. Those motors have a low torque capability and use ON/OFF switch outputs for position control. Each pulse from the command generation system corresponds to an angular positioning, which is defined as a step mechanical increment. The rotating speed varies according to the pulse frequency. The control of this type of motor is related to the sequencing pulse generation system through a personal integrated controller (PIC). Stepper motors can be found in machine tools, typewriters, printers, watches, pointing mechanisms for antennas, mirrors for space applications, telescopes, and so forth.

2.2.4.1 Stepper Motor Modeling

The number of possible steps for each revolution of the rotor S can be estimated:

$$S = 2mN_r \qquad (2.68)$$

with N_r being the number of rotor pole pairs and m representing the stator phases. The angular step or number of degrees a rotor should turn per step is

$$\Delta\phi = \frac{360}{S} \tag{2.69}$$

In the case of an air gap, the magnetic field generates a sinusoidal signal and the motor torque resulting from each phase j, $T_{mj}(t)$, yields

$$T_{mj}(t) = K_m \sin\left[n\phi(t) - \phi_{0j}(t)\right] i_j(t) \tag{2.70}$$

with K_m being the motor constant, $\phi(t)$ the actual rotor position, $\phi_{0j}(t)$ the angular position of coil j in the stator, $i_j(t)$ the current passing over the coil, and n half the number of rotor teeth. The supplied voltage $U_j(t)$ and coil property relationship is given by

$$U_j(t) = E_{bj}(t) + Ri_j(t) + L\frac{di_j(t)}{dt} \tag{2.71}$$

while $E_{bj}(t)$, the emf induced in phase j in each coil, is given by

$$E_{bj}(t) = K_m \sin\left[n\phi(t) - \phi_{0j}(t)\right]\omega(t) \tag{2.72}$$

with $\omega(t)$ being the angular velocity of the rotor, R the coil resistance, and L the coil inductance. The overall torque generated by the stepper motor is given by

$$T_{mj}(t) = \sum_{j=1}^{m} T_{mj}(t) = J\frac{d\omega(t)}{dt} + B\omega(t) + T_f(t) \tag{2.73}$$

The arrangement of the coils has to be considered for more than one phase. An example of the angular position of the coils $\phi_{0j}(t)$ in the stator for various phases m is summarized in Table 2.3.

TABLE 2.3

Coil Location per Number of Stator Phases

Number of phase m	1	2	3
ϕ_{0j} of phase 1	0	0	0
ϕ_{0j} of phase 2	90	60	45
ϕ_{0j} of phase 3	–	120	90
ϕ_{0j} of phase 4	–	–	135

2.3 Electrical Motor Sizing and Selection Procedure

The sizing and selection of an actuating system made of an electric motor and its associated transmission elements consists of (1) finding the electrical power range required to produce the desired process motion and (2) matching it with the existing motor catalog references. The sizing procedure determines the appropriate motor technical characteristics to fulfill load requirements and process operating conditions, such as taking into account a high torque (load increase due to friction) during the start-up, or deriving the motor speed when the generated torque is balanced by the load torque. In order to select the right servomotor, numerous criteria have to be investigated, including the (1) type of power required (DC, AC, number of poles, voltage, and voltage and current levels); (2) deliverable torque range and its resolution under allowable temperature conditions and heat dissipation; (3) torque–speed characteristics; (4) velocity, position range, and accuracy; and (5) full and no-load speed operating characteristics. Overall, the matching is given by the comparison of the motor versus the load torque–speed characteristics. Hence, the motor sizing and selection procedure consists of estimation of the (1) required motor torque, (2) inertia of the attached mechanical transmission structure, and (3) expected motion profile (jerk, acceleration, velocity, and position) required by the operating conditions, all within constraints given by the load torque–speed characteristics. Thus, in order to accomplish the match between the load torque and the estimated motor torque, this procedure should fulfill the following steps:

1. Establish a short list of motor types from
   - Performance requirements for specific applications, such as maximum load torque and velocity profile. For example, a DC motor requires a high starting torque, while for a synchronous motor constant speed is suitable.
   - Environmental factor and safety requirements. For example, in food and pharmaceutical processing, or even aviation, where the environment has to be clean, it may be necessary to encapsulate the DC motor.
   - Energy requirements during operational conditions.
   - Cost of motor and associated accessories.
2. Derive the load torque and desired motion profile (position, velocity, acceleration and even jerk).
3. Select the motor type from the short list obtained, based on desired process type and motion objectives.

4. Compute the motor torque contribution from the rotor inertia of each potential motor being considered for the application, using the derived motion profile.

5. Adjust the motor torque by adding safety factors, in order to cover additional friction forces caused by mechanical transmission components.

6. Integrate motor control drives and accessories based on energy constraints and operating conditions (start and shutdown). This includes taking into account criteria for motion control applications (e.g., energy savings through an intermittent process operating program, cost of ownership evaluation, and reducing downtime).

7. Proceed in the selection of potential motors by matching the torque to the speed curves as follows:

   • Best fit between the required motor torque (including transmission elements) and the load torque

   • Best fit between the estimated motor inertia (including transmission element) and the load inertia

Then, select a motor having at least the required speed. An undersized motor would be unable to displace the load adequately and may face overheating or burning out issues, particularly when the generated heat could not be properly dispersed by the associated ventilation or cooling system. On the other hand, an oversized motor would remain cold and energy would be wasted during inefficient operations. The technical specifications of electric motors are usually presented in the machine nameplate, as illustrated in Table 2.4. Their typical industrial applications are summarized in Table 2.5.

2.3.1 Electric Motor Selection

Industrial applications require either frequent start and stop cycles and reversed or closed-loop positions, a starting torque higher than the running torque, or a transition strategy from the starting torque to the running torque. Based on these requirements, the rationale on motor selection is developed in this section.

2.3.1.1 Electric Motor Operational Conditions (Duty Cycle)

Depending on the industrial processes, electric motors are submitted on either cyclic or random operational conditions, such as timely scheduled machine starts and stops with constant or variable load profiles. Indeed, they can be classified into eight standardized operational conditions, from SS1 to SS8, as specified by the norm NF C51-157 and summarized in Table 2.6.

Here, some motion characteristics specific to industrial applications (e.g., maximum velocity, maximum acceleration, duty cycle time, or dwell time) can be defined. Those characteristics should be determined based

TABLE 2.4

Typical Nameplate of an Electrical Motor

Phase (PH) (1–3)	Voltage (V) 575–115	Horsepower (hp/W)	Physical Size/ Frame (FR)	Rated Speed (rpm)	Frequency. (Hz)	Service Factor (SF)	Nominal Starting Torque	Max. Torque	Protection Index (Thermal, Mechanical, Electrical)	Duty Cycle (dc) Intermittent

TABLE 2.5

Typical Applications of Electric Motors and Their Transmission Elements

Applications	Transmission Systems	Associated Motor Types
1. Elevators	Belt pulley systems	• Induction motor • Motor with ring
2. Grinders, crushers	Belt pulley systems Gear	• Induction motor • Motor with ring
3. Milling machines	Gear	• Synchronous motor • Series-excited DC motor
4. Pumps, compressors, blowers	Gear Crankshaft	• Asynchronous motor • Induction motor • Stepper motor
5. Traction and propulsion instruments	Cylindrical gear Belt pulley systems	• Synchronous motor • Separately excited DC motor
6. Conveyors	Gear Belt pulley systems	• Synchronous motor • Asynchronous motor
7. Machine tools	Gear Belt pulley	• Induction motor • Brushless AC motor • Separately excited DC motor
8. Mixers, vacuum cleaners	Gear	• Motor with ring • Induction motor
9. Medical instruments (e.g., scanners)	Belt pulley Gear	• Stepper motor
10. Compressors	Belt pulley Gear	• Synchronous motor • Asynchronous motor
11. Fans	Gear	• Induction motor • Brushless motor
12. Lifts, cranes	Wheel without end system Belt pulley	• Separately excited DC motor • Synchronous motor • Brushless motor
13. Printers	Gear	• Stepper motor
14. Extrusion train	Gear Wheel without end system	• Induction motor • Motor with ring

on parameters such as velocity and acceleration rate, duty cycle for linear motion within a travel range, type of motion profile (triangular, trapezoidal, or other motion profile), requirements for jerk limitation and thrust load, and characterization (if necessary) of the load change during the duty cycle and application of the holding brake during zero velocity.

2.3.1.2 Motion Profile

A DC motor undergoes a gradual change of angular position, velocity, acceleration, and jerk when it rises or moves to rest. A typical path of the discrete position response is shown in Figure 2.12. In the case of a DC motor shaft ensuring an up-and-down elevator cabin displacement, model

TABLE 2.6

Types of Electrical Motor Operating Conditions

Class	Description of Operating Condition
SS1	Constant load; continuous operation
SS2	Constant load; periods of operation and stop; duty cycles (dc) of 10, 30, 60, and 90 min
SS3	Constant load; periods of operation and stop; identical duty cycles, $dc = (t_f/(t_f + t_a))*100$, where t_f and t_a are operating time and downtime (stop), respectively
SS4	Constant load; periods of start-up, operation, and shutdown/stop; identical duty cycles, $dc = ((t_d + t_f)/(t_d + t_f + t_a))*100$, where t_d, t_f, and t_a are the starting time, operating time, and downtime (stop), respectively
SS5	Constant load; periods of start-up, operation, braking, and stopping; identical duty cycles, $dc = ((t_d + t_f + t_{fr})/(t_d + t_f + t_{fr} + t_a))*100$, where t_d, t_f, t_{fr}, and t_a are the starting time, operating time, braking time, and stopping time, respectively
SS6	Variable or constant load; periods of operating, idling; duty cycle with load, $dc\_load = (t_f/(t_f + t_{fv}))*100$, where t_f and t_{fv} are the operation time and time of idle (preprocessing operation), respectively
SS7	Constant load; frequent reversing of the direction of rotation
SS8	Various constant loads operating at different operating speeds

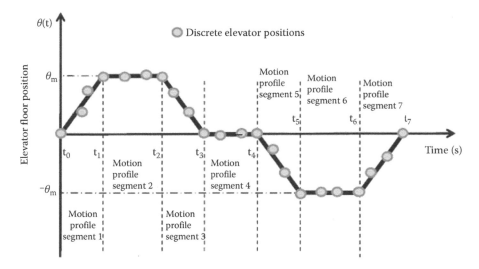

FIGURE 2.12

Linear interpolation of DC motor–driven elevator position profile.

kinematics that are compliant with profile constraints are also expected. Here, the elevator motion between two or more successive floors has the following constraints: smooth and comfortable elevator ride, travel time as short as possible, and if possible, good positioning accuracy. Usually, such command input of a motion profile is represented by discrete values over time. Then, using an interpolation or curve-fitting methods, it is possible to

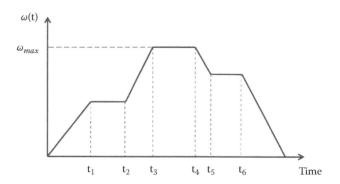

FIGURE 2.13
Generic motion profile.

derive the functions corresponding to the best curve fitting those sequences of data points, as illustrated in Figure 2.12. A typical up-and-down position profile can be decomposed into motion profile segments. As an illustrative example, the motion profile in Figure 2.12 can be depicted in seven segments within three phases: (1) acceleration phase from rest to maximum velocity, consisting of ramping up, holding, and ramping back; (2) constant-speed phase; and (3) acceleration phase from rest to maximum velocity, consisting of ramping up, holding, and ramping back.

With operating conditions such that $a_{mean} = \frac{1}{2}a(t)$ for a ramp-like accelerating and decelerating phase, $a_{mean} = a_{max}$ during the constant-speed phase, $\omega(0)$ is the starting and final position zero velocity, and $\omega_{max} = \omega_m$. It should be recalled that segments t_4, t_5, t_6 and t_7 have the same equations with a negative sign. The rated velocity is given by

$$\omega_{nom} = 2\pi N_M \qquad (2.74)$$

with N_M being the rated speed on the nameplate of the motor. For some applications, the motor speed is constant, while for others it varies by level. Hence, regardless of the applications, the velocity profile is very useful to derive the required motor torque, as the load torque is associated with the dynamic torque. It is possible to distinguish several types of generic motion profiles, as illustrated in Figure 2.13.

2.3.1.3 Load Torque Calculation

The load torque, $T_L(t)$, is the torque required to move the attached mechanical transmission elements and the involved load. This torque can be derived using the estimation of inertia generated by the attached mechanical transmission elements, including the involved load to the motor, $J_{L \rightarrow M}$, and the its motion profile (acceleration or velocity) estimation. The procedure to compute load torque consists of

1. Deriving the inertia of all moving components, J_L
2. Determining the inertia reflected to the motor, $J_{L \to M}$
3. From the motion profile, deriving the velocity, $\omega(t)$, and the acceleration (or deceleration) at the motor shaft, $\alpha(t)$
4. Calculating the acceleration torque at the motor shaft, $T_{acc}(t)$
5. Estimating other external forces, $F_{ext}(t)$, such as gravity, magnetic, and contact
6. Computing their equivalent required torque
7. Adding a computing safety factor of about 10%
8. Computing the torque required to meet the motion profile (velocity and acceleration), such as

$$T_{max_{motor}} \geq T_{acc} \tag{2.75}$$

Example 2.1

Consider torque and speed profiles that are varying with load. In the process to select motor torque and power ratings, it is suitable to evaluate the effective electromagnetic torque developed using the load profile derived from the load cycle (less the rest period from Figure 2.14), such as

$$T_{load\_eff} = \sqrt{\frac{\sum T_i^2 \Delta t_i}{\sum \Delta t_i}} = \sqrt{\frac{T_1^2 t_a + T_2^2 (t_b - t_a) + T_3^2 (t_c - t_b)}{t_a + (t_b - t_a) + (t_c - t_b)}} \tag{2.76}$$

This is different from the average load torque, T_{Lmax}, or the acceleration torque, T_{acc}, which should both be considered.

2.3.1.4 Motor Shaft Torque Calculation

From the nameplate, the steady-state operations or values rated, such as power, P_{nom} (in hp), and speed, N (in rpm), are related as follows:

$$P_{nom} = T_{L \to M} \omega_{nom} \tag{2.77}$$

where $T_{L \to M}$ is the shaft torque, while the rated speed is

$$\omega_{nom} = \frac{2\pi N}{60} \left(\text{in } \frac{\text{rad}}{\text{s}} \right) \tag{2.78}$$

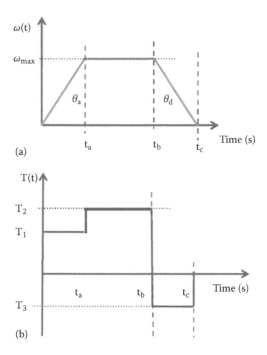

FIGURE 2.14
(a) Speed motion profile. (b) Load torque–operating cycle curve.

Thus, the motor torque developed is given by

$$T_m(t) = T_{L \to M} + T_{friction} = \frac{P_{nom} \times 745.6}{\omega_{nom}} + B\omega_{nom} \text{ (in N} \cdot \text{m)} \tag{2.79}$$

with B being the friction coefficient. If the discrete motion profile and other motor parameters are known, the acceleration torque is given by

$$T_{acc}(t) = J_{total} \frac{d\omega(t)}{dt} \tag{2.80}$$

If the motion profile is trapezoidal or if the acceleration time period is known, using Equation 2.79, the speed variation is such that

$$\alpha_a(t) = \frac{d\omega(t)}{dt} = \frac{\omega_{max}}{t_a} \tag{2.81}$$

Then, the acceleration torque for this time period is given by

$$T_{acc\_1}(t) = J_T \frac{\omega_{max}}{t_a} \tag{2.82}$$

And during the deceleration time period, the speed variation is such that

$$\alpha_d(t) = \frac{d\omega(t)}{dt} = \frac{\omega_{max}}{t_f} \tag{2.83}$$

Then, the deceleration torque for this time period is given by

$$T_{acc\_2}(t) = J_T \frac{\omega_{max}}{t_f(t_c - t_b)} \tag{2.84}$$

Thus, the required motor torque is

$$T_{acc}(t) = max\left\{T_{acc_1}(t),\ T_{acc\_2}(t)\right\} \tag{2.85}$$

The maximum torque is given by

$$T_{motor\_max}(t) = T_{Lmax}(t) + T_{acc}(t) \tag{2.86}$$

2.3.1.4.1 Procedure of Electrical Motor Selection and Sizing

The selected electric motor (P_{motor}, $N_{n_{motor}}$) has to fulfill the following conditions:

$$\begin{cases} P_{motor} = P_{nom} \\ N_{n_{motor}} = N_M \\ T_{motor} \geq T_{L \to M} \end{cases} \tag{2.87}$$

The thermal rating (rated power losses) has to be estimated through the effective motor losses, such as estimation of the effective current for DC motors or the applied voltage for AC motors for load cycle. Then, a safety factor should be applied to ensure that the motor is thermally suitable for operations.

2.3.1.5 Load Torque–Speed Profile Characteristics

The electric motor should operate approximately as described by load torque profiles. Such resistive torques are proportional to mechanical loads, causing variations of speed–torque characteristics. Hence, load torques reported to the motor shaft, $T_{L \to M}(t)$, can be represented by the equation

$$T_{L \to M}(t) = kT_{r0}(t) + \beta_1 \omega(t)^n \qquad (2.88)$$

with k being the proportionality constant, $T_{r0}(t)$ the initial friction torque, and the exponent n characterizing the variation of load torque with speed, $\omega(t)$, and the constant, β_1. Based on the characteristics of the mechanical load, it is possible to derive the exponent n as summarized in Table 2.7 and illustrated in Figure 2.15.

TABLE 2.7

Load Torque Exponent Values

Load Torque Description	Exponent, n
Load torque is not a function of speed (constant load)	$n = 0$
Load torque is proportionally linear to the speed	$n = 1$
Load torque varies proportionally to the square of mechanical speed	$n = 2$
Load torque is inversely proportional to speed	$(n \leq -1)$

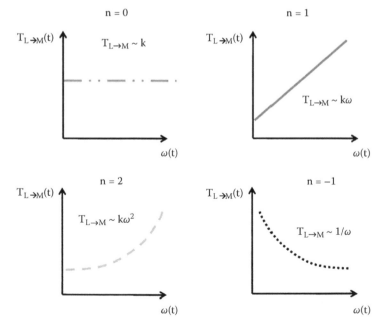

FIGURE 2.15
Typical load torque–speed characteristics.

Hence, the motor torque, $T_m(t)$, yields

$$T_m(t) - T_{L \to M}(t) = J\frac{d\omega(t)}{dt} + B\omega(t) \tag{2.89}$$

with $T_m(t)$ and $T_{L \to M}(t)$ being the motor torque and load torque in (N.m), respectively. During motor operations, there is acceleration when $T_m(t) - T_{L \to M}(t) > 0$ and deceleration when $T_m(t) - T_{L \to M}(t) < 0$, and the motor is moving at constant speed when $T_m(t) - T_{L \to M}(t) = 0$. For example, during the deceleration phase, the motor supplies the load torque and the dynamic torque, and $J\frac{d\omega(t)}{dt} + B\omega(t)$ then has a negative sign.

2.3.1.5.1 Four-Quadrant Operation

Furthermore, the comparative analysis of the motor speed direction with respect to motor torque allows classification of the motor operating modes. Hence, the directional change in armature current flows induced the DC motor shifts between the motoring mode and the generating mode. When it is the case of the power flowing from the motor armature to the DC source, this is called regenerative braking, with the generation of a negative torque as illustrated in Figure 2.16. It has a mirror reflection of the classical torque–speed curve. The first and fourth quadrants have the same forward rotational direction. For a reverse direction of rotation, the third quadrant corresponds to the reverse motoring while the second quadrant refers to the reverse regeneration mode. Based on the load torque–speed profiles of electric motors, their corresponding applications are summarized in Table 2.8.

2.3.1.6 Matching Motor and Load Speed–Torque Curves

Once motor sizing has been achieved based on the determination of the motor torque–speed curve characteristics, the operating point can be estimated graphically, as illustrated in Figure 2.17. Indeed, when a motor is connected to drive a load, the interaction of the load torque with the motor torque determines the point of operation.

Furthermore, initially the motor torque is higher than the required load torque, forcing the motor rotation. Once the motor starts its rotation and increases it speed, the resulting motor torque decreases. The motor finally comes to a stable operating point when the two torques balance each other. Eventually, the motor operating point is derived from the intersection of the motor and load torque–speed curves.

2.3.1.7 DC Motor Parameter Estimation

DC motors are commonly used to drive industrial processes, and their steady-state parameters are highly dependent on the motor mechanical and

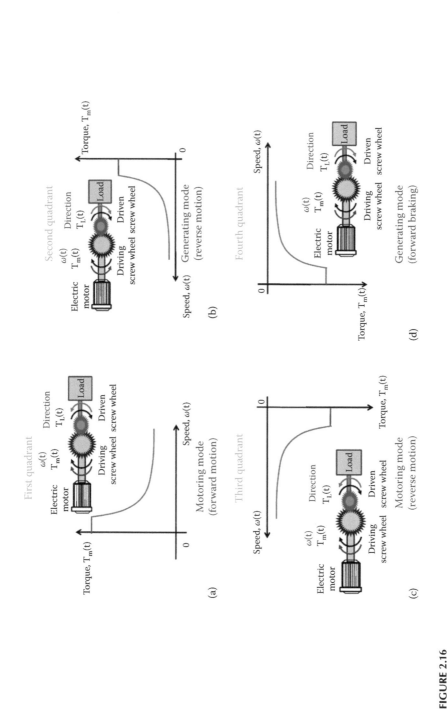

FIGURE 2.16
(a) First of four quadrant operation and torque–speed characteristics. (b) Second of four quadrant operation and torque–speed characteristics. (c) Third of four quadrant operation and torque–speed characteristics. (d) Fourth of four quadrant operation and torque–speed characteristics.

TABLE 2.8

Examples of Load Torque–Speed Profiles of Electric Motors

Industrial Applications	Load Torque Profile Types
Conveyors, displacement pumps, compressors	$T_{L\to M}(t) \propto k$
Pumps (centrifugal), fans	$T_{L\to M}(t) \propto k\,\omega(t)^2$
Lathes	$T_{L\to M}(t) \propto 1/\omega(t)$
Mills, cranes, sawmills, crushers/presses,	Periodic and generic

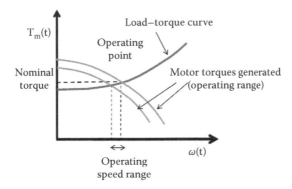

FIGURE 2.17
Interaction between generated motor torque and load torque.

electrical design characteristics. Each motor has its unique values empirically determined. Hence, when designing the position, speed, and current controllers for any electric motor–driven process, it is suitable to have an accurate estimation of the parameters involved, in order to model the dynamics. In this section, estimation techniques for DC motor parameters and process variables are presented and discussed with a theoretical background.

2.3.1.7.1 Estimation of Motor Amplifier Gain, K_a (V/V)

The gain is defined as the ratio of the voltages across the amplifier, and is derived in a straightforward manner by measuring the voltage at the amplifier input and at the output. However, this is only true if signal offsets (biases) exist. Several voltage values could be taken at the amplifier input and output. Then, plotting the voltage readings, such as the input voltage on the *x*-axis and the output voltage on the *y*-axis, would enable us to determine the amplifier gain, which corresponds to the slope of the graph. This could be achieved by using the least-squares error curve fit, which allows capturing significant signal offsets. However, the gain differs when the motor is loaded versus when it is not loaded. To avoid a nonlinearity display around the 0 V range, it is suitable to use fast Fourier transform (FFT) methods to estimate the frequency response function (FRF) and determine the bandwidth limitations of the amplifier.

2.3.1.7.2 Estimation of Current Sensor Gain, K_i (V/A)

The current that travels through the armature motor goes into a current sensor. A voltage is generated proportional to the current passing through the sensor and sent to an output device (e.g., a spectrum analyzer or oscilloscope). Therefore, a gain value associated with the sensor relates the two variables (current and voltage). This value was found by measuring the current and the voltage simultaneously for many different values, using least square estimation (LSE) curve-fitting methods.

2.3.1.7.3 Estimation of Back emf Constant, K_e (V·s/rad)

The rotation of the shaft in a DC motor induces a back emf voltage. The back emf is related to the speed of the motor through a constant known simply as the back emf constant. This value is considered a gain, and it is the ratio of the back emf to the shaft speed. From the measurement through an oscilloscope of the back emf voltages within a motor speed range, this constant can be estimated using LSE curve-fitting methods.

2.3.1.7.4 Estimation of Torque Constant, K_t (N·m/A)

The magnetic torque applied to the motor shaft is generated from the current passing through the coils surrounding the shaft. The torque and current are assumed to be proportional to each other. Therefore, the gain associated with these two variables is calculated by measuring the current passing through the motor and the corresponding torque applied to the shaft. Those measures are obtained by using a multimeter for the current measurement and a device holding the shaft in place for the calculation of the required corresponding torque. After several measurements, the torque constant can be found from the ratio and the LSE curve-fitting methods. Alternative methods could be FFT and FRF on signals gathered through a spectrum analyzer.

2.3.1.7.5 Estimation of Tachometer Gain, K_v (V·s/rad)

The motor shaft acts as a generator of a voltage signal proportional to the angular speed of the shaft. The voltage generated is measured by a tachometer device. Assuming the two variables are proportional to one another, a gain value can be calculated by measuring the speed of the shaft by another device having the same calibration as the tachometer. Then, the voltage from the tachometer is recorded with a multimeter. As usual, the gain value is the ratio of the voltage over the shaft speed using the LSE method for averaging any offset in signal measurements.

2.3.1.7.6 Estimation of the Armature Resistance, R_a (Ω)

Using a PWM servo amplifier or a DC power supply, a current is applied through the motor armature. Hence, values of the current, as well as corresponding voltage drops across the motor armature, are recorded through a static measurement method used to determine the resistance value, R_a. Thus, the armature resistance is derived from Ohm's law by using the LSE curve fit to average any effect of signal offset. An ohmmeter can be used for the winding resistance.

2.3.1.7.7 Estimation of the Armature Inductance, L_a (H)

The inductance of the motor armature, L_a, can be derived from its frequency response plot. On a semilog plot, the graph is nearly two straight lines. At low frequencies, the graph is a horizontal line with a constant gain value of $1/R_a$. The break frequency is found at the point where the amplitude falls to −3 dB. Beyond this point, the graph is a line sloping down. The inductance can be derived from this slope.

Another method to estimate R_a and L_a would consist of coupling two motors (test bed motor and load motor) and varying input voltages, $V_a(t)$, on both motors until $T_{test\text{-}bed}(t) = T_{Load}(t)$. Depending on the DC motor type, the speed, $\omega_m(t)$, over the current, $i_a(t)$, curve can be drawn for each recorded voltage, $V_a(t)$. For a permanent magnet DC motor, using the LSE method would result in a straight line with the slope d and the point intersecting the speed axis c such that

$$\omega_m(t) = c - di_a(t) = \frac{V_a(t)}{K} - i_a(t)\frac{R_a}{K} \tag{2.90}$$

Thus,

$$K = \frac{V_a(t)}{c}$$

$$R_a = Kd$$

The inductance, L_a, is defined as the transient characteristics of the resistance, R_a, to the variation of the current, $i_a(t)$, passing over the motor armature. By blocking the rotor of the motor, and applying a step voltage input to the armature winding, an exponential response of the current is recorded, as illustrated in Figure 2.18.

It can be represented by

$$i_a(t) = \frac{V_a}{R_a}\left(1 - e^{-\frac{t}{\frac{L_a}{R_a}}}\right)$$

This is valid only if at $t = 0$,

$$L_a = \frac{V_a}{\dfrac{I_{ass}}{\Delta t}}$$

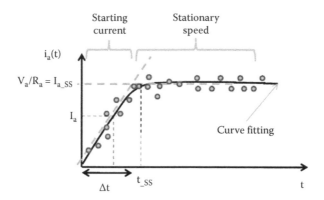

FIGURE 2.18
Typical permanent magnet DC motor armature current recorded data.

2.3.1.7.8 Estimation of Combined Motor Shaft of Inertia J
$\left(N \cdot m \cdot \dfrac{s^2}{rad} \right)$; *Motor Damping Viscous Friction,*
$b_m[N \cdot m \cdot s/rad]$; *and Static Friction (Stiction), $T_T[N \cdot m]$*

Consider an input ramp DC motor velocity resulting in a square-wave shape of the motor shaft acceleration. Hence, the motor inertia, J, is the motor resistance to change into the motor velocity, $\omega(t)$. As such, near motor start-up, the motor moment of inertia is

$$J = \frac{\omega(t \to 0)b_m}{\Delta t} \tag{2.91}$$

In the case of an elevator, as in Section 2.3.1.2, the combined motor inertia is equivalent to

$$J = J_{motor} + \frac{1}{4}r_{dr}^2 M_{cabine} + J_{drum} + J_{gear} \tag{2.92}$$

Thus, plotting the torque and the shaft speed against each other gives the graph drawn in Figure 2.19. Here, the acceleration can be assumed to be jumping between two constant values. And the static (coulomb) friction, the damping coefficient, and even the shaft inertia can be deduced from the resulting graph. It is considered that once the shaft is set into motion, it moves at a constant acceleration until it changes its direction. Since the parameters are sought for the components of a linear system, it is assumed that the damping coefficient is constant. From experimental data,

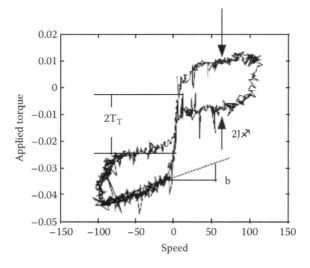

FIGURE 2.19
Typical FFT response of a DC motor.

the torque balance on the shaft gives the following ordinary differential equation:

$$T = sign(\omega(t))T_T + J\frac{d\omega(t)}{dt} + b\omega(t) \tag{2.93}$$

Hence, during the time that acceleration is constant, the slope of the $T_{airgap}\omega$ line will be equal to the damping coefficient (T_T and J terms being constant). Note that the LSE curve fitting could be used to estimate coulomb friction, viscous drag, and to a limited extent, inertia.

2.3.2 Process Dynamics Particularities

Braking electric motor dynamics consists of moving the motor from its operating speed of $\omega(t)$ to zero. It can be done either by cutting off the armature voltage supply to the motor or by transforming the electric motor into a DC generator through a braking system. Commonly used motor braking systems are mechanical (friction) braking, dynamic braking. While cutting off the armature voltage supply could rapidly force the motor speed to zero, the use of a mechanical brake system adds inertia to the motor shaft, thus increasing the motor torque and power required during acceleration and deceleration. The braking effect on the load torque can be important in the case of vertical motion. In addition, the braking system avoids the activation of the engine brake torque during the pause or idling in vertical movement. The dynamic

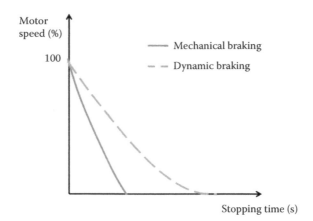

FIGURE 2.20
Speed versus time curves for various braking methods.

braking mode with DC motors brings the motor speed smoothly down to zero, which is immediately followed by acceleration in the opposite direction, or another method consists of modifying two modifies two stator leads, causing the rotor to move in the opposite direction of the magnetic field rotation. Related dynamic braking speed curves are illustrated in Figure 2.20.

In order to simplify the electric motor modeling, it is assumed that (1) the braking system inertia is negligible compared with the load inertia, and (2) the operating conditions include several stop periods. Thus, based on the torque requirements (peak or intermittent), it is possible to select the motor with the smallest torque for any combination of motor and braking system.

2.3.2.1 Summary of Electric Motor Sizing and Selection Procedure

Motor sizing and selection is required to compute the load torque reflected at the motor shaft. Then, the motor performance with regard to its ability to support the load torque and speed requirements has to be checked. If the expected motor performance is not met, another motor or drive should be selected. Motors should not operate over a ratio of 10:1. Hence, when the load inertia is greater than 10 times the rotor, a mechanical transmission element, such as a gearbox, has to be used. Furthermore, the motor torque should be greater than or equal to the load torque at the motor shaft. For safety, a coefficient of 1.05 could be added. Table 2.9 summarizes the equations used to estimate the required motor torque and power train.

2.3.3 Electric Motor Interface and Accessories

Components of electrical-driven actuating systems are (1) the power supply, (2) the power converter, (3) the command and control system with its

TABLE 2.9

Summary of Motor Sizing and Selection Procedure

Procedure Steps	Formula or Method Used	Results
Identify the suitable motor type for the considered industrial applications	See Table 2.6	Type of motor group and power train system
Derive the load torque at the motor shaft, $T_{L \to M}$		$T_{L \to M} = \dfrac{T_L}{N_r \eta}$
Derive the total system inertia, J_{total}	See Table 2.12	$J_{total} = J_M + J_B + J_{B \to M} + J_{L \to M}$
Derive motion profile (speed) and nominal speed, ω_{nom}	Equations 2.76 through 2.86	Using $t_1, t_2, \ldots, t_k$ and t_d, compute $\omega_{nom} = 2\pi N_M$
Derive the acceleration torque, $T_{acc}(t)$	Equations 2.89 through 2.97	$T_{acc}(t) = max\left\{ T_{acc_1}(t), T_{acc\_2}(t) \right\}$ $T_{acc}(t) = max\left\{ J_{total}\dfrac{\omega_{max}}{t_a} ; J_{total}\dfrac{\omega_{max}}{t_d} \right\}$
Derive maximum torque	Equation 2.98	$T_{max} = T_{L\_max} + T_{acc}$
Derive nominal power train	Equation 2.99	$P_{nom} = T_{L \to M}\, \omega_{nom}$

corresponding sensors and detectors, (4) the electric motor or electrical-driven actuating component, and (5) the associated transmission system to the load.

2.3.3.1 Power Source

Typical power sources encountered are AC, single phase or three phase, 50 or 60 Hz, 240/415 V, 220/380 V, 120/90 V, 11 kV/415 V, 3 V direct current (VDC), 12 VDC, and so forth. In addition to the power source, power converter and signal conditioning are the main accessories associated with any electric motor.

2.3.3.2 Power Converter

Power electronic converters are used in the design of variable speed motor drives through phase modulation block, PWM, or even space vector modulation (SVM). Some of them have been described in the section related to the speed control of (AC and DC) electric motors. Using PWM for the design of variable speed motor drives consists of varying the time duration (width) of the pulse to change the average RMS value of a the command input signal. This is achieved by switching the fundamental frequency of PWM through an adjustable speed of the switching. Among PWM commonly used design techniques to vary the RMS voltage magnitude of the output signal there is

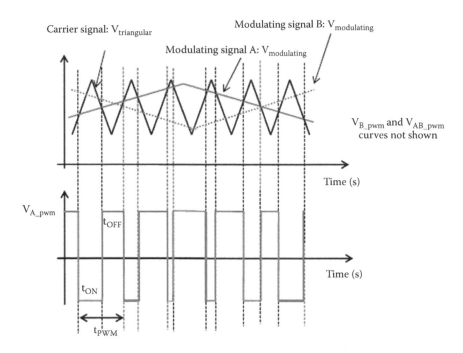

FIGURE 2.21
Generation of PWM signal by comparing carrier and modulating inputs signals.

the sine-triangle PWM, where the switching is done at the intersection of a modulating signal and a high-frequency triangular carrier signal, as illustrated in Figure 2.21 (V_{B_pwm} and V_{AB_pwm} processed signals are not shown). Note that the modulating signal could be a periodic signal, a generic (asymmetrical) signal, or even a straight ramp-like line. Furthermore, this signal can be presampled and held before comparing it with the carrier signal. Depending on the sampling strategy, the resulting sine-wave signals could vary between positive and a zero value or between negative and positive values. In the case of a three-phase AC signal, the switching simultaneously varies the width for all three phases of the AC signal, making it possible to adjust the output voltage through its frequency. This is suitable for realizing constant volts per hertz operation. Naturally sampled PWM faces complexity for its digital modulation system implementation. Another design technique is SVM. Overall, using Figure 2.21, the PWM duty cycle is given by

$$duty\ cycle[\%] = \frac{t_{ON}}{t_{pwm}} 100\% \qquad (2.94)$$

Considering a triangular carrier signal oscillating between –1.0 and +1.0 voltage scale units while the associated modulating wave magnitude is kept at 0.4 voltage scale unit, using Fourier series, it is possible to derive the DC

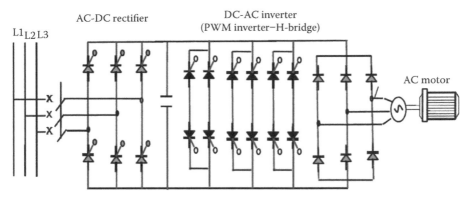

AC-DC rectifier   DC-AC inverter
(PWM inverter–H-bridge)

L1 L2 L3

AC motor

3-phase AC   Constant DC voltage   Variable voltage/frequency
  voltage   magnitude

FIGURE 2.22
Thyristor-based PWM inverter (full bridge) drive for three-phase AC motor.

mean magnitude voltage signal. Considering t_{ON} as the pulses time duration during which the voltage magnitude is at high level (ON), it is given by

$$t_{ON} = \frac{t_{carrier}}{2}\left(1 + \frac{V_{modulating}}{V_{max\_carrier\_pos}}\right) \tag{2.95}$$

with $t_{carrier}$ being the time period of the triangular carrier signal, and $V_{modulating}$, $V_{max\_carrier\_pos}$ being, respectively, the modulating signal magnitude and the maximum carrier signal magnitude (must be positive). Hence, t_{OFF} is the duration of the pulses during which the voltage magnitude is at low level (OFF) and is given by

$$t_{OFF} = \frac{t_{carrier}}{2}\left(1 - \frac{V_{modulating}}{V_{max\_carrier\_pos}}\right) \tag{2.96}$$

The DC magnitude voltage V_A supply to the motor is given by

$$V_A = V_{a\_pwm}\frac{V_{modulating}}{V_{max\_carrier\_pos}} \tag{2.97}$$

Among power electronic devices used to design solid-state sine-wave PWM inverters there are silicon-controlled rectifiers (SCRs) or thyristors, transistors, gate turnoff thyristors (GTOs), and bipolar junction transistors (BJTs). They are selected based on the desired modulation frequency of the PWM. An example of a PWM inverter used to drive a three-phase AC motor is illustrated in Figure 2.22. First, a controlled rectifier converts an input AC voltage signal into a constant-voltage DC using either an SCR based H-bridge or a diode bridge with a DC chopper. This DC signal transformation is followed by a DC three-phase variable-voltage AC inverter consisting of switching transistors or thyristors in six steps to control the AC signal frequency. Usually, a PWM inverter delivers power between 500 and a few thousand horsepower. The nonlinear transformation of the input DC voltage magnitude into the time duration of a pulse is assumed to be a linear operation for motor control purposes. Here, the linear model of the PWM amplifier is a mean voltage equal to the integral of the voltage signal, such as

$$V_{Supply} t_{ON} = V_A t_{PWM} \qquad (2.98)$$

with V_{Supply}, t_{ON}, V_A, t_{PWM} being, respectively, the supply voltage, pulse time duration, average voltage applied into the motor, and switching period. The switching period (t_{PWM}) should be selected to correspond to at least 10 times the motor bandwidth. The duty cycle, t_{ON}/t_{PWM}, must be updated at each sampling instant. Thus, the voltage of a such motor, V_{EQ} H-bridge is dependent on the duty cycle times and supply voltage.

The selection of a power converter is dependent on the motor involved. If a DC motor is used and the power source has an AC nature, the converter should be a rectifier. In the case of a DC source, the converter has to be a chopper. Furthermore, AC variable speed drives, such as those constructed using PWM-type variable v/f converters, are typical speed controllers of induction motors (e.g., squirrel cage induction motors). The compatibility of the AC converter supply power and the expected signals (voltage, current, and frequency) has to be checked based on the motor operating constraints under various load and environmental conditions (humidity, water, temperature, chemicals, dust, etc.). The power converters should be selected in order to improve energy-related efficiency, especially with respect to (1) the type of industrial application and the starting torque operating constraints (benchmarking motor starting current, hence limiting the starting torque during the acceleration phase); (2) the smoothing of the motor start-up; (3) the speed range and variation from zero up to the maximum speed, as well as the motion profile requirements, including acceleration, deceleration, and braking phases, bidirectional or not; (4) the speed control accuracy and dynamic response (speed and torque response) requirements; (5) the torque and power requirements and the torque–speed characteristics; and (6) the ventilation and cooling requirements for the motor and corresponding converter.

2.4 Modeling of Mechanical Transmission Elements

Mechanical transmission elements (e.g., gearbox, pulley, and screw) connected to electrical-driven motors allow us to transfer their rotational speed into the attached load. As such, it is necessary to derive the corresponding load torque model and its inertia at the level of the motor shaft. Hence, the rotor shaft speed, $\omega_m(t)$, for the time interval (t_a, t_c, t_d, etc....), should be estimated from the motion profile. Then, the load torque and required motor shaft speed are

$$T_{req}(t) = J_{total}\frac{d\omega_m(t)}{dt} + B\omega_m(t) \tag{2.99}$$

$$T_{L \to M}(t) = \frac{T_L(t)}{\eta N_r} \tag{2.100}$$

Then, the total torque required is given by

$$T_{m\_max}(t) = T_{req}(t) + T_{L \to M}(t) \tag{2.101}$$

$$\omega_m(t) = N_r \omega_L(t) \tag{2.102}$$

A safety torque, including particular starting and braking torque, should be added. Table 2.10 presents a generic torque equation for some typical mechanical transmission elements, while Table 2.11 lists all parameters used.

TABLE 2.10

Dynamical Equations on Motion and Torque for Some Mechanical Transmission Systems

Mechanical Transmission Types	Equations	
	Motion	Inertia and Torque
Belt pulley (timing belt)		$J_{total} = J_{PL \to M} + J_{L \to M} + J_M + J_{PM}$
		derived from
		$J_{PL \to M} = \left(\frac{1}{N_r}\right)^2 \frac{J_{PL}}{\eta}$
		$J_{PL} = \frac{m_{PL}D_{PL}^2}{8}$
$N_r = \frac{N_{PL}}{N_{PM}} = \frac{D_{PL}}{D_{PM}}$		$J_{L \to M} = \left(\frac{1}{N_r}\right)^2 \frac{J_L}{\eta}$
$\omega_m(t) = N_r \omega_L(t)$		$J_{PM} = \frac{m_{PM}D_{PM}^2}{8}$

(Continued)

TABLE 2.10 (CONTINUED)

Dynamical Equations on Motion and Torque for Some Mechanical Transmission Systems

	Equations	
Mechanical Transmission Types	**Motion**	**Inertia and Torque**

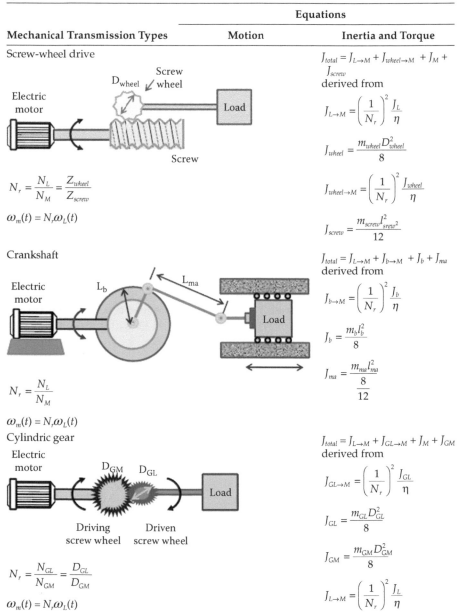

Screw-wheel drive

$$N_r = \frac{N_L}{N_M} = \frac{Z_{wheel}}{Z_{screw}}$$

$$\omega_m(t) = N_r\omega_L(t)$$

$$J_{total} = J_{L\to M} + J_{wheel\to M} + J_M + J_{screw}$$

derived from

$$J_{L\to M} = \left(\frac{1}{N_r}\right)^2 \frac{J_L}{\eta}$$

$$J_{wheel} = \frac{m_{wheel}D^2_{wheel}}{8}$$

$$J_{wheel\to M} = \left(\frac{1}{N_r}\right)^2 \frac{J_{wheel}}{\eta}$$

$$J_{screw} = \frac{m_{screw}l^2_{srew}}{12}$$

Crankshaft

$$N_r = \frac{N_L}{N_M}$$

$$\omega_m(t) = N_r\omega_L(t)$$

$$J_{total} = J_{L\to M} + J_{b\to M} + J_b + J_{ma}$$

derived from

$$J_{b\to M} = \left(\frac{1}{N_r}\right)^2 \frac{J_b}{\eta}$$

$$J_b = \frac{m_b l^2_b}{8}$$

$$J_{ma} = \frac{m_{ma}l^2_{ma}}{8}$$
$$\frac{}{12}$$

Cylindric gear

$$N_r = \frac{N_{GL}}{N_{GM}} = \frac{D_{GL}}{D_{GM}}$$

$$\omega_m(t) = N_r\omega_L(t)$$

$$J_{total} = J_{L\to M} + J_{GL\to M} + J_M + J_{GM}$$

derived from

$$J_{GL\to M} = \left(\frac{1}{N_r}\right)^2 \frac{J_{GL}}{\eta}$$

$$J_{GL} = \frac{m_{GL}D^2_{GL}}{8}$$

$$J_{GM} = \frac{m_{GM}D^2_{GM}}{8}$$

$$J_{L\to M} = \left(\frac{1}{N_r}\right)^2 \frac{J_L}{\eta}$$

TABLE 2.11

Listing of Parameters and Variables Used

Symbol	Definition	Unit
J_M	Motor inertia	$kg \cdot m^2$
J_L	Load inertia	
J_{PL}	Passive pulley inertia	
J_{PM}	Active pulley inertia	
$J_{PL \to M}$	Passive pulley to motor inertia	
$J_{L \to M}$	Load to motor inertia	
J_{GL}	Passive gear wheel inertia	
J_{GM}	Active gear wheel inertia	
J_{ma}	Crankshaft follower inertia	
J_b	Crankshaft cam inertia	
$J_{ma \to M}$	Crankshaft to motor inertia	
J_{wheel}	Wheel inertia	
$J_{wheel \to M}$	Wheel to motor inertia	
J_{screw}	Screw inertia	
J_{total}	Total inertia	
L_b	Lenght of crankshaft cam (m)	
L_{ma}	Lenght of crankshaft follower (m)	
m_b	Mass of crankshaft	kg
m_{screw}	Mass of screw	
m_{PL}	Mass of passive pulley	
m_{PM}	Mass of active pulley	
m_{wheel}	Mass of wheel	
m_{ma}	Mass of crankshaft follower	
η	Efficiency	No unit
N_r	Power transmission ratio	
N_M	Motor rotational speed	$rev \cdot min^{-1}$
N_L	Load rotational speed	
N_{PL}	Passive pulley rotational speed	
N_{PM}	Active pulley rotational speed	
N_{screw}	Screw rotational speed	
N_{wheel}	Wheel rotational speed	
θ_M	Motor shaft angular position	rad
θ_L	Load angular position	
$\omega_m(t)$	Motor shaft angular velocity	$rad \cdot s^{-1}$
$\omega_L(t)$	Load angular velocity	
D_{PL}	Diameter of passive wheel	m
D_{PM}	Diameter of active wheel	
$T_{L \to M}(t)$	Load to motor torque	$N \cdot m$
$T_L(t), T_m(t)$	Load torque, motor torque	
Z_{screw}	Number of screw filets	No unit
Z_{wheel}	Number of wheel teeth	

2.5 Modeling of Electrofluidic Transmission Elements

Electrical-driven fluidic transmission systems, such as pumps, fans, blowers, and compressors, raise the mechanical energy of a fluid, causing an increase of the flow rate and pressure, or an elevation of the fluid. Fluidic transmission elements generally have as input the volumetric flow rate and as output the pressure difference. The categorization of fluidic transmission systems is associated with the type of fluid, the flow rate, and the pressure changes required. While typical fans are used to move gas-based fluid, because they require a high volume and low pressure differential, blowers are suitable for similar volumes but a moderate pressure differential. In the case of fluid requiring a large pressure differential, compressors are used. Usually, fluidic transmission elements offer a higher power-to-weight ratio, higher speed and acceleration responses, and easier direction change than mechanical elements for material handling and transport applications (automotive, aerospace, etc.). Fluids can be characterized by density (mass per unit of volume), viscosity (resistance to deformation), and bulk modulus (compressibility of fluid).

2.5.1 Electric Pumps

Electric pumps are associated with motors to transform electrical energy into potential hydraulic energy by raising the fluid pressure. This is achieved through the reduction of the volume of the inlet-to-outlet pump port travel. Depending on whether the fluid flow can vary, it is either a fixed-flow gear pump or a variable-flow pump. Hence, a pump acts as a fluid flow generator and is usually associated with an electric valve for open directional control, in order to deliver velocity to the fluid and move the load. The volume of fluid being transferred in one revolution is called the pump displacement, D. This pump displacement and the maximum rotational speed, ω_{max}, determine the pump capacity, Q, such as

$$Q = D\omega_{max} \tag{2.103}$$

Typical pumps are either positive displacement pumps or centrifugal pumps. *Positive displacement pumps* are either rotary type or reciprocating type. These pumps generate the fluid motion by mechanically displacing segmented fluid through a discharged nozzle. Reciprocating pumps move the fluid by varying the fluid pressure using a diaphragm. In contrast, a rotary pump adds kinetic energy to the fluid by raising the flow rate, which in turn raises the fluid pressure as it exits the discharged nozzle. Rotary pumps can use other devices to perform fluid compression, including gears, lobes, and screws. Other displacement pumps are volumetric piston pumps, which have either an axial or a radial configuration corresponding to one or more cylinders, with a piston sliding in each of them. Here, the piston displacement determines the fluid flow. *Centrifugal pumps* use a rotating impeller to increase the fluid pressure and velocity. Here, the center of the rotating impeller receives the inflow of fluid, which in turn is mechanically accelerated consequently to its rotation before it leaves through the side of the pump.

The operational characteristics of a pump can be obtained by plotting the curves of the head (H), the power (P), and the efficiency of the pump, according to the flow rate (Q) for a number of constant velocities N, as illustrated in Figure 2.23. Note that the efficiency takes a maximum value and then suffers a reduction. The ideal condition of the operation is achieved when the total head and flow rate have the same values as the maximum efficiency. This point is considered the operating point.

The choice between those pumps depends on the fluid type to be pumped, as well as the expected head and flow rate. The pumps can be set in series such that the overall head is the sum of each pump flow rate. When connected in parallel, the flow rate is increased by adding the total head of each pump working by itself. In this case, the curve H versus Q can be obtained by adding the flow rates of each pump operating by itself with the same total head. The affinity laws or fan laws are as follows.

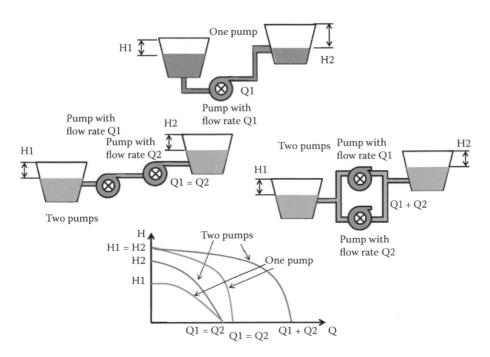

FIGURE 2.23
Serial and parallel pump connections.

The flow rate is a function of the impeller speed (as diameter) and is given by

$$Q_2 = Q_1 \left(\frac{N_2}{N_1} \right) = Q_1 \left(\frac{D_2}{D_1} \right) \tag{2.104}$$

The head is a function of the impeller speed or diameter and is given by

$$H_2 = H_1 \left(\frac{N_2}{N_1} \right)^2 = H_1 \left(\frac{D_2}{D_1} \right)^2 \tag{2.105}$$

The power is a function of the impeller speed or diameter and is given by

$$BHP_2 = BHP_1 \left(\frac{N_2}{N_1} \right)^3 = BHP_1 \left(\frac{D_2}{D_1} \right)^3 \tag{2.106}$$

Electric pumps use drivers such as electric motors to supply power to the fluid. The power requirement of a pump can be derived from the overall head developed, as well as the mass of fluid to be pumped per unit time. This is given by the product of the shaft velocity, ω, and the mass flow rate, Q. Hence, pumps are defined by the relation between the fluid pressure, P; the flow, Q; the shaft torque, T; and the velocity, ω, given by

$$Power = \Delta PQ = T\omega \tag{2.107}$$

If D_v is the volumetric displacement of the pump (m/rad), then

$$\begin{cases} T = D_v \Delta P \\ Q = D_v \omega \end{cases} \tag{2.108}$$

For a fluidic system such as a pump, the load torque of the fluid in the tank can be estimated by

$$T_L = \frac{gQh}{\omega} \tag{2.109}$$

with g being the gravitational acceleration, Q the flow rate, h the tank height, and ω the pump rotational speed. The pump sizing and selection method consists of the alignment of the pump output pressure with the flow in the pipe, and the resistance due to the external load.

2.5.2 Electric Cylinders

Electric cylinders consist of a cylinder, a piston, and a rod. They convert the fluid flow rate and pressure into mechanical force and velocity. Figure 2.24 illustrates a typical cylinder structure that could be unidirectional or bidirectional (single or bidirectional and double acting), with a rod on one or both sides of the piston (single or double rod). With electric cylinders, a pressurized fluid is injected into one side of the cylinder, causing it to expand due to the resulting linear force acting on the piston. This force is proportional to the cross-sectional area of the cylinder. For double-acting and double-rod cylinders, the velocity is given by

$$v(t) = \frac{4q(t)}{\pi(D^2 - d^2)} \tag{2.110}$$

During the retraction, the force is given by

$$F_p = P_{in}\frac{\pi(D^2 - d^2)}{4} - P_{out}\frac{\pi D^2}{4} \tag{2.111}$$

During the extension, the velocity and force are given by

$$v(t) = \frac{4q(t)}{\pi D^2} \tag{2.112}$$

$$F_p = \frac{\pi D^2}{4}(P_{in} - P_{out}) + P_{out}\frac{\pi d^2}{4} \tag{2.113}$$

with P_{in}, P_{out}, A_{in}, and A_{out} being, respectively, the head pressure within the chamber, the rod pressure within the chamber, the head piston area, and the rod piston area. Based on the orifice equation, the flow rate is given by

$$q(t) = CA\sqrt{\frac{2}{\rho}(P_p - P_{in})} = kx\sqrt{\frac{2}{\rho}(P_p - P_{in})} = A_{in}v(t) + \frac{V_1}{\beta}\frac{dP_{in}}{dt} \tag{2.114}$$

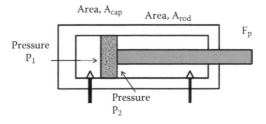

FIGURE 2.24
Single-acting, pressured air-powered cylinder.

2.5.3 Electric Fans and Blowers

Centrifugal fans raise the speed of a gas-type fluid through a rotating impeller. Similar to centrifugal pumps, fluid enters through the center of a centrifugal fan and increases its speed as it moves progressively toward the end of the blades. The blade shape defines the resulting fluid pressure. Table 2.12 summarizes the types of fans and their corresponding blade shapes.

With *axial fans*, the fluid flows along their axis. Among axial fans, there are propeller, tube-axial, and vane-axial fans, as summarized in Table 2.13. Both centrifugal and axial fans are driven by electric motors.

Blowers generate negative pressures in the case of vacuum systems. Commonly encountered blowers are centrifugal blowers and positive displacement blowers, both driven by electric motors. Centrifugal blowers are similar to centrifugal pumps. The gear-driven impeller can rotate at a speed up to 15,000 rpm. The positive displacement blowers mechanically move fluid, similar to positive displacement pumps. By this mechanism, it enables the movement of constant fluid volume at various pressures.

2.5.4 Fluid Flow–Controlled Electric Valves

Electric valves dictate the fluid transmission pressure and flow rate by (1) fixing whether the fluid is flowing (ON/OFF valves) or fixing its direction of flow (bidirectional sliding valves), (2) maintaining a desired pressure valve outlet independently of inlet flow pressure variations (pressure regulator valves), (3) regulating the flow rate using an orifice with a variable area (flow rate regulator valves), and (4) continuously controlling a solenoid displacement, speed, and force through the flow rate or pressure difference in order to ensure the actuating system positioning (flow proportional valves, pressure control proportional valves, or servo valves). Those valves could be electric

TABLE 2.12

Centrifugal Fans

Fan Category	Blade Shape
Radial	Flat
Forward curved	Forward curved
Backward inclined	Flat, curved, or airfoil

TABLE 2.13

Axial Fans

Fan Category
Propeller fan
Tube-axial propeller fan
Vane-axial fan

powered as well as hydraulic or pneumatic, depending on the driven fluid state (liquid or gas). They could be fully or partially closed or opened in a position dictated by signals transmitted from the controlling unit. Basic valve types are ball valves, butterfly valves, diaphragm valves, gate valves, plug valves, globe valves, and eccentric valves. Globe valves are suitable for large flow size and high pressure drop, while gate valves are used in the case of fluid operating at high temperature and pressure drops, as well as requiring intermittent operations. Ball valves are suitable for a high flow rate with just a quarter turn to operate, while butterfly valves require a high motor torque to operate. Similar to ball valves, diaphragm valves offer a tight shutoff.

The three-way control valves have inlet, outlet, and exhaust ports, which could be closed or open, as illustrated in Figure 2.25. When there is no energy supplied, the outlet port and exhaust port are interconnected. When the solenoid is energized, the inlet and outlet ports are interconnected. There are also two-way control valves (one inlet port and one outlet port) such that the valve is considered open when the energy is not ON, and a three-way control valve (three ports, such as one inlet and two outlets), as illustrated in Figure 2.26a and b. There are three types of ports (ways): P for pressure or pump-connected port, T for tank-connected port, and A or B for working port.

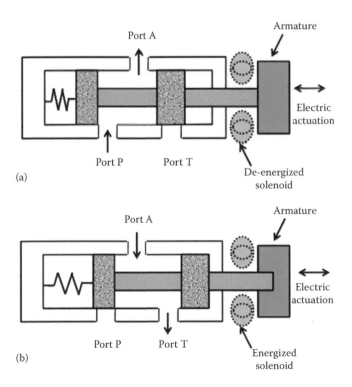

FIGURE 2.25
Two-position, three-way (2/3) directional solenoid-actuated control valves (open and closed).

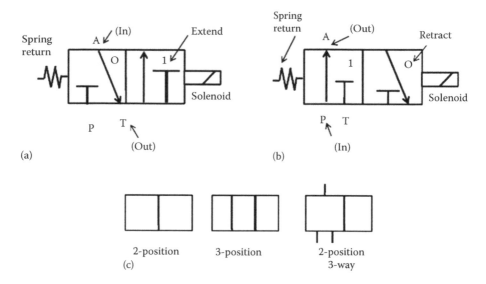

FIGURE 2.26
(a) Equivalent closed two-position, three-way (ports) (2/3) directional control valves.
(b) Equivalent open two-position, three-way (ports) (2/3) directional control valves. (c) Some
symbolic directional control valves.

The fluid flowing through the valve can be controlled by a solenoid mounted
on the side. Here, n-way is related to the number of connections for inlet and outlet ports, while normally open or closed terminology is related to the valve status when the power is turned off. Commonly encountered valves are designed
as two-position valves, three-position valves, and even four-way-position
valves, as shown in Figure 2.26c. They are used for the operating functions
of ON/OFF fluid flow control, unidirectional and single-acting cylinders, and
double-acting and bidirectional cylinders. Among valve design mechanisms
there are also direct-operated mechanical (e.g., pneumatic and hydraulic) actuations and indirect actuations, when the cylinder is not directly connected to the
valve. A key factor when selecting the type of actuated valve is the switching
time (delay), which is comparatively less for electric actuation. Either 12, 24, 115,
or 230 V (AC/DC) voltage could be applied to energize the solenoid.

2.5.4.1 Valve Dynamic Equations

Considering that the flow pressure P is produced in the fluid flow, the valve
behavior is described by the following nonlinear equation of the flow and
valve trim:

$$q(\theta) = Cf(\theta)\sqrt{\frac{\Delta P}{g}} \qquad (2.115)$$

where θ $(0 \le \theta \le 1)$ is the valve position, g is the fluid density (kg/L), C is the valve coefficient decided by the valve size, and $f(\theta)$ is the valve trim type for different plugs, given by

$$f(\theta) = \begin{cases} \theta & \text{for linear} \\ \sqrt{\theta} & \text{for quick opening} \\ R^{\theta-1} & \text{equal percentage} \end{cases} \tag{2.116}$$

where R is the rangeability or ratio between the lowest and highest fluid flow rates, such that a larger value of R indicates better accuracy of the valve position, which is given by

$$R = \frac{Q_{max}}{Q_{min}} \tag{2.117}$$

An equal percentage is suitable for a large pressure drop, while linearity is used where a fairly constant pressure drop is expected. Quick opening is used for intermittent service requiring a large and instant flow rate. The opening angle, θ, depends on the opening sectional area, A, and the flow pressure, β (including a constant factor given by fluid viscosity characteristics), such that

$$\theta = \frac{A}{\beta} P \tag{2.118}$$

Thus, the valve coefficient is given by

$$C = Q_{max} \sqrt{\frac{G_t}{\Delta P_v}} \tag{2.119}$$

with Q_{max} being the maximum flow through valves (L/min), $\Delta P = P_1 - P_2$ the upper and downstream pressure drop across the valve (bar), and G_t the liquid specific gravity. Using the flow velocity across the sectional area of the valves, v, the maximum flow through those valves, Q_{max}, can be estimated from

$$Q_{max} = Av \tag{2.120}$$

The airflow rate (for a pneumatic valve) is given by

$$q(t) = K_q \theta(t) \tag{2.121}$$

Considering that the overall flow pressure in a system ΔP_{total} is delivered by a pump or a compressor, the variation in opening angle of the valve causes a flow change, which in turn produces a variation in pressure difference, ΔP_v, across the valve. The overall required pressure difference for pumping, ΔP_{total}, could be estimated from the ΔP at the maximum flow such that

$$\Delta P_{total} = \Delta P_v + \Delta P_s = \left(\frac{Q_{max}}{C}\right)^2 g + kQ^2_{max} \qquad (2.122)$$

where

$$\Delta P_v = \Delta P_{total} - \Delta P_s = \left(\frac{Q_{max}}{C}\right)^2 g + kQ^2_{max} - kq^2 \qquad (2.123)$$

As a rule of thumb, ΔP_{total} should be around 1/3 to 1/4 ΔP_{total} at a nominal flow rate. In order to control a motorized valve, it is required to control its electric motor in revolutions per minute, position, acceleration, and torque. The sizing of flow control valves for incompressible fluid must be based on the following information requirements: (1) the flow through the valve, (2) the pressure drop across the valve, and (3) the specific gravity of the liquid. For a compressible fluid flow, additional information on inlet pressure and temperature, as well as the average molecular weight of the fluid, is required. Once the valve coefficient is derived, an appropriate valve can be selected from the catalog ($C_{v\_catalogue} \geq C_{v\_calculated}$). Hence, the selection must take into account safety considerations. For example, the valve injecting the fuel into the furnace is closed in case of utility failure, for example, upstream pipe oil leaks.

2.6 Modeling of Electrothermal Transmission Elements

Electrical heating elements, such as heaters, heat pumps, and furnaces can convert electrical energy into heat. Based on the Joule principle, the current passing through a resistor converts the electrical energy into heat energy. A typical example is the immersion of a heater in the top of a hot water cylinder. Inversely, heat pumps extract the heat energy from the ambient air or from the ground, and raise its temperature via the heat exchanged with a fluid boiling at low temperature. Then, the resulting vapor is compressed and condensed to a liquid form, in a condenser inside the building for space heating. A commonly encountered industrial heating system is furnaces (induction, electric, and muffle). Among the advantages of electric heating methods over

TABLE 2.14

Dynamics Equations Governing Heat Transfer

Conduction	$$q''_{cond} = -K\frac{dT}{dx}$$	q_{cond} = heat flux (W/m^2) K = thermal conductivity
Convection	$$q''_{conv} = h(T_s - T_\infty)$$	h = heat transfer coefficient $\left(\dfrac{W}{m^2\ K}\right)$ T_s = surface temperature (K) T_∞ = fluid temperature (°K)
Radiation	$$E = \varepsilon\sigma T_R^4$$	E = emissive power (W/m^2) T_R = surface temperature (K) σ = Stefan–Boltzmann constant $\left(5.67\times\dfrac{10^{-8}\,W}{m^2\,K^4}\right)$ ε = radiative emissivity $(0 < \varepsilon < 1)$
Conductive thermal resistance, R_{th} (the ability of an object to transfer heat between two points)	$$q_{cond} = q''_{cond}\,A = \frac{-KA\Delta T}{L}$$ $$R_{th} = \left|\frac{\Delta T}{q_{cond}}\right| = \frac{A}{k}\frac{L}{A} = \rho_{th}\frac{L}{A}$$ in $\left(\dfrac{K.s}{W}\right)$	ΔT = temperature change (K) q = heat flow (m^3/s) L = length of longitudinal rod (m) A = heat transfering pipe cross-sectional (m^2) q_{cond} = heat conduction (W/s)
Heat capacity and constant time for heating and cooling	$$\tau = R_{th}C_{th}$$ $$Q = sh \cdot m \cdot \Delta T = C_{th}\Delta T$$	Q = stored thermal energy (W) C_{th} = heat capacity (J/kg) sh = specific heat (J/(kg K)) m = mass (kg) τ = time constant (s)

other forms (gas-based heating system) there are the (1) better distribution of heat energy than chemical combustion, (2) refined temperature control, (3) near-real-time heating from a cold start-up, and (4) immediate and instantaneous shutdown. The typical heat transfers encountered are summarized in Table 2.14 with the formulas governing their behavioral dynamics.

2.6.1 Electrical Heating Element Sizing and Selection

There are many factors to take into account for the proper sizing of a heat transmission device, including (1) the heat transfer ratings and (2) the relationship between the heat exchanger's coverage area and its speed, as well as its efficiency rate. Here, during the process of electrical heater sizing, the following technical device data are required:

- Minimum and maximum operating intervals of the heat mass flow rate
- Inlet and outlet temperatures for each fluid

- Heat exchanger category
- Physical properties of the fluid heat transfer (specific heat, viscosity, and density)
- Pressure drop limitations

2.6.2 Control Requirements of Electrical Heating

Generally, there are two heater control techniques: contactor control and thyristor control. Contactor control is a device that enables switching of the power supply (current ON/OFF) toward the heater. Even if though this method has longer response time to the input signal, it is suitable for high-temperature fluids. Another control technique is based on a thyristor device that enables switching as much as several hundred times a second and is capable of providing a much faster response to the desired heat level. This method allows refining of the temperature control strategy. It is suitable for high-temperature fluids, such as fuel gas, or fluids burning at higher temperatures, such as glycol.

2.7 Electrical Binary Actuators

Discrete actuators are used to activate or deactivate some process operations. Binary actuators can be (1) bistable actuating systems when designed around material with bistable properties, such as electromechanical relays and solid-state devices, or (2) discrete actuated systems when associating switching power electronic devices with actuators, such as electric motors and hydraulic-powered cylinders. While the latest has been largely covered, bistable devices are discussed in depth in the next section.

2.7.1 Bistable Actuating Systems

A typical binary actuating system consists of actuators flipping between one or two discrete states. Some binary actuators are lightweight (e.g., smart material changing its elasticity properties when heated). Hence, those actuators are embedded in a load structure in order to change its dynamic behavior autonomously (e.g., vibration attenuation and noise cancellation). Furthermore, those devices do not need power to maintain each stable state. Among bistable actuators there are

- *Dielectric elastomer actuators* (DEAs), which have a ratcheting transmission that acts as a power spring. High-speed switching DEAs are suitable for robotic and mechatronic systems.

- *Shape memory alloys* (SMAs), which display alloy metal phase contractions in front of thermal variations. Those actuators are very sensitive to environment conditions (e.g., temperature, humidity, and dust).

- *Ionic polymers*, which vary their volume once when some ions are absorbed by their polymer microstructure.

- *Electroactive polymer actuators*, which generate a decreasing force as their deformation increases.

- *Piezoactuators*, which have a stacked or laminar design configuration. Laminar design actuators consist of piezoelectric strips with electrodes bonded onto them, while stacked actuators consist of some thin wafers of piezoactive material between metallic electrodes in parallel connection. Those actuators are suitable for (1) *suppressing oscillations*, thanks to piezoactive materials that convert the mechanical oscillations into electrical energy; (2) *microrobot* applications, with legs being piezoactuators, which can be lengthened, shortened, or bent in response to an applied voltage at the electrodes; (3) *micropump* applications, where their diaphragms are actuated by piezoactuators, allowing I/O check valves to be opened for fluid pumping; (4) *micromanipulators*, which convert contractions into gripping operations and are suitable for positioning applications; (5) *microdosage* devices using piezoactuators that allow high-precision dosage of liquids within a range of nanoliters, and (6) *piezomotors* that convert oscillations into a continuous motion, resulting in an elliptical motion in the contact area. Various oscillations offer possibilities to develop different kinds of piezomotors: longitudinal, transversal, shear, and torsional.

There are also micro- and nanodevices (less than 15 mm in size) with embedded electronic circuitry, such as electrostatic motors, which use the electrostatic principle to generate actuating forces.

2.7.2 Solid-State-Based Switching Power Electronics

In order to ensure the energy conversion within actuating systems, some devices are required to supply power amplification and modulation. This can be achieved by converting the low-power signal from the command signal into a high-power signal to be sent to the actuating system. Amplification and modulation are performed using power electronics. The resulting electronic circuitry (e.g., power converters) is used to drive electromechanical motion actuating devices, such as motors and solenoid valves. Those devices are usually made of silicon (Si), germanium (Ge), and cadmium sulfide, and their ON/OFF switching defines the operation of the converter. Hence, in contrast to naturally bistable devices, such as diodes and thyristors, transistors have to be biased fully ON in order to have similar behavior. During the bistable mode, two conditions are possible: (1) the blocking phase (i.e., switch turned OFF), where the voltage across the device is at the highest level while the current flows through the device at its lowest level, or (2) the conducting phase (i.e., switch turned ON), where the voltage across the device is at the lowest level while the current through the device is at the highest level. The description and operating phases of those semiconductors are revisited in the following sections.

2.7.2.1 Power Diodes

A power diode is a two-terminal electronic PN device that is constructed by joining p-type (anode) and n-type (cathode) semiconductors together to form a p-n junction, as illustrated in Figure 2.27a. The diode allows the current to run across the p-n junction for the positive terminal connection of the p-layer and the negative terminal connection of the n-layer. When the supply voltage, V_{diode}, exceeds (around 0.7 V in the case of silicon and 0.3 V in the case of germanium) V_{fb}, the diode is called forward biased, similarly to a closed switch. Inversely, when $V_{rb} < V_{diode} < V_{fb}$ (V_{rb} being the reverse breakdown voltage of the diode), the diode behaves as an open switch and is called reverse biased. These diodes are suitable for signal rectification and peak.

The Zener diode acts as a reverse-biased voltage regulator, with the supply voltage being higher than the rated reverse breakdown voltage. The diode resistor is used to build AND or OR logic gates, as shown in Figure 2.27b and c.

There is also the diode–transistor logic (DTL) gate, which is used to design NAND and NOR logic gates, as illustrated in Figure 2.27d and e. Diodes are suitable to design OR gates, while the transistors are suitable for NOT gates.

2.7.2.2 Power Transistors

The power transistor is a three NPN or PNP device that modulates current and voltage signals using controlled switching (closing and opening

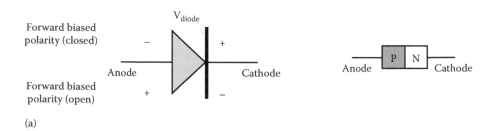

(a)

FIGURE 2.27
(a) Diode schematics.   (*Continued*)

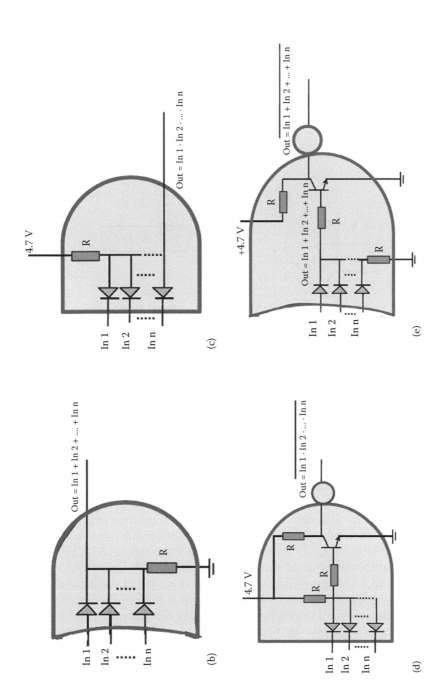

FIGURE 2.27 (CONTINUED)

(b) AND and OR gates designed using a diode resistor. (c) AND and OR gates designed using a diode resistor. (d) n-Input DTL NAND gate. (e) DTL NOR gate.

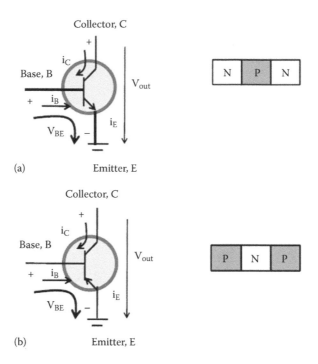

FIGURE 2.28
(a and b) Power transistor schematics.

connection) between two terminals. Figure 2.28a and b illustrates a device with three flowing currents (semiconductor emitter [E], collector [C], and base [B]) and two across voltages, V_{BE} and V_{CE}. Among common types of transistors there are resistor–transistor logic (RTL), transistor–transistor logic (TTL), BJT, and field-effect transistor (FET).

RTL is suitable for implementing a high-switching-speed NOT and NAND gate design, as shown in Figure 2.29a and b. It is operated on the principle that a low-voltage drop across the collector–emitter junction and a low current in the base–emitter loop could bring the transistor into saturation.

TTL is suitable for NAND gates, as illustrated in Figure 2.30. Here, the base–emitter is forward biased during conduction, while the collector–base is reversed biased. The base–emitter conducts when one of the inputs is low, while the emitter (first transistor at 0.2 V) output is high at the inverter. The emitter is cut off when all inputs are at logic high, allowing all other transistors to conduct at logic low. Symbols of some power electronic devices are shown in Figure 2.31.

There are also FETs, which are suitable for high-speed switching applications where the conducting or blocking is activated by the voltage instead of the current. Among FETs there are junction FET (JFETs), insulated gate bipolar transistors (IGBTs), and MOSFETs. MOSFETs have three terminal devices

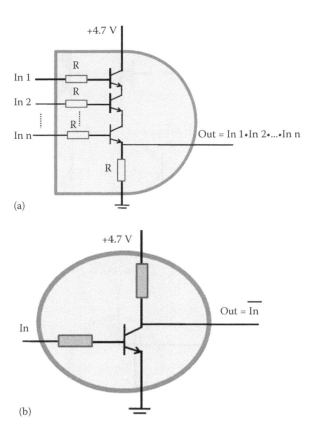

(a)

(b)

FIGURE 2.29
(a) RTL NOT schematics. (b) RTL AND gates schematics.

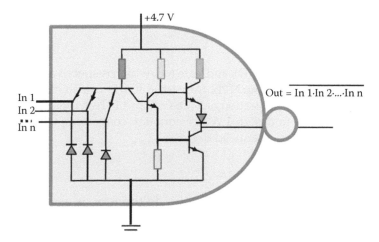

FIGURE 2.30
TTL implementation of NAND gate.

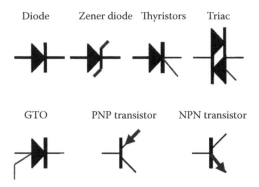

FIGURE 2.31
Symbols of some electronic devices.

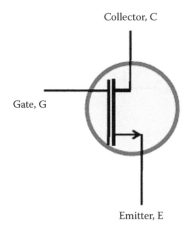

FIGURE 2.32
MOSFET schematic.

called the source (S), drain (D), and gate (G), which respectively are the emitter, collector, and gate of the NPN transistor. IGBTs use a voltage-controlled transistor. IGBT is an integration of MOSFET and BJT, as illustrated in Figure 2.32. FETs are used to design PWM frequency converters, while MOSFETs are used to design variable speed drives.

2.7.2.3 Thyristors

The SCR, also called reverse blocking triode thyristor, is a controllable time conduction diode. It comprises a four-layer silicon wafer PNPN device with three junctions. It also contains two power terminals (the anode A and the cathode K) and one control terminal (the gate G), as illustrated in Figure 2.33a–c.

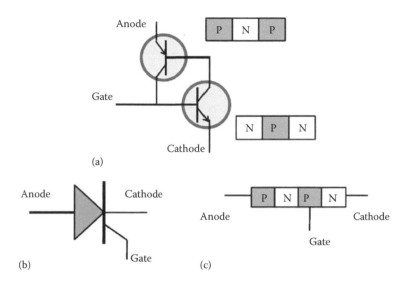

FIGURE 2.33
(a–c) Thyristor schematic.

The thyristor is ON when the voltage across the anode is higher than the one across the cathode while allowing the gate current to flow into it for a few microseconds. It switches from a blocking phase to a conduction phase by a suitable gate pulse, as illustrated in Figure 2.34. Only an external positive pulse on the gate activates the forward conduction. Among thyristor family devices there are the power thyristor, the GTO, the field-controlled thyristor (FCT), and the triac. A *GTO thyristor* switches OFF when a negative current pulses through the gate. A *triac* is a controlled switch operating similarly to a pair of thyristors. A triac can be turned ON in both the reverse and forward directions. Thyristors are suitable for rectifiers of AC sources.

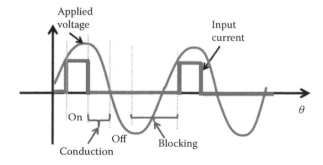

FIGURE 2.34
Thyristor switching due to voltage and current variations.

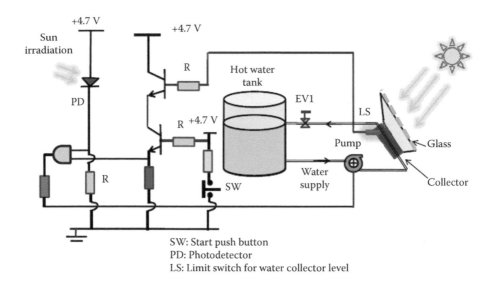

SW: Start push button
PD: Photodetector
LS: Limit switch for water collector level

FIGURE 2.35
Solid-state circuit for control of solar thermal heating system.

They modulate the average output voltage by varying the firing timing of the voltage or current gate, which is derived from

$$V_D = 1.35 \times V_{RMS} (RMS - Phase\ Voltage) \tag{2.124}$$

Example 2.2:

A logic function for motor pump activation of the solar heating system is illustrated in the circuitry shown in Figure 2.35. Here, the motor pump is activated by a pulse-type start push button (SW) and two digital sensors: irradiation from the sun toward a photodiode (PD) and a limit switch (LS) turned ON (closed) when the collector is not full, all connected to the two-input AND gate and corresponding to the logic Boolean function given by

$$Pump = SW \cdot PD \cdot LS$$

This circuitry corresponds to the truth table shown in Table 2.15.

2.7.2.4 Logic Integrated Circuit and Programmable Logic Devices

Programmable logic devices (PLDs) are made of integrated circuits (several hundred logic gate structures). Among PLDs there are (1) programmable logic array (PLA), which has an AND layer in the middle and an OR layer

TABLE 2.15

Truth Table for Motor Pump Activation

PD	SW	LS	Motor
1	1	1	1
1	0	0	0
0	1	0	0
0	0	0	0
0	1	1	
1	1	0	
1	0	1	
0	0	1	

level at the output, and (2) programmable array logic (PAL), which has a programmable AND layer constructed along with a fixed OR layer.

2.8 Solenoids

Commonly encountered solenoids operate by moving an iron core inside a wire coil as depicted in Figure 2.36a. Initially (no voltage is applied), the iron core is maintained outside the coil by a spring. When a voltage is applied across the coil, the current flowing through it generates a magnetic field surrounding the coil to produce a magnetic force capable of moving the core inside the coil. Those inductive devices can create voltage spikes. Typical electromechanical switches, such as relays and contactors, operate based on this principle, as well as pneumatic valves, car door openers, and so forth. Due to their low voltage and current requirements, such devices can be connected to logic control unit (e.g., PLC) outputs.

A typical example is when a current is going through the coils, inducing a magnetic force that can balance the force of gravity and cause the train (which is made of a magnetic material) to be suspended. Then, the system model around the equilibrium (the train is suspended) could be given by

$$
\begin{cases}
m\dfrac{d^2h(t)}{dt^2} = mg - \dfrac{Ki(t)^2}{h(t)} \\[2mm]
V(t) = L\dfrac{di(t)}{dt} + i(t)R
\end{cases}
$$

This is illustrated in Figure 2.36b.

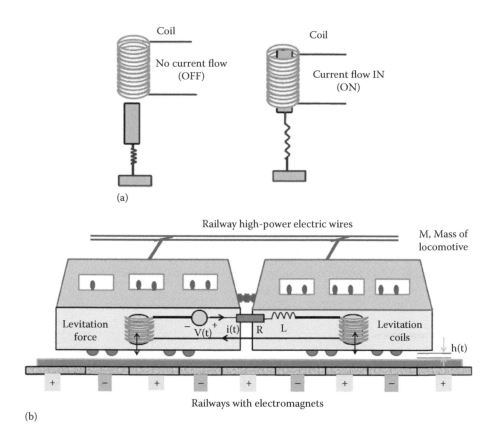

FIGURE 2.36
(a) Typical energized and unenergized solenoids. (b) Magnetically lifted high-speed train.

Exercises and Problems

2.1. Find the suitable motors among the applications listed in Table 2.16 (justify your choice).

2.2. a. For the following systems or processes, list the types of electric actuating devices encountered:

    i.   Electric bicycle for grocery delivery

    ii.   Artificial hearth

    iii.   Roll paper mill

    iv.   Beer fermentation tank

    v.   Automatic car airbag

    vi.   Japanese high-speed electric train

TABLE 2.16

Application versus Type of Electric Motor

Application	Electric Motor
1. Electric garage door	1. Permanent magnet DC motor
2. Constant-speed baking conveyor oven	2. Brushless DC motor
3. Window drive in vehicle	3. Synchronous AC motor
4. Air conditioning fan for five-story building	4. Stepper motor
5. Kuka robot for spray-painting vehicle	5. Asynchronous AC motor

    vii. Dialysis blood treatment

   viii. Washing machine

  b. Identify typical examples of loads with variable torque among the systems below.

    i. Belt conveyor

    ii. Hoisting gear

   iii. Pump and fan

   iv. Wind turbine

2.3. A three-phase PWM inverter uses a 600 VDC source. If the modulation is 1.0, select the corresponding RMS voltage magnitudes for the fundamental frequency.

  a. 0.6 kV

  b. Around 380 V

  c. Around 480 V

  d. Around 570 V

  e. None of the above

    What is the rpm speed of a two-pole, three-phase AC induction motor with an input frequency of 60 Hz?

  a. 900 rpm

  b. 1200 rpm

  c. 1500 rpm

  d. 1800 rpm

  e. None of the above

    From the torque–speed curve, the motor torque is in the opposite direction as the motor rotation in which of the following quadrant combinations?

  a. II as well as III quadrants

  b. II as well as I quadrant

  c. I as well as IV quadrant

    d. IV as well as III quadrant

    e. None of the above

2.4. a. List five applications of stepper motors requiring a motor position and/or speed control. Describe a method to derive the stepper motor speed and the corresponding motion resolution.

    b. Identify three DC motors found in a vehicle and specify the types of DC motors encountered.

    c. List three piezoelectric actuating systems and determine the supply voltage required. Identify the types of low-inertia motor used in a robot design.

2.5. A 47 kg electric motorized garage door is lifted 3.1 m at 650 rpm constant speed in 16 s.

    a. Determine the motor torque required if all friction torques are neglected.

    b. For a 380 AC input supply voltage to the motor, determine the amount of current required.

2.6. A 6 × 1.5 V battery package is used to energize the motor of a 4 kg unmanned vehicle.

    a. Compute the minimum electric motor power to be supplied to move up the unmanned vehicle along a 5% inclined road at an average speed of 2.1 km/h.

    b. Derive the DC motor specifications for such applications. (Hint: See Table 2.4.)

2.7. Consider an electric motor connected to a DC/DC chopper (380 VDC and switching frequency of 440 Hz). Here, the motor nominal power and velocity are 35.5 kW and 1250 rpm, respectively. At steady state, the current resistance and impedance are 0.13 Ω and 0.015 H, respectively. Determine the expected duty ratio cycle of the converter in order to achieve a motor speed of 700 rpm such that the load and friction torques are proportional to the speed.

2.8. From Figure 2.37, identify the corresponding motor torque–speed curve (if it exists):

    a. Induction motor

    b. DC shunt motor

    c. DC series motor

    d. Synchronous motor

2.9. Consider two 5-ton fully loaded underground gold mining carts moving along an 8% inclined hill, as illustrated in Figure 2.38. An electric motor attached to the drum (with 0.5 m diameter) of the first cart is moving at a constant speed of 50 km/h.

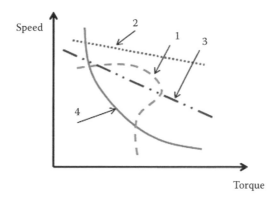

FIGURE 2.37
Different motor torque–speed curves.

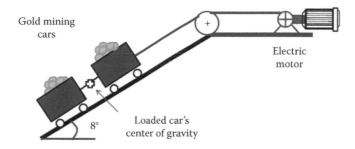

FIGURE 2.38
Electrical-driven car moving up a hill.

    a. The friction coefficient to the road surface is 0.4 kg/s. Determine the minimum motor torque required and the power consumed (in watts). Assume the weight of both carts acts at their center of gravity, which is at the middle of the junction.

    b. When the carts are going downhill at a constant speed of 65 km/h, derive the power delivered by the motor.

    c. What speed and power rates are required to move the carts uphill along 75 m of travel distance and within a 3 min time interval?

2.10. As illustrated in Figure 2.38, a motor-driven gold mining cart is moving upward uniformly as follows: (1) during the acceleration phase, it rises from zero to a maximum power of 1500 kW in 15 s; (2) during the full load period, it requires a power of 1000 kW for 50 s; and (3) during the deceleration period. when the regenerative braking takes place, its power is uniformly reduced from 500 W to 0 in 12 s. The interval for decking before the next load cycle starts is

25 s. Derive the suitable power (kW) rating of the motor based on the RMS power.

2.11. You are given an electric DC motor with the following and nameplate:

DC Power (kW)	DC Voltage (V)	Current (A)	Duty Cycle (dc)	Nominal Speed (rpm)	Brush Voltage Drop (V)	Field Power Input (kW)	Armature Resistance (Ω)	Armature Inductance (mH)
700	60	4.8	Permanent	600	2	50	0.0046	15

a. Draw the torque–speed characteristic for the interval 0–5.6 N·m (Hint: assuming a starting torque of 150 of full torque at rated speed, use power and torque equations).

b. For a friction torque (assume the torque constant [in Nm/A] to be 1) given by

$$T_f(t) = 0.2\omega(t) + 0.04$$

derive the motor torque for a maximum efficiency, as well as the corresponding steady-state motor speed (in this case, the load torque is 100 N·m, the motor shaft inertia is 0.145 kg · m², and the damping coefficient is 0.01 N · $\dfrac{m}{rad}$ /s).

c. Derive the electromagnetic torque and the nominal motor power (field and armature) (i.e., at its operating point).

2.12. An electric motor without load displays a starting torque of 6 N·m and rotates at a steady speed of 200 rpm. If the friction torque is such that

$$T_f(t) = 0.2\omega(t)$$

draw the torque–speed characteristic curve and determine the operation point of the maximum motor power (hint: use power and torque equations).

2.13. Consider a variable (0–18 VDC) voltage supply to both field and armature windings independently such that the rated flux is obtained when the field voltage is 120 V. Assume that the field voltage can be safely taken to a minimum of 18 V.

Draw the torque–speed characteristics of a separately excited DC motor having parameters as shown in Table 2.17.

2.14. Consider a PWM used to drive a DC motor at a speed of 750 rpm with a duty ratio (cycle) of 100%.

a. Determine how to adjust the PWM duty ratio in order to reduce the speed to 400 rpm.

TABLE 2.17

Separately Excited DC Motor Parameters

Supply Voltage (V)	Nominal Speed (rpm)	Motor Inertia (kg·m²)	Torque Constant (V·s/rad)	Damping Ratio (Nm/rad/s)	Armature Resistance (Ω)	Field Resistance (Ω)	Field Inductance (H)	Armature Inductance (H)
230	1400	3	0.564	0.0105	0.74	98	1.7	0.0024

b. What is the minimum switching frequency required of this PWM when designing a PWM (DC/DC chopper) with a 5 VDC input voltage and a current varying from 20 to 400 mA with an inductance of 7 μH?

2.15. Some 2 × 12 VDC motors (with PWM-based variable speed drives) are used for a single-arm robot handling unit and have the capability to change its direction. If the PWM has a 1.4 kHz switching frequency, what would the RMS voltage range be?

2.16. An electric motor (MTR1) drives up and down the spindle of a drilling machine bidirectionally, as illustrated in Figure 2.39. For such machine operation, explain the motion of the power flow using the four-quadrant drive concept.

2.17. For an induction motor with the load torque–speed characteristics depicted in Figure 2.40, do the following:

a. Determine the number of poles and the rated power.

b. Find the steady-state speed and the slip.

2.18. It is desired to determine the power rating (kW) of a 550 rpm motor used to drive equipment that has the load torque curve depicted in Figure 2.41.

2.19. Consider a motor rotating bidirectionally at a constant speed of 180 rpm and driving a screw wheel connected to a gear mechanical transmission element, as illustrated in Figure 2.42. Here, the drive motor shaft has 24 teeth, while the intermediate gear has 32 teeth.

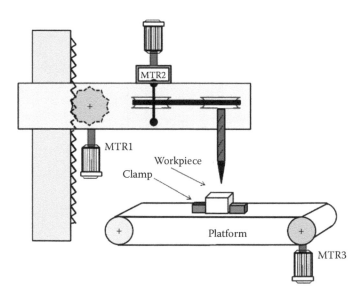

FIGURE 2.39
Drilling machine.

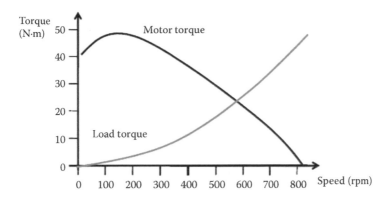

FIGURE 2.40
Motor and load speed–torque curves.

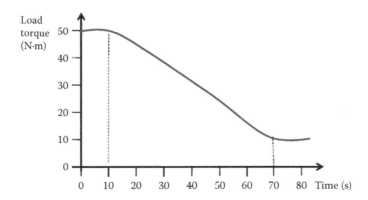

FIGURE 2.41
Load torque profile.

This motor moves through a 24-tooth pinion along a 96-tooth rack for 9 cm pitch.

a. Select a suitable DC motor for such application.

b. Derive the transmission elements (screw, gear, and wheel system) configuration that could triple the motion speed along a travel distance of 25 cm with the same selected motor.

c. If a stepper motor with 1.8° per step is used, what is the number of steps required to cover the same distance?

2.20. A hoist gripper of a seaport gantry system has to unload a 20-ton container from a ship cargo, as depicted in Figure 2.43.

a. Determine a suitable power rating of hoisting motor (group motor 1).

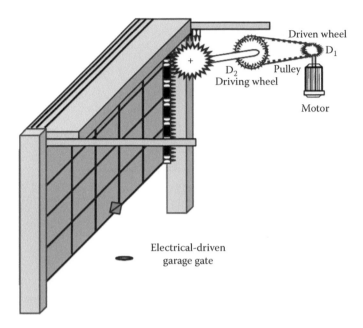

FIGURE 2.42
Screw-based gear wheel motion for the garage gate.

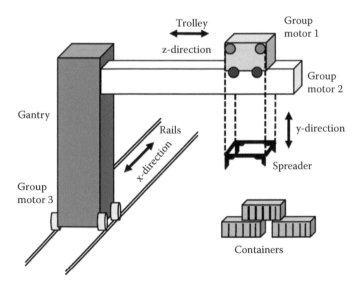

FIGURE 2.43
Seaport gantry.

b. Derive the suitable motor for this operation.

The duty cycle of the motor driving this gripper is indicated in Table 2.18.

2.21. An eight-pole, 50 Hz three-phase induction motor has an inertia of 1200 kg · m². The load torque of an oil well (upstream) is 200 kg·m for 10 s, and the motor has a slip of 6% at a torque of 100 kg·m. Assuming that (1) the torque–speed curve of the motor is linear and (2) the motor pump never gains its full speed during the oil loading period, based on the pipeline schematic depicted in Figure 2.44,

a. Derive the maximum torque developed by the motor.

b. Estimate the speed at the end of the deceleration period.

TABLE 2.18

Operating Characteristics of Seaport Gantry

Gantry Operations	Closing of the Gripper	Hoisting	Opening of the Gripper	Lowering of the Gripper	Rest
Duration (s)	8	15	5	15	20
Power required (W)	45	95	35	55	0

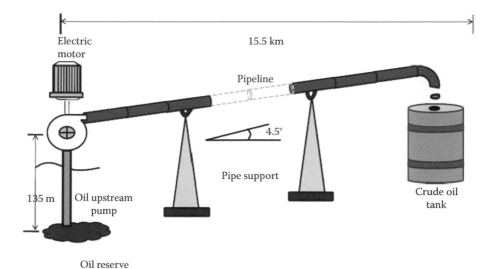

FIGURE 2.44
Upstream oil processing.

Bibliography

Alciatore D.G., M.B. Histand, *Introduction to Mechatronics and Measurement Systems*, 2nd ed., McGraw-Hill, New York, 2003.

Baker B., *Real Analog Solutions for Digital Designers*, Elsevier, Amsterdam, 2005.

Bishop R.H. (ed.), *The Mechatronics Handbook*, CRC Press, Boca Raton, FL, 2002.

Cetinkunt S., *Mechatronics*, Wiley, Hoboken, NJ, 2006.

de Silva C.W., *Mechatronics: A Foundation Course*, CRC Press, Boca Raton, FL, 2010.

Fraser C.J., Electrical and electronics principles, in *Mechanical Engineers Reference Book*, pp. 2.1–2.57, ed. E.H. Smith, 12th ed., Elsevier, Amsterdam, 1994.

Gottlieb I., *Electric Motors and Control Techniques*, 2nd ed., McGraw-Hill, New York, 1994.

Herman S., *Industrial Motor Control*, 7th ed., Cengage Learning, Boston, 2013.

Hughes A., *Electric Motors and Drives: Fundamentals, Types, and Applications*, 3rd ed., Newnes, Boston, 2005.

Isermann R., *Mechatronic Systems: Fundamentals*, Springer, Berlin, 2003.

Johnson J.L., *Basic Electronics for Hydraulic Motion Control*, Penton IPC, Cleveland, OH, 1992.

Jury E., Y.Z. Tsypkin, *On the Theory of Discrete Systems*, Automatica, Elsevier, Amsterdam, 1971.

Karnopp D.C., D.L. Margolis, R.C. Rosenberg, *System Dynamics: Modeling, Simulation, and Control of Mechatronic Systems*, 5th ed., Wiley, Hoboken, NJ, 2012.

Kazmierkowski M.P., R. Krishnan, F. Blaabjerg, J.D. Irwin, *Control in Power Electronics: Selected Problems*, Academic Press, Cambridge, MA, 2002.

Krishnan R., *Electric Motor Drives: Modeling, Analysis and Control*, Prentice-Hall, Upper Saddle River, NJ, 2001.

Ljung L., *System Identification: Theory for the User*, 2nd ed., Prentice-Hall, Upper Saddle River, NJ, 1999.

Lyshevski S.E., *Nano- and Micro-Electromechanical Systems: Fundamentals of Micro- and Nano-Engineering*, CRC Press, Boca Raton, FL, 2000.

Manring N., *Hydraulic Control System*, Wiley, Hoboken, NJ, 2005.

Manring N., *Fluid Power Pumps and Motors: Analysis, Design and Control*, McGraw-Hill, New York, 2013.

Moran M.J., H.N. Shapiro, B.R. Munson, D.P. Dewitt, *Introduction to Thermal Systems Engineering: Thermodynamics, Fluid Mechanics and Heat Transfer*, Wiley, Hoboken, NJ, 2002.

Novotny D.W., T.A. Lipo, *Vector Control and Dynamics of AC Drives*, Oxford University Press, Oxford, 1996.

Onwubolu G.C., *Mechatronics Principles and Applications*, Elsevier, Amsterdam, 2005.

Petruzella F., *Electric Motors and Control*, McGraw-Hill, New York, 2009.

Pinsky M., S. Karlin, *An Introduction to Stochastic Modeling*, Academic Press, Cambridge, MA, 2010.

Ripka P., A. Tipek, *Modern Sensors Handbook*, ISTE Ltd., London, 2007.

Sashida T., T. Kenjo, *An Introduction to Ultrasonic Motors*, Oxford University Press, Oxford, 1993.

Singh R.P., D.R. Heldman, *Introduction to Food Engineering*, 3rd ed., Academic Press, Cambridge, MA, 2001.

Shumway R.H., D.S. Stoffer, *Time Series Analysis and Its Applications: With R Examples*, Springer, Berlin, 2010.

Shumway-Cook A., M.H. Woollacott, *Motor Control: Theory and Practical Applications*, Williams & Wilkins, Philadelphia, 1995.

Soloman S., *Sensors Handbook*, McGraw-Hill Handbooks, McGraw-Hill, New York, 1998.

Stephen L., S.L. Herman, *Industrial Motor Control*, 6th ed., Delmar Cengage Learning, Boston, 2010.

Sul S.-K., *Control of Electric Machine Drive Systems*, Wiley-IEEE Press, Hoboken, NJ, 2011.

Suzuki Y., K. Tani, T. Sakuhara, Development of new type piezoelectric micromotor, *Sensors and Actuators*, 83, 244, 2000.

Uchino K., *Piezoelectric Actuators and Ultrasonic Motors*, Kluwer Academic Publishers, Norwell, MA, 1997, p. 349.

Yeaple F., *Fluid Power Design Handbook*, 3rd ed., Dekker, New York, 1996.

Zhou B., Y. Liu, Z. Wang, J.-S. Leng, A.K. Asundi, W. Ecke, Modeling the shape memory effect of shape memory polymer, *Proceedings of the SPIE*, 7493, 2009.

3

Logic Controller Design

3.1 Introduction

Many control systems involve a finite set of values (e.g., binary variables such as 1 or 0, ON or OFF, and open or closed) rather than continuous process variables. These systems operate by switching between ON and OFF position motor starters, valves, and other devices according to operating sequences (event occurrence in process) or time-synchronized events. Such systems are called logic-based controlled systems. Here, switched conditions of inputs (e.g., limit switches, relay contacts, and push buttons) dictate the output states of process devices (e.g., electromechanical relay coils and indicator lights) through logic control systems. Thus, a process operating sequence can be directly defined by successive output states of the devices involved. Hence, the sequential logic controller converts the switching input conditions into energized or de-energized state outputs of each device. This is usually achieved through hardwired relay networks, solid-state logic devices, logic programmable devices (e.g., programmable logic controller [PLC] and field-programmable gate array [FPGA]), microcontrollers, and digital signal processing.

In the logic controller design, the challenge is to determine the logical linkages in such a way as to always validate any combination of switching input conditions with changes in state outputs. This can be achieved by modeling the process operations. Such formal modeling requires a functional description and analysis of the operating sequences to ensure the execution of predictable operational sequences. For processes with a single cycle of operation, the modeling consists of converting the operating sequences into the corresponding truth table or process switching sequence table. For the modeling of more complex processes with several cycles of operations and concurrent operating sequences, state diagrams can be used. But in the case of multiple cycles of operations and parallel operating sequences, petri net or sequential function chart–based modeling is preferred. Hence, from those formal models it is possible to derive the Boolean functions relating process state outputs with state input transition conditions.

The main objective of this chapter is to lay out a methodology for the design of logic controllers, resulting in input/output (I/O) Boolean functions. First, Boolean algebra is reviewed, along with combinatorial and sequential logic tools. Then, some logic controller design methods based on functional analysis and various formal process modeling techniques (e.g., switching theory, state diagram, and sequential function chart) are presented. Thus, a Boolean-based logic controller is implemented by using solid-state electronic devices or programming languages, such as a ladder diagram (LD) and instruction list (IL). This is covered, along with the procedure to develop schematic electrical wiring diagrams for automation design projects. Finally, automation applications in breweries, manufacturing processes, traffic management, and handling systems are described to illustrate the design and implementation of logic controller programs and circuitries.

3.2 Logic System Design Preliminaries and Methods

Logic systems have a two-state condition, which can be represented by distinct levels (e.g., +V and 0 V voltage levels, light ON or OFF, open or closed limit switch, and motor running or stopped). By standards, binary 1 (or logic 1) corresponds to ON, TRUE, HIGH, and so forth, while binary 0 (or logic 0) corresponds to as OFF, FALSE, LOW, and so forth. Behavioristic characteristics of logic systems can be captured using this two-state concept, which is manageable by the binary number system and powerful Boolean algebra tools. Considering each state of a logic system as an event, logic systems can be described as discrete-event systems evolving from changes within the input conditions. Hence, the logic relationship between switched input conditions and system output states can be either combinatorial or sequential. This can be captured in the form of Boolean functions resulting from the analysis of their corresponding truth table and sequence table, as well as from the switching theory. For simple logic systems, direct conversion from those tables into I/O Boolean functions is possible, while for more complex logic systems, there are some methods (e.g., switching theory and state diagram) to derive and simplify I/O Boolean functions. The subsequent step is to derive a logic algorithm or solid-state circuitry enabling to us solve those I/O Boolean functions. In this section, a review of methods to design logic systems is presented.

3.2.1 Combinatorial and Sequential Logic Systems

In combinatorial logic systems, the system output is only derived from current inputs. Examples of combinational logic systems are solid-state electronic devices, such as gates, decoders, and multiplexers. In contrast, sequential logic systems derive their output from past inputs, as well as current inputs, implying that a memory is required. Typical sequential logic systems are latches (e.g., SR, SR latch with enable, and D) that change states at any time due to input change, or flip-flops (e.g., edge-triggered D flip-flop, edge-triggered SR, edge-triggered JK, and T flip-flop) that change states only when a clock edge is applied. Latches and flip-flops are summarized in Appendix A. The formal modeling tool for logic systems is Boolean algebra with variables having a value of either 1 or 0 and using logic gates AND(), OR(+), NOT, EOR(+), NAND, or NOR to describe the logical links between input conditions and past or current state outputs. In short, the design of combinational or sequential logic systems would consist of deriving I/O Boolean functions representing the operating combination or sequence. Modeling methods widely covered in digital system design are reviewed in subsequent sections.

3.2.2 Sum of Product and Product of Sum Methods

Sum of product (SOP) is a method used to derive I/O Boolean relationships for combinational logic systems by using their corresponding truth table. Here, for each ith row of the truth table, the logic input combination activating any output is given by

$$Product(i) = \prod inputs \tag{3.1}$$

This could be implemented by connecting all involved relays in series (AND gate). Hence, the SOP method defines the Boolean relationship between all input combinations activating a jth output by

$$Output(j) = SOP = \sum_j Product(i) \tag{3.2}$$

The hardwired implementation of this Boolean function is required to use the AND gate and OR gate (parallel collection of series-connected relays).

Example 3.1

Consider a cement rotary kiln, where the intense heat from the fire is expected to remove a minimum of 85% humidity from the pozzolana before being mixed with gypsum, in order to form ready-to-use cement.

As long as a flame is maintained in the kiln flame chamber, it is suitable to inject the pozzolana to be dried. Otherwise, it would be undried, and consequently useless. It is desired to design a flame detection logic circuit to dictate whether the pozzolana should be injected. The presence of a flame is monitored using three different flame detection technologies: optical detector (OD), thermal detector (TD), and gas analyzer (GA),

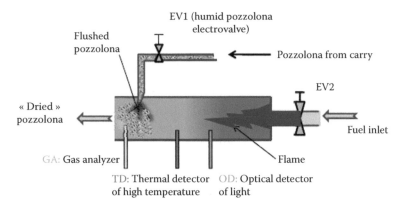

FIGURE 3.1
Pozzolana-based cement rotary kiln.

as depicted in Figure 3.1. The humid pozzolana electronic valve (EV1) is energized if any of the detectors are actuated. All detectors are normally closed contact switches connected in series. Hence, they maintain the EV1 output energized status even in the event of a detector electric circuit wiring failure (open switch contacts, broken wire connections, open relay coils, burned fuses, etc.). In short, the pozzolana injection valve (EV1) should be open if and only if two of three detectors (OD, TD, and GA) acknowledge good flame intensity. This is considered to be a three-input and one-output combinational system, which is described by the eight truth table combinations in Table 3.1A.

Here the I/O relationship is given by

$$Output(EV1) = OD \cdot TD \cdot \overline{GA} + OD \cdot \overline{TD} \cdot GA$$
$$+ \overline{OD} \cdot TD \cdot GA + OD \cdot TD \cdot GA$$

TABLE 3.1A

Truth Table When Two Out of Three Detectors Are Checked

	OD	TD	GA	Output (EV1)
	0	0	0	0
	0	1	0	0
	0	0	1	0
	1	0	0	0
$Prod\,1 = OD \cdot TD \cdot \overline{GA}$	1	1	0	1
$Prod\,2 = OD \cdot \overline{TD} \cdot GA$	1	0	1	1
$Prod\,3 = \overline{OD} \cdot TD \cdot GA$	0	1	1	1
$Prod\,4 = OD \cdot TD \cdot GA$	1	1	1	1
				$SOP = Prod\,1 + Prod\,2 + Prod\,3 + Prod\,4$

In case there are more output conditions than inputs, it is suitable to use the product of sums (POS) method to derive the I/O Boolean function. Here, for each ith row of the truth table, the logic input combination activating any output is given by

$$sum(i) = \sum inputs \tag{3.3}$$

Hence, the POS method captures the Boolean relationship between all input combinations activating a jth output by

$$Output(j) = POS = \prod_j sum(i) \tag{3.4}$$

This could be implemented using the OR gate feeding into the AND gate (series collection of parallel connected relays).

Example 3.2

Consider a cement rotary kiln process having as outputs a flame detector failure and pozzolana valve (EV1) activation. The POS method could be summarized as in Table 3.2.

Recall that if two of the three detectors (OD, TD, and GA) acknowledge a good flame intensity, the EV1 is open and the failure of one of three detectors is assumed a detector disagreement. When detector outputs are all in logic level 0 or 1, all detectors agree. From the truth table in Table 3.1B, there are two outputs (but six output activation combinations) and three inputs. The implementation of this detector disagreement is expected to offer a safer logic execution of the process control.

TABLE 3.1B

Truth Table When Using a Virtual Disagreement Detector

	OD	TD	GA	Output (EV1)	Detector Disagreement
	0	0	0	0	0
$Sum\,1 = \overline{OD} + \overline{TD} + \overline{GA}$	0	1	0	0	1
$Sum\,2 = \overline{OD} + \overline{TD} + GA$	0	0	1	0	1
$Sum\,3 = OD + \overline{TD} + \overline{GA}$	1	0	0	0	1
$Sum\,4 = \overline{OD} + TD + GA$	0	1	1	1	1
$Sum\,5 = OD + TD + \overline{GA}$	1	1	0	1	1
$Sum\,6 = OD + \overline{TD} + GA$	1	0	1	1	1
	1	1	1	1	0
					$POS = Sum\,1 \cdot Sum\,2 \cdot Sum\,3 \cdot$ $Sum\,4 \cdot Sum\,5 \cdot Sum\,6$

3.2.3 Karnaugh Maps

It is usually difficult to derive and simplify the resulting I/O Boolean expression from a truth table with more than two inputs and one output by using Boolean algebra and theorems. A reduction technique tool has been proposed based on Karnaugh maps (K-maps) consisting of a two-dimensional representation of the truth table where columns and rows correspond to each logic input variable. Here, adjacent row and column designations differ by only 1 bit. This is called gray code sequence. For example, with three inputs, the gray code sequence format as a change of one binary between sequence would be 000 -001 -011 -010 -110 -111 -101 -100, as shown in Table 3.1C. In the case of n-input variables, the truth table would correspond to 2^n boxes in Karnaugh maps. Hence, developing a reduced I/O Boolean function would consist of

- Constructing the truth table.
- Selecting an appropriate K-map with 2^n boxes for n inputs from the truth table.
- Applying the gray code sequence for the row and column designations of the K-maps.
- Copying the 1s and 0s from the output position within the truth table to the corresponding boxes in the K-maps. If the variable is undetermined, an output value X should be assigned in that box.
- Encircling all adjacent 1s in the same column, as well as in the same row, of the K-maps.
- Writing the Boolean function and the product term for each circle (group).
- Simplifying the resulting I/O functions by retaining only the common variables. Undetermined logic output variables X can be discarded in the minimization process. In that case, the resulting set

TABLE 3.1C

Truth Table for Three Inputs and One Output

A	B	C	Y
0	0	0	0
0	0	1	0
0	1	0	1
0	1	1	0
1	0	0	1
1	0	1	0
1	1	0	1
1	1	1	1

TABLE 3.1D

Corresponding K-Map for Three Inputs

		BC			
		00	01	11	10
A	0	0	0	0	1
	1	1	0	1	1

of product terms would be combined using an OR logic function to derive the minimized I/O Boolean relationship.

- Writing SOPs for all Boolean expressions above to derive I/O Boolean expressions (sum of the common variables from each group).

Karnaugh maps can be used for up to eight input variables. Above that, the simplification can be readily programmed by a computer-aided design tool, such as Quine-McCluskey's technique, in order to derive a minimum SOP expression as the sum of minterms, or Petrick's technique to derive all possible minimum SOP solutions.

From Table 3.1C, the corresponding K-maps would be given as shown in Table 3.1D. Then, using the grouping method for simplified SOP, the Boolean equation for each group is

Group (in green) = $\bar{A}(B\bar{C}) + A(B\bar{C})$ with $(B\bar{C})$ as common variables

Group (in blue) = $(A)B(\bar{C}) + (A)\bar{B}(\bar{C})$ with $A\bar{C}$ as common variables

Group (in red) = $(AB)(\bar{C}) + (AB)(C)$ with AB as common variables

The simplified SOP Boolean expression is

$$Y = B\bar{C} + A\bar{C} + AB$$

3.2.4 Moore and Mealy State Diagrams

A state diagram is a graphical technique to capture the behavior of discrete-event systems characterizing the state output change under specific activation of input transition conditions. These diagrams are classified into either (1) Moore type, when the state output value is derived from the present state, or (2) Mealy type, when the state output value is derived from the present state and state input transition conditions. In order to sketch the state diagram, the following assumptions are considered: (1) state transition conditions are forced only by system inputs; (2) a system is always in one and only one state; (3) states are described as the combination of state output values; (4) states are equivalent if the same system inputs produce an identical combination of state outputs; (5) with n system inputs, there must be at least 2^n outgoing arrows per circle (state); and (6) system input variables are classified into either (a) level-type operating mode, when the input transition condition (value) is maintained active until the state output value changes (e.g., light switch with a spring to hold position), or (b) pulse-type operating mode, when the input transition condition actuates the state output and returns to its initial condition immediately after (e.g., emergency stop of a machine tool). From the Moore or Mealy state diagrams, it is possible to derive I/O Boolean functions based on the type of solid-state devices chosen (e.g., flip-flops and latches). There is also a simplified version of state diagrams suitable for discrete-event systems that can easily be converted into programming languages using the transition equation method. All those state diagrams are presented in this section. First, a general procedure to sketch state diagrams is summarized as follows:

1. Identify all system inputs that cause process transitions (e.g., sensing devices and control panel push buttons), as well as all system outputs (actuating devices, control panel light indicators, etc.).
2. List all system operating sequences in a sequence table or all possible input combinations that activate system outputs in a truth table.
3. Define a state as a combination of system outputs that are activated by a set of process inputs.
4. Construct a state (excitation) transition table with at least three columns: present state (the value for each possible state at time t), system input (all possible input transition conditions), and next state (the value for the possible state at time $t + 1$). Based on step 3, a column capturing system output may be added, as illustrated in Table 3.2A. The next state is defined using solid-state devices.
5. Convert the present state variable values, the system input, and the next state into binary values using the gray code sequence format.

TABLE 3.2A

State Table Based on Mealy Machine

Present State	System Input	Next State	System Output

TABLE 3.2B

State Table Based on the Mealy Machine

$x_i y_i$ q	00	01	11	10
q_0	$q_0, 0$	$q_0, 1$	$q_1, 0$	$q_0, 1$
q_1	$q_0, 1$	$q_1, 0$	$q_1, 1$	$q_1, 0$

TABLE 3.2C

State Table Based on the Moore Machine

$x_i y_i$ q	00	01	11	10	Z
q_{00}	q_{00}	q_{01}	q_{10}	q_{01}	0
q_{01}	q_{00}	q_{01}	q_{10}	q_{01}	1
q_{10}	q_{01}	q_{10}	q_{11}	q_{10}	0
q_{11}	q_{01}	q_{10}	q_{11}	q_{10}	1

6. From the finite machine type chosen (Mealy or Moore), construct either (a) a state table based on the Mealy machine with four columns— (i) present state, q; (ii) process inputs $(x_i y_i)$; (iii) next state, for example, illustrated by $(q_0, 0)$; and (iv) system output (value of the system output)— or (b) a state table based on the Moore machine with the particularities of q_{xx} and Z being, respectively, the state (present or next) values and the system output values. This is illustrated in Table 3.2B and C.

7. Sketch the state diagram by representing the system states by circles, with each arrow corresponding to each possible input transition condition. An example of a Mealy state diagram is illustrated in Figure 3.2, where the process state changes from state 1 to state 2 due to the activation of the input transition condition causing the process output to be set at 0, as illustrated in Figure 3.2. In the case of a Moore state diagram, the process outputs are indicated within the state, as they are independent of the input transition condition.

8. Apply the binary coding on system input transition conditions and state outputs. They are usually written above each arrow and separated by a slash (/), as depicted in Figure 3.2b and d.

Example 3.3

Consider an automatic gate for accessing a public transportation innercity train (metro) requiring a 75-cent ticket with the input and output devices summarized in Table 3.3A. It can only accept a 25-cent coin. Figure 3.3a illustrates a typical automatic metro gate.

Table 3.3A can be converted into the process formal model using a Moore state diagram, as shown in Figure 3.3b, or a Mealy state diagram equivalent to that shown in Figure 3.3c.

Using the gray sequence coding of system state outputs and input transition conditions as summarized in Table 3.3B, it is possible to describe the process using a Moore state diagram with binary coding, as shown in Figure 3.4a and b.

Alternatively, from the system I/O binary coding, it is possible to describe the process using the state transition table and D flip-flop depicted in Table 3.3F. Therefore, K-maps can be used to derive the corresponding I/O Boolean functions, as shown in Table 3.3.

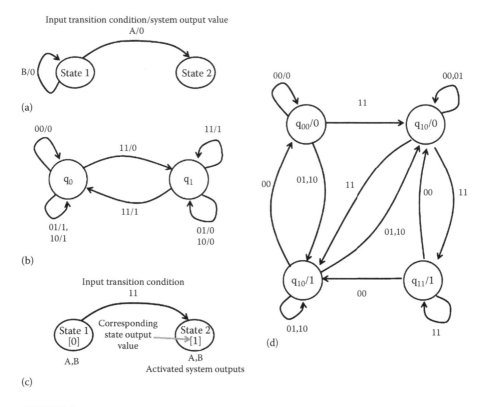

FIGURE 3.2
(a) Mealy state diagram. (b) Mealy state diagram with binary coding. (c) Moore state diagram.
(d) Moore state diagram with binary coding.

TABLE 3.3A

Listing of Input and Output Devices Involved in an Automatic Metro Gate

Input Equipment	Symbol	Output Equipment	Symbol
25-cent (quarter) coin sensor	CN	Motor gate	M1
10-cent (dime) coin sensor	D	Light indicators (green/red)	L1/L2
Reset push button	RT	Horn	HR
		Money return	R

TABLE 3.3B

Binary Coding of Input Transition Conditions and Vending Machine States

Coin entered (cents)	00	25	50	75
State	S00	S25	S50	S75
Binary coding (CN)	00	01	10	11

TABLE 3.3C

D_1 Karnaugh Table

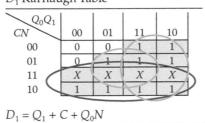

CN \ Q_0Q_1	00	01	11	10
00	0	0	1	1
01	0	1	1	1
11	X	X	X	X
10	1	1	1	1

$D_1 = Q_1 + C + Q_0 N$

TABLE 3.3D

D_0 Karnaugh Table

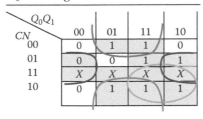

CN \ Q_0Q_1	00	01	11	10
00	0	1	1	0
01	0	0	1	1
11	X	X	X	X
10	0	1	1	1

$D_0 = \bar{Q}_0 N + Q_0 \bar{N} + Q_1 N + C$

TABLE 3.3E

Open Output Karnaugh Table

CN \ Q_0Q_1	00	01	11	10
00	0	0	1	0
01	0	0	1	0
11	X	X	X	X
10	0	0	1	0

$Open = Q_1 Q_0$

3.2.5 Simplified State Diagram

For any sequential operating system, the number of possible state changes per state could be reduced. Therefore, a simplified state diagram can be derived directly from the corresponding sequence table. Figure 3.5 depicts the generic sketching rules of a simplified state diagram. Here, through the construction of arrows leaving and arriving, each state S_i could be sketched

TABLE 3.3F

State Transition Table with D Flip-Flop Gate

Present State (cents)	Present State		Inputs		Next State (cents)	Next State		Output Open
	Q1	Q0	C	N		D1	D2	
00	0	0	0	0	00	0	0	0
			0	1	25	0	1	0
			1	0	50	1	0	0
			1	1	75	X	X	X
25	0	1	0	0	00	0	1	0
			0	1	25	1	0	0
			1	0	50	1	1	0
			1	1	75	X	X	X
50	1	0	0	0	00	1	0	0
			0	1	25	1	1	0
			1	0	50	1	1	0
			1	1	75	X	X	X
75	1	1	0	0	00	1	1	1
			0	1	25	1	1	1
			1	0	50	1	1	1
			1	1	75	X	X	X

as such: (1) arrows arriving to S_i correspond to those input transition conditions that turn ON state S_i and keep it ON while deactivating a preceding state of S_j; (2) all arrows leaving S_i correspond to input transition conditions, T_input_conditions that turn OFF S_i and activate a following state of S_j. The state diagram is converted into I/O Boolean functions using the transition equation method. Hence, the Boolean functions for any present state S_i and any state output O_m using the state diagram shown in Figure 3.5 are given by

$$S_{present} = \left(S_{present} \cdot \sum S_{previous} \cdot T_{input\ conditions} \right) \cdot \prod S_{next}$$

$$O_m = \sum S_{states\ where\ O_m\ is\ energized}$$

(3.5)

3.2.6 Logic Controller Design Methods

An automation system aims to ensure the execution of a number of discrete-event process operations. Discrete events occurring in a given order are

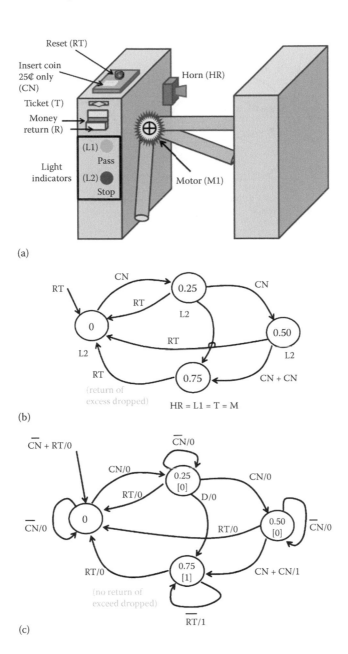

FIGURE 3.3
(a) Automatic metro entrance gate. (b) Moore state diagram. (c) Equivalent Mealy state diagram.

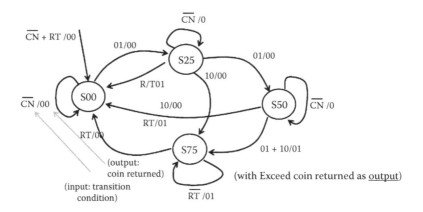

FIGURE 3.4
Equivalent Mealy state diagram with binary coding.

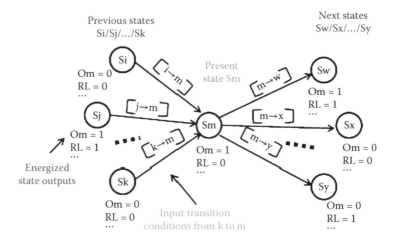

FIGURE 3.5
Generic state diagram sketching.

called sequential events, while those occurring randomly based on a defined I/O relationship are called combinational events. Thus, the design of a logic controller would require establishing the logic model of a combination of system state outputs based on a set of system inputs and present system states. In this review on the design of logic systems, subsequent sections present some techniques to design a logic controller enabling the execution of repetitive process operations, while circumventing the time-consuming trial-and-error approach commonly used. A design methodology ensuring the automatic execution of process operations by logic controllers is summarized in Table 3.4A.

TABLE 3.4A

Step-by-Step Logic Controller Design Methodology

Design Step	Item Description
Process Description and Functional Analysis	
1. List all the devices involved in the process operations.	List process equipment (analog/digital) related to the operating sequence execution, such as the start-up/shutdown of actuating, sensing devices, as well as inputs/outputs from the human–machine interface (e.g., AutoPB, StopPB, mode indicator lamp, and top LS) with their technical specifications.
2. Identify all devices' operational conditions and constraints, operating threshold values, and initial conditions.	Bidirectional motor, n-multispeed drive requiring n contactors, etc. Operating cycle conditions: Simultaneous or concurrent operating sequences, level type or pulse type switch operating mode, etc. Assign a symbol for each I/O device.
3. Classify them into analog/digital, I/O system variables, and (level or pulse) types.	Inputs: For example, push buttons, and limit switches. Outputs: For example, solenoids, lights, and motors.
4. Draw piping and instrumentation diagram (P&ID) and process flow diagram (PFD).	If possible.
5. Develop functional decomposition and analysis.	Using decomposition methods (e.g., FAST and SADT), organize each process combinational or sequential event in the equivalent process operation and input transition conditions.
6. Derive the chronological activation order of process operations (operating cycle) for a group of sequential events or for all possible input conditions and combinations of event groups.	For all sequential events, regroup all correlated process operations into successive sequences (operating cycle). List each sequence with the system outputs and inputs causing the transition between sequences. For all combinational events, summarize within the truth table all conditions of activation for each process operation (output). For remaining conditions with any activation, mark them as undetermined.
7. For each operating cycle, identify and describe the type of system operating cycle.	Single operational cycle. Multiple operational cycles: Mutually exclusive or concurrent. Cyclic or acyclic (auto/manual/semiautomatic). Synchronization of the system operations (time triggered, event triggered, counter triggered).
8. Perform a system output activation analysis: Either on each sequence for each operating cycle or on each combination of input conditions for each group of combinational events.	Describe the initial condition of each operating cycle (activation). Identify duplicate and priority (emergency) sequences. Identify the system operating cycle deactivation condition.

(Continued)

TABLE 3.4A (CONTINUED)

Step-by-Step Logic Controller Design Methodology

Design Step	Item Description
9. Perform safety analysis of process operational execution.	Identify the potential defaults and develop interlocks to maintain the normal system operating in case of their occurrence; specify other safety constraints and remedy actions.
	Identify process hardware and software operational defaults.
	Develop interlocks and forcing action design.
10. Derive logic operating sequences for each process operation, including safety measures in the table in a diagram format.	Construct either truth, sequence, or stable transition tables or state diagrams containing information on all inputs and the corresponding process outputs activated, as well as the current/next state for each operating sequence.
11. Develop process start and stop mode graphical analysis.	Classify within a coordinated cycle hierarchy all cycles of process operations in a normal automatic production cycle: manual, semiautomatic, or maintenance cycle; safety cycle; etc.

Formal Modeling of Discrete-Event System

12. Sequential or combinational formal modeling of discrete-event system.	In the case of a single operating cycle, construct a truth table (combinational event) or a sequence table (sequential event).	From the sequence table, identify and separate each binary process output activation Output[1]/deactivation Output[0]; for the input combinations within the truth table, use Karnaugh maps.
	In the case of multiple operating cycles with mutually exclusive sequences, construct the corresponding state transition diagram (sequentially structured event) or several truth tables (combinational event).	Derive an I/O Boolean equation for each state (even of state output) from the state transition diagram; from input combinations within the truth table, use Karnaugh maps and then combine the resulting state outputs with POS and SOP methods.
	In the case of multiple concurrent exclusive operating cycles, construct a sequential function chart.	

Sequential Logic Controller Circuit Design

13. Select solid-state devices (latches, flip-flops).

(Continued)

TABLE 3.4A (CONTINUED)

Step-by-Step Logic Controller Design Methodology

Design Step	Item Description
14. From the truth table or the sequence table, derive the state table associated with the logic devices selected.	
15. Use K-maps to derive I/O Boolean functions for device inputs.	
16. Simplify and check the consistency of I/O Boolean functions.	Use a timing diagram.

Wiring Diagram and Logic Controller Programming Languages

17. Develop the wiring diagram between the sensing and actuating devices, as well as the human–machine interface and the logic programmable controller unit.	Electrical power supply diagram. I/O control unit wiring diagram (in/out of control unit).
18. Set a table of mnemonics for the I/O system variables based on the physical addresses in the programming environment.	Use wiring diagram to assign addresses to system variables. This table should have information on the type of variables (analog/digital), address of I/O, range of values, etc.
19. Choose the programming language and develop logic controller application.	Convert I/O Boolean equations into a classic programming language, such as ladder, function block, SFC, IL, ST, or even assembly language.
20. Apply first-scan subprogram.	
21. Sizing and selection of the computing hardware (control unit) and the detection/sensing equipment.	
22. Verify the cycles of system operation and check if logic meets the safety specifications.	Use sequential logic software analysis tools and timing diagrams for all input transition conditions and system outputs.

Automation Project Documentation

23. Establish wiring and electrical connection diagrams integrated into the control unit schematics with physical addresses.	Compile P&ID, PFD, and electrical wiring diagram.
24. Documentation	Detailed system design and implementation, I/O wiring diagram, I/O address assignments, data memory address assignments, mnemonic and process variable specifications, control program printout, maintenance disaster recovery plan.

3.3 Process Description and Functional Analysis

The functional analysis aims to characterize all process operating cycles by decomposing the process into major operating tasks or functions and identifying their activation and deactivation conditions. Then, it is possible to establish the chronological order of the process operations' execution, also called process sequencing. This is achieved by gathering overall process information, such as present system state conditions (e.g., initial condition, condition of action termination, amd operating time constraints), input transition conditions (level or push type), and even resulting activated (next) system state outputs. The procedure to perform function analysis of a discrete-event system can be summarized as follows:

1. Perform the functional decomposition of system operations using FAST or Structured analysis and design technique (SADT) methods, which consist of breaking down into primary and secondary levels some process operating functions. FAST is presented in Section 4.5.3. With SADT, a function is defined as the process operation to be carried, while system operations are lists of actions to be performed by system actuating devices. Hence, in the function table (FT), for each action, there is an activation condition required prior to its execution (precondition). Information on all actions assigned to each function is presented in an FT in chronological order, as depicted in Table 3.4B.

2. Once each function has been structured in a chronological order of involved actions, this table can be converted into a dependency chart to present those functions in a sequence. Here, the dependency chart illustrates the relation over time between the different functions through rectangles, as depicted in Figure 3.6. The sequential execution of functions is illustrated by arrows, which could describe either a parallel execution or an alternative execution of functions. The dependency chart and FT should be refined while taking into account all design specifications. Subsequently, the FT and dependency chart could be transformed into a sequential function chart, as presented later in this chapter.

3. Perform the safety analysis of the process operation's execution from expected defaults and specify other safety constraints (e.g.,

TABLE 3.4B

Module Component from FT Using SADT Method

Function (from Module)	Activation Condition	Activation Device	Action Description (sequence)	Execution Device
Action 1				

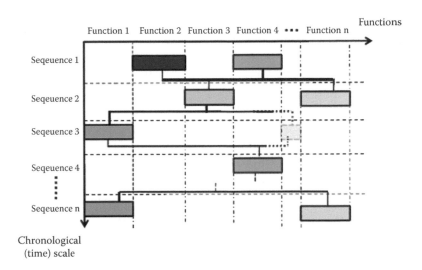

FIGURE 3.6
Dependency charts showing the sequence of process events.

maximum operating values). Then, for each expected system hardware and software operational default, develop interlocks and design forcing action to restrict the execution of some process operations (by adding process inputs and outputs, as well as process states). For each process operation with safety constraints, the FT and dependency chart should be refined according to newly identified process operating sequences, as well as input transitions.

4. Derive all process operating sequences, including those related to safety measures, for each process operating cycle. This should lead to gathering all input conditions and their corresponding system outputs activated for each process operating sequence.

5. Present information representing the characterization of the process operations in either a truth table (combinational structured event); sequence table (sequentially structured event); state transition table, even an excitation table; or state transition diagram.

3.4 Formal Modeling of Discrete Systems

Discrete-event systems have switching system outputs resulting from a change in the set of binary input variables, sometimes in a specific timely order. Thus, the formal modeling of such discrete-event systems consists of capturing the logic relationship between all input transition conditions causing the activation of output variables, in terms of Boolean functions. Hence, among formal modeling methods there are (1) the truth table and K-maps; (2) the switching theory in sequence table analysis; (3) the state diagram, which is a graphical representation of process operations based on expected input transitions; and (4) the sequential function chart (SFC). The resulting I/O Boolean functions can be implemented using hundreds of relays or logic programmable devices in the same way as a general-purpose computer is used to solve algebraic equations. Those methods are presented in the subsequent sections.

3.4.1 Process Modeling Using Truth Table and K-Maps

In the case of combinational discrete-event systems having few inputs (i.e., less than three inputs), the I/O Boolean relationship can be derived using the following steps: (1) construct the truth table, (2) derive the stable transition table, and (3) convert it into K-maps by using selected solid-state devices, such as flip-flops or latches, to derive the output Boolean function.

Example 3.4

Consider a system with three logic input variables (push buttons START and STOP) that are used to activate and deactivate the system output variable MOTOR_STARTER. This defines the process state of the motor ON or OFF. The relationship between those variables is given through the truth table operating conditions, as presented in Table 3.5A. By choosing a D flip-flop, the corresponding state transition table can be derived as depicted in Table 3.5B.

Hence, the resulting K-maps can be obtained as illustrated in Table 3.5C. Applying the SOP method, the I/O Boolean yields

$$MOTOR_{STARTER} = START \cdot \overline{STOP} + MOTOR_{STARTER} \cdot \overline{STOP}$$
$$= (MOTOR_{STARTER} + START) \cdot \overline{STOP}$$

Using the D flip-flop and considering N as the number of system states (system output combination) and n as the number of flip-flops, the design rule is given by

$$2^n \geq N$$

TABLE 3.5A

Truth Table of Motor Starter

START	STOP	MOTOR_STARTER
0	0	MOTOR_STARTER
0	1	0
1	0	1
1	1	0

TABLE 3.5B

State Transition Table of Motor Starter

MOTOR_STARTER Present State ($Q(t)$)	Process Inputs		MOTOR_STARTER Next State ($Q[t+1]$)	D
	START	STOP		
MOTOR_STARTER	0	0	MOTOR_STARTER	0
0	0	1	0	0
0	1	0	1	1
0	1	1	0	0
1	0	0	1	1
1	0	1	0	0
1	1	0	1	1
1	1	1	0	0

TABLE 3.5C

K-Maps of Motor Starter

		MOTOR_STARTER			STOP
		00	01	11	10
START	0	0	0	0	1
	1	1	0	0	1

Here, with $n = 1$, the D flip-flop gate output can be summarized as

$$Q_1(t+1) = START \cdot \overline{STOP} + Q_1(t) \cdot \overline{STOP} = \left(Q_1(t) + START\right) \cdot \overline{STOP}$$

As such,

$$D = \left(Q_1(t) + START\right) \cdot \overline{STOP} = \left(MOTOR_{STARTER}(t) + START\right) \cdot \overline{STOP}$$

The equivalent logic controller circuit is given in the Figure 3.7.

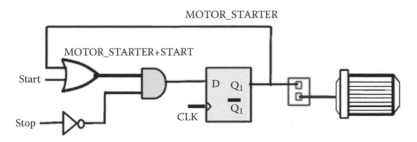

FIGURE 3.7
Logic circuit of a motor starter with a D flip-flop.

3.4.2 Process Modeling Using Sequence Table Analysis and Switching Theory

In the case of sequentially structured discrete-event systems, it is possible to directly express the sequential execution of process operation in Boolean statements using the switching theory. For all events within each operating cycle, the sequence of operations must be compiled into a sequence table containing each sequence, information related to the input transition conditions, and the switched system output value. A procedure for obtaining I/O Boolean functions from the sequence table is as follows:

1. List in a table all input and output devices involved in process operations with their respective logic variables. If possible, add information regarding the type of operating modes (level or push) and their initial conditions.

2. From the process functional analysis, consequently construct the sequence table for each operating cycle such that the operating sequences are listed in chronological order in the first column. In the following columns, list their corresponding activated system inputs and the logic value for each system output, respectively.

3. From the analysis of the sequence table, search for any change of a set of system output values that have the same combination of input values but result in different system output values. Those identical sequences (i.e., where the same previous sequence output values, along with the same system inputs, cause different system outputs) lead to a confusing situation where identical input variables from similar previous sequence output values can set more than one set of system output values.

4. In order to differentiate those identical sequences from all other sequences in the operating cycle, some virtual output variables can be created—as many as necessary.

5. For each system (virtual and real) output variable, derive the I/O Boolean function by identifying the input transition conditions and the output and virtual variables required to be activated or deactivated (i.e., turn those variables ON and OFF) such that

$$Output_i [1] = \sum all\ input\ transition\ conditions\ activating\ Output_i \qquad (3.6)$$

while

$$Output_i [0] = \prod all\ input\ transition\ conditions\ deactivating\ Output_i \qquad (3.7)$$

Then,

$$(Output_i = (Output_i + (Output_i [1]) \cdot \overline{(Output_i [0]}) \tag{3.8}$$

Recalling Example 5.4 with a motor starter, $Motor_{starter} [1] = START$ and $Motor_{starter} [0] = STOP$, the I/O Boolean function would result in

$$Motor_{starter} = \left(Motor_{starter} + Motor_{starter} [1] \right) \cdot \overline{Motor_{starter} [0]}$$
$$= (Motor_{starter} + START) \cdot \overline{STOP}$$

This logic process modeling method allows us to have a system output turned or maintained at zero (OFF), especially when there is a conflict between input transition conditions (e.g., pushing start and stop at the same time).

Example 3.5

Consider a double-tank water pretreatment system (Figure 3.8) whose devices are listed in Table 3.6A.

The operating sequences for the water treatment tank should ensure that PUMP 2 is turned ON ($CR2 = 1$) when the tank 2 level limit switch

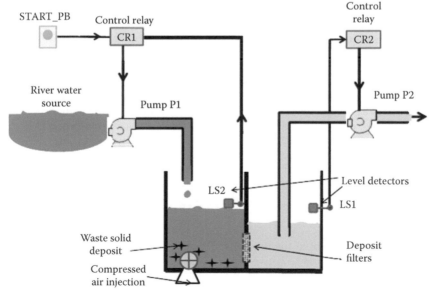

Potable water pretreatment stages

FIGURE 3.8
River-based water treatment system.

TABLE 3.6A

Equipment Involved in a Water Treatment Tank

Input Equipment	Symbol	Output Equipment	Symbol
Start push button	START_PB	Control relay 1 for outflow PUMP 1	CR1
Level tank 2 limit switch	LS1	Control relay 2 for inflow PUMP 2	CR2
Level tank 1 limit switch	LS2		

TABLE 3.6B

Sequence Table Analysis for Water Treatment with Double Tank

	Input Devices			Output Devices		Analysis	
Sequence	START_PB	LS1	LS2	CR1 (Pump 1)	CR2 (Pump 2)	Changing	Virtual
1. Tank empty	1	0	1	1	0	X	
2. Filling tank	1	0	0	1	0		
3. Tank full	1	1	0	1	1	X	
4. Emptying tank	1	0	0	1	1		

(*LS1*) is reached (activated by *LS1* = 1) and should stay ON until the water-level limit switch (*LS2*) is reached in tank 1 (activated by *LS2* = 1). Then, PUMP 2 should be turned OFF (*CR2* = 0) and stay OFF until filtered water rises up again to the limit switch *LS1* and activates it (*LS1* = 1). It is assumed that the flow rate out of PUMP 2 is higher than the inflow from PUMP 1. This is summarized in Table 3.6B. Here, PUMP 1 for the inflow of water from the river is always activated once the start push button (*CR1* = *START_PB*) is pushed. Limit switches (*LS1*) and (*LS2*) are operating on the level mode. A memory function could be required to maintain the pump ON until *LS2* is reached. From the sequence table analysis, only *CR2* changes its logic output value (activated or deactivated), such that *LS1* = 1 is required for *CR2* = 1, resulting in *CR2*[1] = *LS1*, while *LS2* = 1 is required for *CR2* = 0, causing *CR2*[0] = *LS2*. Hence, using the equipment listed above, the I/O Boolean function for *CR2* yields

$$CR2 = (CR2 + CR2[1]) \cdot \overline{CR2[0]}$$

This is equivalent to the relationship between output variable PUMP 2 and inputs *LS1* and *LS2*, given by

$$CR2 = (CR2 + LS1) \cdot \overline{LS2}$$

$$CR2 = PUMP2$$

Example 3.6

Recall Example 3.5 of the first tank filled by water from the river through PUMP 1. After the solar-based water pretreatment process, the water is removed through PUMP 2 once a tank 1 limit switch detects the upper water level. In order to ensure proper timing of treatment, the pump is alternatively activated and deactivated every 15 s for a total period of 200 s, as depicted in Figure 3.9. The resulting I/O Boolean equations are given by

$$CR1 = (CR1 + START\_PB) \cdot \overline{LS2}$$

$$PUMP1 = CR1$$

$$CR2 = (CR2 + LS2) \cdot \overline{TIMER[15\,sec]}$$

$$TIMER_{200\,sec} = CR2$$

$$PUMP2 = TIMER[200\,sec]$$

$$TIMER_{15\,sec} = TIMER[200\,sec]$$

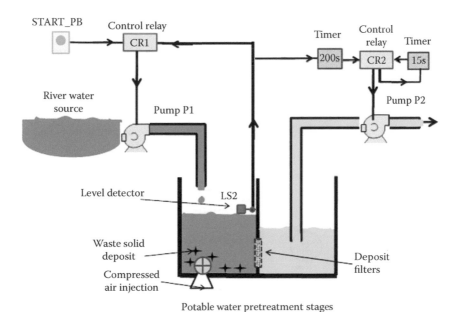

Potable water pretreatment stages

FIGURE 3.9
Water pretreatment and pumping system.

3.4.3 Process Modeling Using Simplified State Diagram

In the case of sequentially structured discrete-event systems with multiple operating cycles that are mutually exclusive (e.g., an alternative sequence exists), it is suitable to use a simplified state diagram. Hence, the sketched state diagram described in Figure 3.5 can be converted into an I/O Boolean expression using

$$S_i = \left(S_i + \sum_{j=1}^{n} (IT_{j,i} \cdot S_j) \right) \cdot \prod_{k=1}^{m} \left(\overline{IT_{i,k} \cdot S_{i+1,k}} \right) \tag{3.9}$$

with $\prod_{k=1}^{m} \left(\overline{IT_{i,k} \cdot S_{i+1,k}} \right)$ being all next states and involved input transition conditions deactivating the S_i, while $\sum_{j=1}^{n} (IT_{j,i} \cdot S_j)$ is all previous states and involved input transition conditions that activate the S_i. Recall that S_i represents the state of the system at the ith operating sequence characterized by all the output logic values, n is the overall number of input transitions leading to the ith state, m is the total number of input transitions out of the ith state, $IT_{j,i}$ is the logical input conditions of a transition from the jth state to the ith state, and $IT_{i,k}$ is the logical input conditions of a transition from the ith state to kth state. If it is not possible to reach a state, a virtual intermediary state with output values identical to those of the preceding state should be created and inserted between the two states. Their state I/O Boolean function yields

$$Output_i \ of \ S_i = \sum_{i \ [among \ m...to...n]} S_{i \ [among \ m...to...n]} \ where \ Output_i \ activated \tag{3.10}$$

Example 3.7

Recall Example 3.5 on the motor starter system. The corresponding state diagram is illustrated in Figure 3.10 such that MOTOR_STARTER = 0 in state 0 (S0) and MOTOR_STARTER = 1 in state 1 (S1). A system state

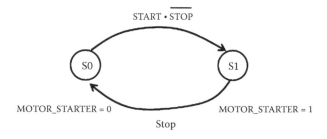

FIGURE 3.10
Two-state motor starter state diagram.

change occurs only when the start push button is activated, while simultaneously the stop push button is not activated $(START \cdot \overline{STOP})$. The system returns to its initial system state 0 only when $STOP = 1$.

Here, applying Equation 3.9 for two states $S0$ and $S1$ leads to Boolean equations describing $S0$ and $S1$, such as

$$S0 = (S0 + S1 \cdot STOP) \cdot \overline{\left(S1 \cdot START \cdot \overline{STOP} \right)}$$
$$S1 = (S1 + S0 \cdot START \cdot \overline{STOP}) \cdot \overline{\left(S0 \cdot STOP \right)}$$

From the equations above, it can be noted that to solve those Boolean equations, it is necessary to have the initial conditions of each state and the order of execution. Equation 3.5 can only be applied if such information is not available; otherwise, once the state I/O Boolean functions are described, the system outputs can be defined as functions of the states. Using Equation 3.10, it can be found that the system output ($MOTOR\_STARTER$) is ON in state 1, such that

$$MOTOR\_STARTER = S1$$

Substituting $MOTOR\_STARTER$ into $S1$ ($S0$ being discarded, as there is no output variable activated), the equation above yields

$$MOTOR\_STARTER = (MOTOR\_STARTER + START \cdot \overline{STOP}) \cdot \overline{STOP}$$

which could be simplified to

$$MOTOR\_STARTER = (MOTOR\_STARTER + START) \cdot \overline{STOP}$$

With this method, the I/O Boolean functions are directly obtained from the state diagram with little effort. Instead of hardwired-based solid-state device implementation, as in the previous example, this I/O Boolean function can be implemented using a software-based logic programmable device. However, some safety measures should be inserted to prevent overlapped sequence execution. This could be done by ensuring enough elapsed transition time between multiple new and old states during the activation and deactivation of output devices.

Example 3.8

A T-junction is controlled by traffic lights, as depicted in Figure 3.11a. In this example, the following are not considered: road surface detectors to acknowledge any vehicle presence, pedestrian push buttons and traffic lights ("Walk" and "Don't Walk" lights), law enforcement for manual control of the intersection, priority lights, traffic signal preemption for emergency vehicles, and coordinated signals with other street lights.

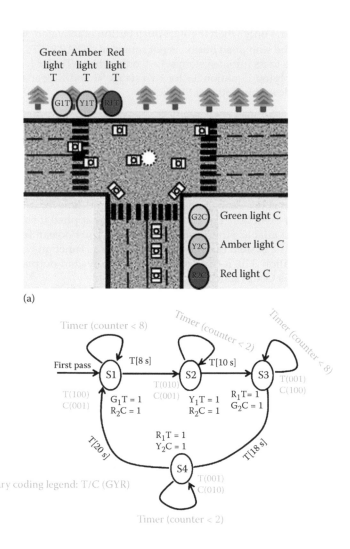

FIGURE 3.11

(a) T-junction type of vehicle traffic. (b) Moore-like state diagram of traffic light (with incremental timer).

Here, the objective is to design a car traffic light logic controller. In that regard, the corresponding timing diagram depicting the red–green–yellow sequence is shown in Figure 3.12.

There are six traffic lights: red (R1T), yellow (Y1T), and green (R1T) lights in the straight direction and similar red (R2C), yellow (Y2C), and green (R2C) lights in the crossing direction. The traffic lights operate based on a timer ON (TON) instruction that resets after every cycle of 20 s, as presented in Figure 3.11a. From the functional analysis, the input transition conditions are timer (8 s) (T[8]), timer (10 s) (T[10]), timer (18 s)

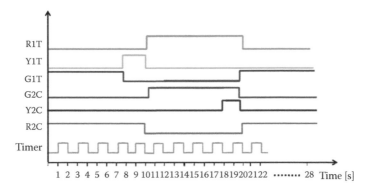

FIGURE 3.12
Timing diagram of traffic light at junction.

(T[18]), and timer (20 s) (T[20]). The system outputs are the six lights (R1T, Y1T, G1T, R2C, Y2C, and G2C) and the timer. Note that there are four states. Table 3.7A and Figure 3.11a present the operational characteristics.

From the state diagram depicted in Figure 3.11b, the subsequent I/O state Boolean equations are

$$S1 = (S1 + S4 \cdot T[20]) \cdot \overline{S2 \cdot T[8]} + First\ pass$$
$$S2 = (S2 + S1 \cdot T[8]) \cdot \overline{S3 \cdot T[10]}$$
$$S3 = (S3 + S2 \cdot T[10]) \cdot \overline{S4 \cdot T[18]}$$
$$S4 = (S4 + S3 \cdot T[10]) \cdot \overline{S1 \cdot T[20]}$$

The output Boolean functions are then

$$G1T = S1 + S3$$
$$R2C = S1 + S2$$
$$Y1T = S2$$
$$G2C = S3$$
$$Y2C = S4$$
$$R1T = S3 + S4$$

Revisit the T-junction of five routes for vehicles, with no collision-free crossing, as previously covered. It is desired to develop a collision-free traffic light with six routes instead. As such, another traffic light, F, is added to ensure collision-free left crossing from the T-road direction to the C-road direction. Here, there are five states or light patterns. Their tables of sequence are presented in Table 3.7A and B, respectively. All four states required for collision-free crossings are illustrated within the resulting state transition diagram shown in Figure 3.13.

TABLE 3.7A

Switching State Table of Traffic Junction

Sequence	System Inputs								System Outputs						
	T[8]	T[10]	T[18]	T[20]	G1T	Y1T	R1T	G2C	Y2C	R2C	Timer
GT1 and R2C ON	1	0	0	0	1	0	0	0	0	1	1
Y1T and R2C ON	0	1	0	0	0	1	0	0	0	1	1
R1T and G2C ON	0	0	1	0	0	0	1	1	0	0	1
R1T and Y2C ON	0	0	0	1	0	0	1	0	1	0	1

TABLE 3.7B

Sequence State Table of Traffic Junction (with Left Crossing)

Sequence	System Inputs							System Outputs					
	T[8]	T[10]	T[18]	T[20]	T[22]	F	G1T	Y1T	R1T	G2C	Y2C	R2C	Timer
R1T, G2C ON	1	0	0	0	0	0	0	0	1	1	0	0	1
R1T, Y2C ON	0	1	0	0	0	0	0	0	1	0	1	0	1
G1T, R2C ON	0	0	1	0	0	0	1	0	0	0	0	1	1
Y1T, R2C ON, F ON	0	0	0	1	0	1	0	1	0	0	0	1	1
R1T, R2C ON, F ON	0	0	0	0	1	1	0	0	1	0	0	1	1

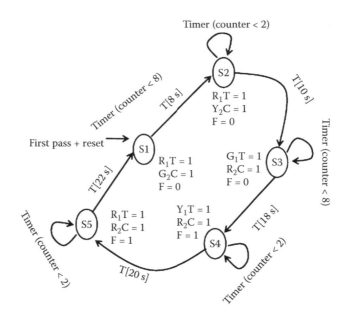

FIGURE 3.13
State diagram of traffic light with left crossing signaling.

3.4.4 Process Modeling Using Sequential Function Chart

In the case of a sequentially structured discrete-event system with multiple concurrent operating cycles, a graphical-based modeling method called the SFC can be used. Similar to the state diagram, with SFC, the state is replaced by a step and the system output by an action. A typical example is a lock that is activated through two 3-digit codes. Here there are two parallel operating sequences that have to be validated within the same time frame. Because it is also a logic programming method, it is presented later in Section 3.6.6.

3.5 Logic Controller Circuit Design

From any I/O Boolean function derived, the digital system or logic controller circuit could be implemented to solve it recursively using solid-state devices. This could be designed using a Moore or Mealy logic controller circuit, as illustrated in Figure 3.14. Electronic devices such as flip-flops or latches are commonly used to implement such logic controller circuits. As illustrated in Figure 3.14, with Moore implementation, the system logic state is a combination of switching outputs only when inputs change. It is suitable for applications not requiring "memory." In the case of the time-triggered Moore implementation,

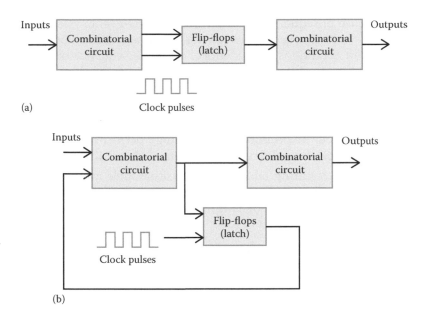

(a)

(b)

FIGURE 3.14
(a) Moore implementation of logic controller circuit. (b) Mealy implementation of logic controller circuit.

the input changes only on the rising clock edge such that the switched outputs would remain at this state until the next clock cycle. In the case of the Mealy implementation, the output logic is a combination of switching outputs when changes of inputs and previous system states occur. Here solid-state electronic device outputs are considered a present system state. When a clock signal is used as an input signal, it is a time-triggered Mealy implementation. The logic controller circuit can be represented as a state diagram.

Hence, the procedure to design the logic control circuit can be summarized as follows:

1. List all system input and output devices, operating characteristics, and sequences per operating cycle, if possible.
2. From the functional analysis, derive the relationship between the combination of activated system inputs and their corresponding system outputs for all sequences per operating cycle.
3. Summarize the system input and outputs relationship in a truth or sequence table.
4. Construct the state transition tables using either a truth or sequence table.
5. Convert all present states, system inputs, and next states into binary gray code sequence format.

6. In the case of sequentially structured discrete-event systems, identify any operating sequences having the same combination of input transition conditions, and identical previous sequence output values, leading to different current sequence outputs. A virtual system output could be used to differentiate those identical operating sequence activation conditions, in order to have a unique system I/O relationship.

7. Otherwise, per inspection (row matching), perform on the truth or sequence table the reduction of numbers of identical sequences.

8. Convert the resulting table into a transition table and perform the solid-state device (i.e., flip-flop) assignment to the system state using the rules of the number of assignments (N_D) given by

$$N_D = \frac{(2^{N_{FF}} - 1)}{(2^{N_{FF}} - N_S)! N_{FF}!} \qquad (3.11)$$

where N_S is the number of states from the resulting state transition table and N_{FF} is the number of solid-state devices (i.e., flip-flops) to be used. This is also equivalent to $2^{N_{FF}-1} \le N_S \le 2^{N_{FF}}$. As a rule of thumb, for states in the state transition table, there are solid-state devices that could be used given by $2^{n-1} \le m \le 2^n$.

9. Choose the solid-state device type (flip-flops or latches, e.g., SR, D, T, or JK flip-flops) to be implemented accordingly.

10. Design the logic controller circuitry using the switching theory for level mode operations. Based on equivalence between the states of selected solid-state devices (flip-flops or latches) and each state from the resulting state transition table, derive the output equations as functions of system input and present state (Q^-). For each selected device, the output equation (next state equation [Q^+]) is given using either the state transition table (Table 3.8A) or the characteristics Equation 3.12.

11. Substitute the assigned flip-flop states for each state in the state transition table and construct the upgraded state transition table by

TABLE 3.8A

Generic State Transition Table and Equivalent Solid-State Device I/O

Q– (Present State)	Q+ (Next State)	S	R	D	T	J	K
0	0	0	d	0	0	0	d
0	1	1	0	1	1	1	d
1	0	0	1	0	1	d	1
1	1	d	0	1	1	d	0

specifying the next states of the solid-state devices (flip-flops and latches), as well as each output in terms of the present states and the input.

12. Derive the Boolean equation of the I/O functions using equivalence from the table and the logic equations below.

$$Q_{SR}^+ = S + \bar{R}Q$$
$$Q_D^+ = D$$
$$Q_T^+ = T\bar{Q} + \bar{T}Q \quad (3.12)$$
$$Q_{JK}^+ = J\bar{Q} + \bar{K}Q$$

13. Simulate circuit realization and plot the timing diagram of system input and output variables to check the results achieved.

Example 3.9

Reconsider the water treatment process in Example 3.6. With $S2 = CR2$ and $S1 = CR2$, using Equation 3.9, the state Boolean function would lead into

$$S2 = (S2 + LS1 \cdot S1 + LS1 \cdot \overline{LS2} \cdot S2) \cdot \overline{(LS2 \cdot S1)}$$

which is equivalent to

$$CR2 = (CR2 + LS1 \cdot \overline{CR2} + LS1 \cdot \overline{LS2} \cdot CR2) \cdot \overline{(LS2 \cdot CR2)} = CR2 + LS1 \cdot LS2$$

The equivalent Mealy and Moore state diagrams are depicted in Figures 3.15 and 3.16. The transition table could be derived as summarized in Table 3.8B.

Using the current state (Q1Q0) and LS1 and LS1 from the state transition table above, it is possible to fill in the corresponding value of the next state corresponding output value (CR2) in the excitation table. This is equivalent to the Karnaugh table summarized in Table 3.8C.

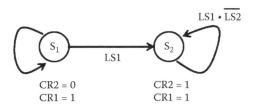

FIGURE 3.15
Mealy state diagram of the water treatment process.

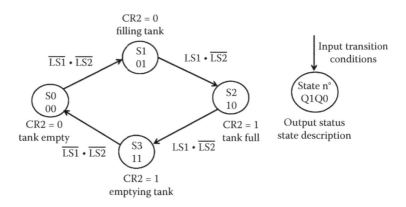

FIGURE 3.16
Moore state diagram of water treatment process for tank 2.

TABLE 3.8B

Water Treatment State Transition Table

State (Q1Q0)	LS1	LS2	Next State (Q1Q2)	Corresponding Output CR2
00	0	0	01	0
01	1	0	10	0
10	0	0	11	1
11	0	1	00	1

TABLE 3.8C

Corresponding K-Maps of Water Treatment Process

		LS1 LS2			
		00	01	11	10
Q1Q0	00	1	X	X	X
	01	X	X	X	0
	11	X	1	X	X

Example 3.10

Consider the sequence table for the water treatment process whose operating sequences are summarized in Table 3.9A; the corresponding state transition table is given in Table 3.9B.

Using the sequence table, the corresponding Karnaugh map is summarized in Table 3.9C.

TABLE 3.9A

Sequence Table Analysis for Water Treatment Process

	Input Devices			Output Devices	
Sequence	Start_PB	LS1	LS2	CR1 (valve)	CR2 (pump)
1. Tank empty	1	0	1	1	0
2. Filling tank	1	0	0	1	0
3. Tank full	1	1	0	1	0
4. Emptying tank	1	0	0	1	1

TABLE 3.9B

Water Treatment Process State Transition Table

CR2(t)	LS1	LS2	CR2($t + 1$)
0	0	0	1
1	1	0	1
1	0	0	0
0	0	1	0

TABLE 3.9C

Corresponding K-Maps

		LS1 LS2			
		00	01	11	10
CR2	0	1	0	X	X
	1	0	X	X	1

Using the D flip-flop, the state Boolean function is given by

$$CR2(t+1) = LS1 + \overline{CR2(t)} \cdot LS2 = \overline{CR2(t) \cdot LS2} \cdot \overline{LS1} + CR2 \cdot LS1 \cdot \overline{LS2}$$

3.6 Logic Controller Programming Languages

Embedded devices such as FPGA and μController use assembly language programming instructions to implement I/O Boolean functions. Among those instructions used to transfer programs or data to or from the controller processing unit are (1) register-to-register operations, such as ADD; (2) memory-to-register operations, such as LOAD and STORE; and (3) memory-to-memory operations, such as STRCMP (string comparison). There is also a set of read input and write output instructions indicating whether to transfer data and memory address buses to or from I/O devices and the controller processor. Those instructions have similar assembly language, with some variations based on a vendor-specific compilation environment. There are some methods to implement control logic, known as the five International Electrotechnical Commission (IEC) 1131 standards. They consist of two graphical languages, such as ladder or relay diagrams (LDs) and function block diagram (FBDs) (a derivative of data flow graphs), and two text-based languages, such as Instruction lists (IL) (equivalent to assembly language), structured text (ST) (a derivative of Fortran), and an object-oriented programming framework called SFCs. The graphical languages are symbol-based program control instructions, while the text-based languages are text-based program control instructions. SFC is actually an organizational structure that can integrate ladder, ST, IL, and function block programming languages. All these programming languages are presented in this section.

3.6.1 Synchronizing Mechanism of Computer Program Execution

Logic controller programs are embedded within processing devices that interface with other I/O devices involved in the execution of system operations. Hence, the execution of such programs has to be synchronized with lower-pace system I/O devices for data reading and writing operations. The synchronizing mechanism could be done using clock signals (synchronous mode) by associating next state output variable occurrences with input clock signal changes (from logic level 0 to 1, or vice versa). This method requires prior knowledge of a system's time delays, such as the controller processor computing time or system I/O data processing and transfer times. Another synchronizing mechanism uses interrupt programs (asynchronous mode). It relates I/O device status (1/true or 0/false) to its readiness to read or write process data to or from the controller processor. This interrupt program temporarily stops the control logic program and activates the data transfer to or from the specific I/O device. In contrast to the other method, it allows an update of command input signals without interfering with the control process. For example, emergency push buttons are typical interrupt signals activating contingencies such as control panel alarm lights.

3.6.2 Ladder Diagram

A ladder circuit diagram consists of a compound of individual circuits that are executed sequentially. Each circuit is a contact terminology known as a rung (or network). Each rung can only activate or deactivate only one system output device or individual virtual system state variable. The utilization of virtual state variables should be reduced to maintain the simplest logic relationship between system inputs and outputs.

The LD program embedded within the controller devices (e.g., PLC) establishes the logic relationship between the system input and output field devices. Similar to a hardwired relay ladder, it represents the system input

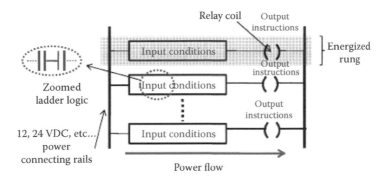

FIGURE 3.17
Typical structure of relay-based logic program.

devices assembled in series and parallel configurations to control several connected system output devices. When activated (e.g., ON or closed position), current can flow through input devices, eventually energizing the system output, as well as stopping (e.g., OFF or open position) the current flow. Hence, depending on the configuration, this allows switching the output devices ON or OFF. This language is similar to that for the electrical relay diagram. Figure 3.17 depicts the general structure of a ladder circuit diagram.

3.6.2.1 Elements of the Ladder Diagram

Typical LD elements are control relay coils, relay contacts, inputs (e.g., push button, electromechanical limit switch, selector switch, proximity switch, and timer contact), and outputs (e.g., solenoid, indicator light, horn, and timer). In addition, there are counting and timing logic functions that are used to set a system output device ON or OFF after the expiration of a time delay or a preset value is reached. Table 3.10A illustrates some symbols of those devices.

Time delays are used to set a pause between two events (e.g., machine operations) or to define the time when an event is to take place. Among timer functions there are (1) the delay turning ON action timer (TON) function, where the timer should switch ON (contact closed) after a countdown to the current time (similarly, there is the delay turning OFF action timer [TOFF] function to switch OFF the timer); (2) the retentive ON action timer (RTO) function, where the timer turns and remains ON even if the power supply is turned OFF (similarly, there is the retentive OFF action timer [RTF] function); and (3) the retentive reset action timer (RTR) function, which, for example, is suitable to track the time required before any elevator preventive maintenance. Timer functions contain information on timer resolution or accuracy, called time base (TB), and the number of counts, called ticks. For example, with a 30 s event, when the timing counter has a time base of 1 s, the timing counter must count 30 times (ticks) before it setting its output ON. Those timing functions use a

TABLE 3.10A

List of Common LD Elements

Contact Relays		Relay Coils	
⊣⊢	Normally open	⊣( )⊢	Coils
⊣/⊢	Normally closed	⊣(/)⊢	Negating coils
⊣P⊢	Edge contact, positive edge	⊣(S)⊢	Setting coils
⊣N⊢	Edge contact, negative edge	⊣(R)⊢	Resetting coils
⊥	Normally open push button	⊣(P)⊢	Edge coil, positive edge
⊥	Normally closed push button	⊣(N)⊢	Edge coil, negative edge
—◠—	Normally closed limit switch	—⋀—	Solenoid
—◟—	Normally open limit switch		

TIMER control output signal to activate the timer (TIMER is changed from 0 to 1 or vice versa) and a controller timing (delay, Δt) input signal to indicate whether the desired time delay (TIMER [Δt] is reached (changing from 1 to 0 or vice versa). Similarly the counter functions use a preset value to be reached and an accumulated count value that is varied during the counting process. Among counter functions, there are (1) the count up (CTU) function, to count until an accumulated count value reaches a preset value; (2) the count down (CTD) function, when the accumulated count value decreases equal to the preset value; and (3) the counter reset (CTR) function, which resets the accumulated count value of the up and down counters to zero.

Whether using I/O Boolean functions or state diagrams, it is possible to convert them into LD programs. Indeed, using basic logic Boolean functions, some configuration of contacts in the ladder rung can be easily realized, as illustrated in Table 3.10B for AND, OR, EXCLUSIVE OR, and NOT functions.

From the state diagram, it is possible to directly convert each state and each system output into Boolean functions using Equations 3.5 or 3.9 and 3.10. Then, a translation of those state Boolean functions into an LD program consists of

- Constructing a ladder rung for each state $State_i$ Boolean function using the table of equivalence given in Table 3.10. This should be achieved through successive order such that each state preceding $State_i$ is constructed first.
- Similarly constructing a ladder rung for each system output O_m Boolean equation.

A typical generic ladder rung for a state S_m is illustrated in Figure 3.18.

TABLE 3.10B

Table of Ladder Equivalence between AND, OR, EXCLUSIVE OR, and NOT Functions

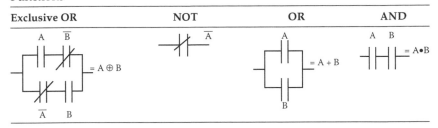

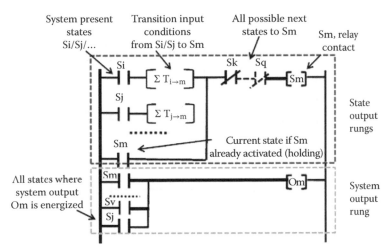

FIGURE 3.18
Generic equivalent LD relating states to system output.

Example 3.11

Consider the system formal model given by the state diagram in Figure 3.19.
From the simplified state diagram depicted in Figure 3.19, each Boolean state function is given by

$$S_1 = (S_1 + S_3 \cdot STOP) \cdot \overline{S_2}$$
$$S_2 = (S_2 + S_1 \cdot \overline{PD1 \cdot PD2}) \cdot \overline{S_3}$$

while each system output Boolean function results in

$$KM_2 = S_3$$
$$KM_1 = S_2$$
$$TIMER = S_2$$

The equivalent to the LD is given in Figure 3.20.

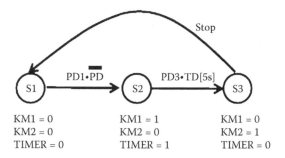

FIGURE 3.19
Equivalent state diagram.

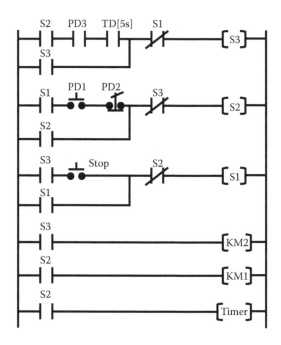

FIGURE 3.20
Equivalent LD.

3.6.3 Function Block

The FBD is a symbol-based programming language derived from data flow graphs that allows us to program logic controllers in a way similar to that of wiring electrical circuits. FBDs are either customized or vendor-designed function blocks to be modified according to control program requirements. This is function block encapsulation. Usually, FBDs are represented by rectangles, with input and output types specified as illustrated in Figure 3.21. Typical function blocks are listed in Table 3.11.

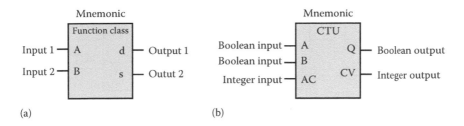

FIGURE 3.21
(a) Generic function block sketching. (b) Function block of incremental counter.

TABLE 3.11

Standard Function Blocks

Symbol	Definition
SR	Set bistable block
RS	Reset bistable block
CTU	Counter up block
CTD	Counter down block
TP	Pulse time block
TON	Timer on delay block
TOF	Timer off delay block
R_TRIG	Rising edge trigger block
F_TRIG	Failing edge trigger block

The declaration of a system output variable allows them to be assigned and used in other instructions throughout the program by mapping their addresses. Other examples of function block programs are presented in Figure 3.22a and b. Also, with the FBD language, some loop structures can be constructed using a feedback path, as shown in Figure 3.22.

Figure 3.22c represents another example of using a function block where a gate would be closed after successive integer and Boolean comparisons.

Example 3.12

Figure 3.23 illustrates the equivalent hardwired start and stop circuit implementation using LD in a customized FBD to activate both MOTOR ON and LED ON. Note that there are two rungs for the logic activation of the two system outputs by both the input and output variable conditions. The mnemonic should be compliant with the I/O address in the hardwired circuit. This LD is programmed and encapsulated within a FBD, as shown in Figure 3.23.

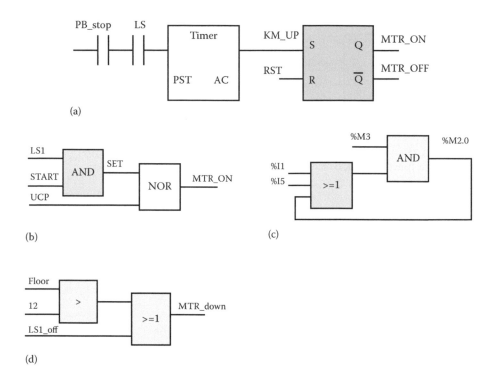

(a)

(b)   (c)

(d)

FIGURE 3.22
(a) FBD associated with LD. (b) FBD language. (c and d) Comparison function block.

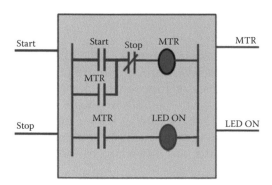

FIGURE 3.23
Motor start and stop function block.

3.6.3.1 Invocation and Incorporation of Functions and Function Blocks

LD allows the utilization and invocation of an infinite amount of functions and function blocks. The only requirement to incorporate functions and function blocks within an LD is the declaration of corresponding binary input and output variables of the involved block. If this is not the case, it is to

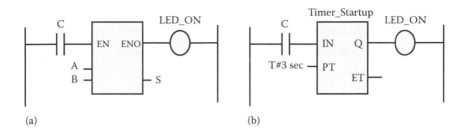

FIGURE 3.24
(a) Incorporation of function blocks. (b) Invocation of function blocks.

add a binary input with a parameter EN (enable), as well as a binary output with a parameter ENO (enable OK), at the involved functions or function blocks. This ensures the power flow through the block.

An example of incorporation commonly used for reading input signals A and B is shown in Figure 3.24a. Here, if a logic level 1 signal is sent to the binary input variable EN, then the data socket is opened and a Boolean addition operation is performed between input variables A and B, with the result assigned in variable value S. Simultaneously, the value of output ENO is updated if the addition operation is properly executed and correct (ENO = 1) based on the activation of the variable S = 1. When the block is not properly processed, the output ENO remains at the 0 logic level signal.

Figure 3.24 (b) depicts another example of invocation is when the time delay is a function block within the LD without additional EN and ENO parameters. Here, the timer delay function block is connected to its Boolean input IN and output Q variables. If a 1 logic level signal is sent to the input variable IN, the timer starts with a preset time (PT) of 3s within the function block Timer_Startup. This can be assigned for the duration that the value of output Q is at the 1 logic level to the Lamp.

3.6.4 Structured Text

ST is a text-based programming language similar to FORTRAN and BASIC-type computer languages. ST uses variable declarations according to the physical addressing of input- and output-connected devices, as well as virtually defined variables. Similar to conventional programming, ST iterative or conditional programs can be built using code such as WHILE...DO and REPEAT...UNTIL, FOR....END, or IF...THEN...ELSE. ST uses English-like instructions, which are summarized in Table 3.12. How to use them is presented in examples below.

ST is also used to create function block routines to perform the data processing of real-time I/O process devices, such as the process data monitoring function depicted in Figure 3.25. ST can be useful in forcing system output action. ST is suitable (1) for automation applications related to process data handling and sorting, and complex computational loads requiring values such as floating point, and (2) for advanced control-oriented applications,

TABLE 3.12

List of Statements and Instructions Possible in the ST Language

Statement/Operation/Instruction	Symbol
Exponentiation	**
Sign	–
Complement	NOT
Multiplication	*
Division	/
Comparison	<, >, ≤, ≥
Modulo	MOD
Addition	+
Substraction	–
Equality	=
Inequality	<>
Boolean AND, OR, XOR	AND, OR, XOR
Function processing	Function name (i.e., LOG(X))
Selection statement	IF, CASE
Interaction statement	FOR, REPEAT, WHILE…END
Assignment	:= (i.e., Y: = cos(X))

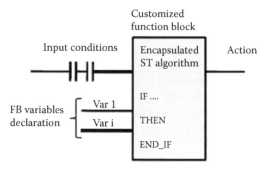

FIGURE 3.25
Function block (FB) created using ST.

including artificial intelligence (AI) computations, fuzzy logic, and distributed decision making.

ST statements (IF, CASE, FOR, REPEAT, WHILE, and EXIT) are described in the following examples of generic program segments.

3.6.4.1 IF Statement (IF THEN or IF THEN ELSEIF)

IF Boolean function 1 THEN output action1;
[ELSIF Boolean function2 THEN output action2;]
[ELSE output action;]
END_IF;

```
CASE selector OF:
      Value 1: condition 1;
         .
         .
         .
      Value k: condition k;
         .
         .
         .
      Value n: condition n;
ELSE condition(s);
END_CASE
```

3.6.4.2 CASE Statement (in the Case of Multiple Selections)

The CASE statement consists of a selector among possible statement groups. Each group is assigned an integer value type.

In the CASE statement, conditions of activation are grouped according to that selector value and are executed based on the value of the selector.

The *FOR loop statement* is used in the case where an output action should be repeated.

For Var:= Mathematical expression or Boolean function DO

Actions(s);
END_For

The REPEAT loop statement is used in the case where a system output action should be repeated based on an input transition condition.

Repeat
   Actions;
   UNTIL Mathematical expression or Boolean function
END_Repeat;

The *WHILE loop statement* is used in the case where a system output action should be activated or deactivated based on an input transition condition.

While Mathematical expression or Boolean function DO

Actions
END_While;

Other statements include the *EXIT statement*, which is to be is used in the case where a system output action should be terminated.

3.6.5 Instruction List

The IL is another text-based programming language corresponding to the machine or assembly language or C for programmable automation

TABLE 3.13A

Example of Described IL Programs for Automatic Conveyor Oven

Operator	Command	Comment
LD	temp	Read sensor temperature measurement
GE	185	Compare if greater or equal to 185 degrees
OR	Heater_off	Heater deactivated
ST	Conveyor_on	Conveyor motor activated

TABLE 3.13B

Example of Described IL Programs for Automatic CNC Machine Tool

Operator	Mnemonic	Comment
S	%I2.3	"Activate switch"
OR	%M4.1	"Turn on motor"
ST	%Q2.5	"Advance cutting saw"

devices (e.g., µController). It is useful for customized applications requiring optimization of the processing speed in the program execution or a specific routine in the program. IL instruction contains information on an operator, optionally a modifier, and at least one IL operand, such as LD, ST, AND, or LE, corresponding to, respectively, read accumulator value, store operation current, Boolean AND expression, and less than equal to instructions. Once the jump address is specified, all instructions are separated by commas.

In the examples listed in Table 3.13A and B, the logic control of a conveyor oven is used in order to cut off the heater when the temperature is above 185°C. In the second example, the band saw cutting tool advances toward the wood workpiece only after the limit switch positioning is uploaded into the register in the detector flag %I2.3 of the accumulator unit of the µController. This is an OR instruction to the logic value of the motor status (which is logic level 1 when ON), and it is setting the output %Q2.5 to 1.

3.6.6 Sequential Function Chart

SFC is a flowchart-like programming language. It is used to sequentially structure the execution of process operations. It can encapsulate LD, FBD, IL, and ST instructions into control subroutines. Commonly encountered symbolic instructions of SFC are steps, transitions, and actions that are organized as

1. Initial step, representing the first active step of the system operating cycle.
2. Simple step, characterizing the state of process operation and described by the output action. There are three types of actions:
   a. Activating action, when the corresponding step is passing from an inactive to active state
   b. Deactivating action, when the corresponding step is passing from an active to inactive state
   c. Continuous action, when the action is performed for the same duration the corresponding step is active
3. Macrostep, a group of steps characterized by an input (simple) step and an output step that is not associated with an system output.
4. Transitions, allowing us to move from one step to another. They are defined by a specific logic condition necessary to cross the transition.

When an input transition condition is changed and all the steps preceding the transition are active, then the SFC program can execute the next step(s). Hence, each SFC program consists of steps and transitions (step enabling conditions). Table 3.14 presents some elements of SFC programming. Since SFC requires the storing of information on the status of the process execution (e.g., the active steps at a given time), it is structured into programs and function blocks, as illustrated in the text below.

A *step* can be represented graphically by a block, as depicted in Figure 3.26a and b. Each step has a symbolic name and should differ from other steps in a program or function block names. For the step in Figure 3.26a, an equivalent textual representation of an SFC program could be as follows:

```
Step Pump_ON
(*program content to activate or deactivate action *)
END_Step
```

The *transition* is a step represented by a horizontal line across a vertical line between two steps containing the logic condition allowing the transfer from one step to another. Hence, the transition is the binary evaluation of an I/O Boolean function. The execution of the logic control program or the FBD continues when the condition is TRUE, ON, or 1. When the transition condition is always TRUE, it is done by setting the transition to be the number 1, as

TABLE 3.14

Typical SFC Graphical Instruction

Element	Symbol

Step with identifier

Initial step with identifier *** is drawn by a double line

Action block, containing the actions assigned to a step
    Field a: Action qualifier
    Field b: Name of action
    Field c: Feedback variable
    Field d: Action content

Transition with identifier *** or transition condition ***

Alternative branch

Junction of alternative paths

Parallel (simultaneous) branch

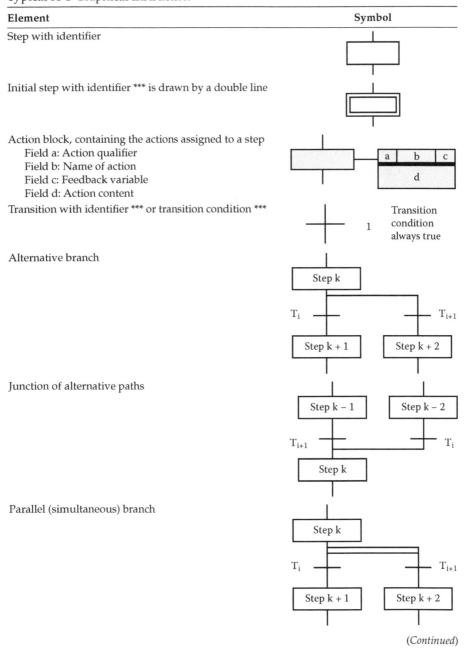

(Continued)

TABLE 3.14 (CONTINUED)

Typical SFC Graphical Instruction

Element	Symbol
Junction of parallel paths	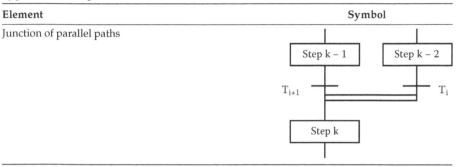

illustrated in Figure 3.26c and d (similarly with the condition always FALSE or OFF).

Transition conditions can be programmed in the LD, function block, ST, and IL languages, as depicted in Figure 3.26e–g. Those are examples of two steps connected via a logic AND function. The power flows from left to right. Each transition condition can be assigned a name, which could be used within the formulation of the I/O Boolean function.

    Transition in IL
    Transition k-1_to_k
        LD (load) A
        AND   B
        END_Transition
    Transition in ST
    Transition k-1_to_k
        := A & B
        END_Transition

The *alternative branch (OR and AND)* connects alternative step transition conditions to be executed. It is represented by a number of transition conditions after a horizontal line, as illustrated in Figure 3.27a for the case of three steps and three alternative branches. Each alternative should contain a sequence of steps and transitions. However, those transitions can be concurrent or mutually exclusive, as well as the resulting step. Here, transition conditions are usually executed from left to right, but this depends on the vendor's configuration.

The *parallel branch (AND)* contains all connected sequences of step transition conditions to be activated simultaneously, even if they are

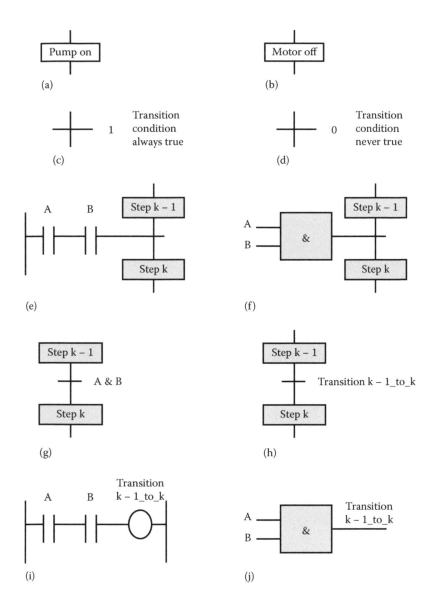

FIGURE 3.26
(a) Example of pump-activated step. (b) Example of motor-deactivated step. (c) Example of a transition condition that is always true. (d) Example of a transition condition that is never true. (e) LD and SFC. (f) Function block diagram and SFC. (g) ST and SFC. (h) Transition name. (i) Transition into LD. (j) Transition into function block.

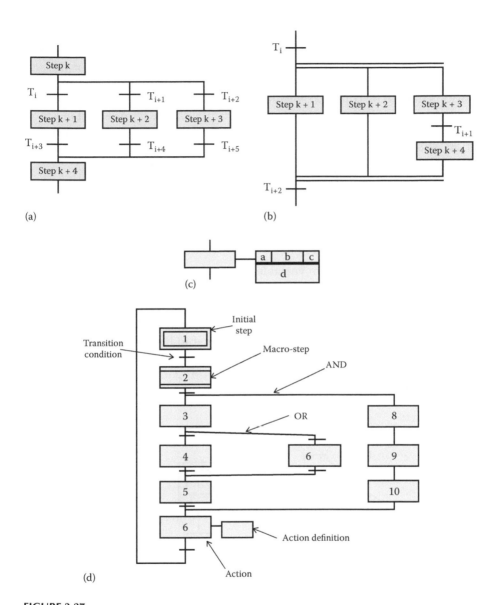

FIGURE 3.27
(a) Triple alternative branches. (b) Parallel branches. (c) Generic action block structure. (d) Generic sequential function chart.

interindependent. Here, the joining parallel branches of sequences of step transition conditions are represented by a horizontal double line, as depicted in Figure 3.27b. Prior to the synchronized execution of the step transition sequence within parallel branches, all preceding steps and transitions have to be completely executed.

Associated with each system output, *action blocks* are represented in a tabular format containing information on the action name, the qualifier specification, and the content, as depicted in Figure 3.27c. In addition, the action is a Boolean variable corresponding to a real internal system output.

The fields can be filled as

a: for a data processing action definition (store, set, store at predefined time, etc.)

b: for a given name of action

c: for an expected return variable

d: for a data action value

An illustrative SFC program with parallel and alternative branches is presented in Figure 3.27d. There is a relationship between the sequences of events using the SFC program and the functional decomposition using the FAST method. Figure 3.28a and b shows a comparison of a flowchart diagram and the equivalent SFC program.

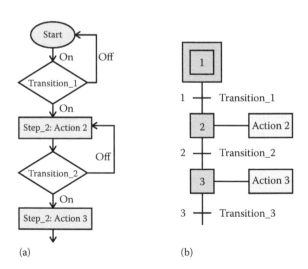

FIGURE 3.28
(a) Example of flowchart diagram. (b) Equivalent SFC diagram.

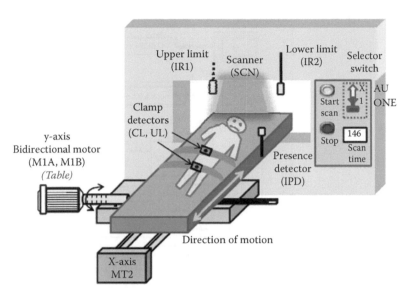

FIGURE 3.29
Scanning machine.

Example 3.13

Consider a scanner, as illustrated in Figure 3.29. A simplified process operating sequence is (1) the scanner power is turned on, (2) the operator pushes the start button, and (3) a presence detector (IPD) checks for a patient's presence.

If the patient is not present, the process is idle. When a part is present, (4) a clamp (CL) locks the part in place, which is detected by the sensor (SW1), and (5) the motor moves the table over the scanner box (M1A). Hence, by moving the table from an upper position given by the limit switch (IR1) to a lower position given by the limit switch (IR2), the scanning is processed (SCN). (6) At the IR2 position, the direction of motion of table change (M1B is activated), while the scanning is complete. (7) At the IR1 position, the clamp is unlocked (UL) (SW1 is deactivated) and the patient can be moved out of the table. (8) The process is interrupted if the machine is in the one-cycle mode, and it continues if it is in automatic mode (AU).

The start push button (START), the patient presence detector (IPD), the cycle switch selector (AU/ONE), the limit switches (upper [IR1] and lower [IR2]), and the patient clamping (locked [SW1]) are considered overall scanner system inputs. And the table motor contactor (M1A, M1B), the scanner contactor (SCN), and the clamp contactors (locked [CL] and unlocked [UL]) are classified as overall system outputs. Following the functional analysis, two mutually exclusive cycles are identified; the resulting state diagram and the equivalent SFC diagram are presented in Figure 3.30a and b, while Table 3.15 summarizes the corresponding sequence table for one scanner operating cycle.

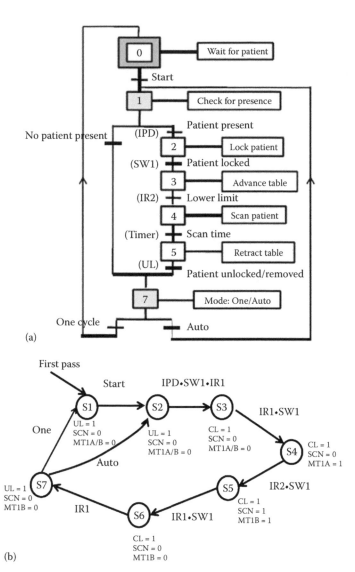

FIGURE 3.30
(a) Example of an SFC program for logic control of the scanner. (b) State diagram of scanner machine.

3.7 Wiring Diagram and Automation Project Documentation

An automation system design project requires sketching electrical wiring diagrams to connect system components with automation system devices. There are two types of wiring diagrams: (1) the power supply schematic and (2) the command system diagram. The power supply wiring schematic

TABLE 3.15

Table of Scanner Processing Sequence for One Cycle Operation

Sequence	System Inputs						System Outputs				
	Start	IPD	AU/ONE	IR1	IR2	SW1	CL	UL	MT1B	MT1A	SCN
Wait patient	0	0	0	1	0	0	0	1	0	0	0
One cycle	0	0	0/1	1	0	0	0	1	0	0	0
Start	1	0	0	1	0	0	0	1	0	0	0
Check presence	0	1	0	1	0	0	0	1	0	0	0
Locked patient	0	0	0	1	0	1	1	0	0	0	0
Advancing table	0	0	0	1	1	1	1	0	1	0	0
Scan patient	0	0	0	0	1	1	1	0	0	1	1
Retracting table	0	0	0	1	0	1	1	0	0	1	0
Unlocked patient	0	0	0	1	0	0	0	1	0	0	0
Removing patient	0	0	0	1	0	0	0	1	0	0	0

illustrates the physical connections between the electrical elements used in the process operations and power supply units. The command system diagram illustrates the detailed connection between each process automation field device involved (control panel elements, logic controllers, actuators, and detectors or sensors). Table 3.16 summarizes commonly encountered binary system automation switches, and Table 3.17 presents some electrical components and their electrical representation. Binary switches are actuated by changes in system physical properties (e.g., temperature, flow, and weight) causing a contact between two conductors. This can result into a large flow of electric current. Hence, it is required to size the power supply in order to limit switch contact ampacity and avoid (1) melting contact due to heat generated by the current passing through, (2) sparks during opening or closing switch contacts, or even (3) a jumping current across open switch contacts due to the high voltage.

As such, the normal switch condition is defined by default when the switch is set OFF (0 logic level) unless it is defined otherwise. In the case of a *speed switch*, it is when the motor shaft is not turning; in the case of a *pressure switch*, it is when any pressure is applied. With the *temperature switch*, it is calibrated to the normal processing temperature. With the *level switch*, it is set for the case of an empty tank. And in the case of the *flow switch*, it is when there is no liquid flow.

TABLE 3.16

Typical Encountered Binary Switching Devices

Symbol	Description
	Toggle switches whose activation is controlled by angular positioning of the contactor.
	Push-button switches are buttons activating two-position devices with a spring ensuring the return mechanism.
	Select switches whose activation is performed at a fixed-angle rotary level to select at least two positions. Some of them are constructed with a spring to ensure a return mechanism.
	Lever actuator limit switches are fitted such that a lever is tipped or by a roller bearing after contact with the moving component.
	Rotary speed switches use a scale mechanism or optical-based detection mechanism to sense the angular speed of a motor shaft.
	Pressure switches capture fluid pressure sensitivity when they are in contact with a piston, diaphragm, or bellow.
	Liquid flow switches detect any fluid flow rate above a specific threshold.
	The level switch is actuated when the liquid level rises over a specific level, allowing the height detection of solid materials.
	Heat or cool variation causes thermal expansion between the two metals, bending them and actuating a switch contact mechanism.

TABLE 3.17

Electrical Components Encountered in Process Automation

Solenoid	Thermocouple	Transformer
Relay time delay	Resistor	Relay contact
Bell	Horn	Capacitor
Lamp	Relay coil	Motor (3-phase AC)
Circuit breaker (3-phase)	Thermal relay	Fuse

Table 3.17 presents some electrical actuating devices, such as *solenoids*, which consist of a magnetic object moving bidirectionally while energizing its surrounding coil. The motion direction of the object depends on the flowing electric current direction (from power supply unit). Then, among other electromagnetic, electromechanical, and electrostatic actuators encountered in the process automation there are solenoid-based binary actuating elements, such as *control relays* or *relay coils*, when a solenoid is used to activate a set of switching contacts or contactors, as depicted in Figure 3.31a, and large *relays* acting as a binary (on or off) amplifier, when the relay is energized and an electric current running through the conductor produces a magnetic field along the length of the coil. The strength of the magnitude field and the current magnitude are proportionally related. Once the coil is energized or de-energized, the core is pushed inside or pulled outside the coil by the attached spring. Then, the relay coil produces the magnetic field, closing or opening various contacts able to conduct to a load hundreds of times that amount of power. The relay also contains switching contacts to open and close as a function of the core displacement. Figure 3.31b illustrates a power supply schematic for energizing the coil relay coil by 24 V direct current (VDC) voltage supply in order to close or open a 380 V alternating current (VAC) power supply once the start button is activated.

Similarly, there are *contactors* whose contact relays are used to control flow in a large-scale electrical power grid through circuit breakers. They are constructed with multiple contacts, which are usually open to avoid the power flow to the load when the coil is de-energized. In Figure 3.32,

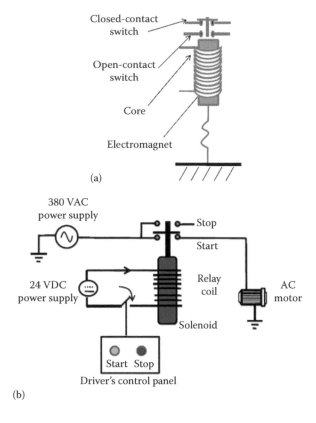

(a)

(b)

FIGURE 3.31
(a) Typical electromagnetic control relay. (b) De-energizing and energizing relay coil circuit with 380 VAC power supply.

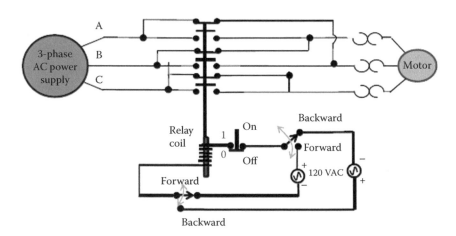

FIGURE 3.32
Bidirectional motor control power supply schematic.

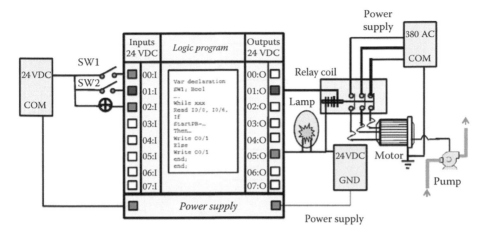

FIGURE 3.33
Power supply and command wiring diagram for AC motor control.

there is an example of contact switches used to change the phases of the three-phase power supply, such as the lowest contact having a current rating that can actuate a bidirectional induction motor through the current flow over its armature. In Figure 3.33, a bidirectional AC motor connected to two contactors, to switch its polarity, is presented. Such a power wiring design aims to avoid the forward and backward contactors being energized at the same time.

An example of the command wiring diagram of a bidirectional AC motor is shown in Figure 3.33. Here 24 VAC is supplied to the start and stop push buttons of the 8-bit I/O PLC, and 24 VDC and 120 VAC are supplied to the control relay and the motor, respectively. A time-based activation of interlocks can be used with motor contactors, such that simultaneous energizing of other contactors is avoided during the motor running phase, and energizing of the same contactor is prevented for a specific time after motor shutdown. Here, the connection addresses with the logic controller unit are sketched to ease the variable declaration and addressing.

Command wiring diagrams present the electrical connection between the control unit and the control panel as the command component and the electrical wiring between the control unit, the process, and the associated sensing system as the operative component. An illustration of a detailed command wiring diagram for the control of a bidirectional motor and a pump is depicted in Figure 3.34.

Example 3.14

Consider that it is desired to implement the following I/O Boolean function:

$$MOTOR = (MOTOR + START) \cdot \overline{STOP}$$

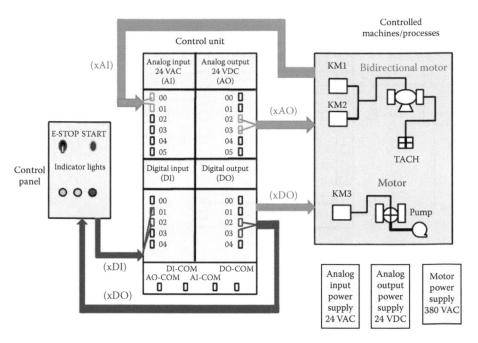

FIGURE 3.34
Example of a command wiring diagram with a PLC interface.

Using the command wiring diagram of the 380 VAC motor with a personal integrated controller (PIC)-based PLC as depicted in Figure 3.35, the corresponding assembly language control program could be

```
Load #0 % for memory stored process data
Store Motor
Port 01:I %
Store Start
Port 06:I %
Store Stop
Load Start
Or Motor
Store Motor
Load Stop
Not
And Motor
Store Motor
Port 01:O %
Jump Loop
```

Similar results could be obtained using the solid-state logic circuit shown in Figure 3.36.

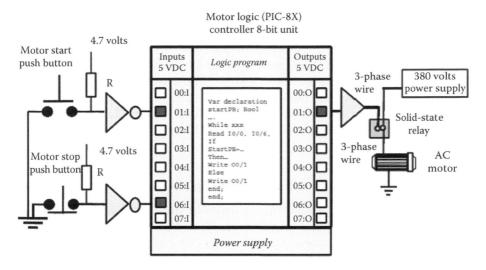

FIGURE 3.35
PIC-based logic control of an AC motor.

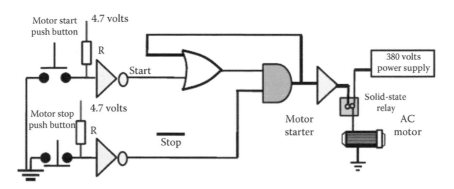

FIGURE 3.36
Equivalent solid-state logic control circuitry of AC motor.

3.8 Sizing and Selection of Automation Systems

For automation system design using a computer-based device, the following functions are expected: (1) monitoring system output signals; (2) executing logic instructions, sequencing, timing, and counting functions for control or diagnostics of process operations; (3) driving actuators or indicators; and (4) communicating with other automation field devices. Therefore, for the implementation of large-scale automation systems (e.g., robot-based man-ufacturing transfer line), as well as low-scale automated systems (e.g., automated guided vehicle), it is required to select among potential logic con-troller devices (e.g., FPGA, μController, and PLC) which one can achieve an optimized overall performance. The sizing and selection of automation sys-tems should be done based on criteria such as the computing and processing capacity (CPU memory size) and the number of I/O ports. Table 3.18 presents a checklist of information to be gathered for the assessment of logic control-ler technical characteristics.

3.8.1 Memory Structure and Capacity Estimation

The memory of the controller unit consists of storage cells, each storing in binary format (1 or 0) word length (8, 16, 32, and 64 bits) data to be handled simultaneously. For example, an OFF could correspond to storing informa-tion 0 and could be estimated as 1 bit of information stored. Usually, the logic control program memory size is specified as K units, each corresponding to 1024 word locations to store each type of control logic control instruction. As summarized in Table 3.19, for example, a 4 K logic control program size cor-responds to a device with a memory requiring a maximum of 4096 storage areas for program instructions. In the case where process data and program instructions are stored and executed in different memory locations, it is suit-able to double the memory storage size required.

Example 3.15

It is desired to derive the memory requirements for the logic control program of a process having the following characteristics: 70 system outputs where each output results from a logic composed of 5 contact elements, 11 timers connected to 6 contact elements and 3 counters con-nected to 5 contact elements, and 15 Boolean instructions connected to 5 contact elements.

An estimation of memory size could be derived from the sum of the size estimation from each individual device. It would be given from

- The number of system outputs. There are a total of 350 (70×5) contact elements and a total of 70 (70×1) system outputs, cor-responding to a total of 420 ($350 + 70$) words.
- The number of timers. There is a total of 66 (11×6) contact ele-ments and a total of 33 (11×3) timers, corresponding to a total of 99 ($66 + 33$) words.
- The number of counters. There is a total of 15 (3×5) contact ele-ments and a total of 9 (3×3) timers, corresponding to a total of 24 ($15 + 9$) words.
- The number of instructions. There is a total of 75 (15×5) contact elements and a total of 15 (15×1) Boolean instructions, corre-sponding to a total of 90 ($15 + 75$) words.

TABLE 3.18

Information Required for the Sizing Logic Control System (Memory and I/O Modules)

Consideration	Information to Record		Description
1. System migration and compliance issues	Designed system	Current system	Perform compliance system check to determine the technical compatibility between design and current system components, interfaces, and accessories.
2. Environmental and safety issues	Environmental issues related to devices and process	Operating and personal safety issues (legal compliance)	Consider any environmental issues capable of affecting the system (temperature, dust, vibration, etc.), as well as operational safety issues. This will require the system to have operating conditions to support any extreme environmental conditions through the design of protective installation.
3. Sizing and characterization of digital devices	AC input DC input	AC output DC input	Derive the number of discrete devices involved in the system and the power types and range (AC, DC, etc.). Hence, size digital I/O modules according to the identified signal types.
4. Sizing and characterization of analog devices	Type inputs: Voltage Current Thermo RTD	Type outputs: Voltage Current	Derive the number of analog devices involved in the system and the signal types and range (voltage, current, temperature, etc.). Hence, size analog I/O modules according to the identified signal types.
5. Specialty modules or features (application specific)	High-speed counter Positioning Servo/stepper BASIC programming Real-time clock		Decide if the system requires any customized functions or features, such as (1) high-speed counting or positioning and (2) real-time clock requirement. List them and adapt the system accordingly.

(*Continued*)

TABLE 3.18 (CONTINUED)

Information Required for the Sizing Logic Control System (Memory and I/O Modules)

Consideration	Information to Record	Description
6. Identify CPU requirements and sizing	Estimate the memory size required by careful estimation of (1) program memory size, (2) data memory size, (3) minimum scan time allowable, and (4) redundant energy source required (type and power) Software/special function requirements: PID	Determine the type of CPU data memory required, the size of the program instructions, and consequently, the program memory. Also, define the program scan time needed. Here data memory is the size of process data to be manipulated and stored (data table size). For a high scan time, a high processor speed should be considered to perform quick instruction execution speed. Boolean logic is quicker at executing instructions than mathematical operations. To estimate the program memory required, a rule of thumb consists of using a 5-word bit memory size for an I/O digital device and a 25-word bit memory size for an I/O analog device.
7. I/O locations and communication issues	Remote locations Specific I/O protocol Ethernet, ControlNet, wireless, ASCII serial, etc., required? Local	Determine local or remote I/O device locations. Also determine the communication requirements—whether the system is communicating to other networks or field devices and each I/O data format. If necessary, choose a system capable of supporting remote I/O devices and estimate allowable distances and speeds suitable for the remote control application or select an appropriate protocol.
8. Programming and RAM/flash memory issues	Process, controller data size (integer, character) and type, scan time Special features: PIDs MPC Optimal Interrupts Etc.	Derive the programming requirements: classic logic control programming instructions or special instructions (e.g., built-in PID functions to perform closed-loop control of process operations and counters). Choose a controller unit able to support all instructions, including specific application.

TABLE 3.19

Typical Memory Utilization Requirements for Logic Controller Unit

Program Instruction Type	Words of Memory Required
ON or OFF contacts	1
Output coil signal	1
Boolean instructions (add/compare, etc.)	1
Timer	3
Counter	3

Thus, a memory size capable of handling a total of 633 words (420 + 99 + 24 + 90) would be required to store program instructions, as well as outputs, timers, and counter data. This is below 1 K of memory. For safety measures and additional I/O table requirements, it is suitable to add 30% to the memory size estimate.

3.8.2 Power Supply and I/O Unit

Power supply is expected to provide the voltages and current levels required by the automation system components (I/O analog and digital devices, as well as memory and CPU subcomponent). Common voltage levels required by logic control devices are 5 VDC, 24 VDC, and 220 VAC. For input devices, typical devices, such as push buttons, switches, photodetectors, timers, and relay contacts, send ±5 to ±10 V signals.

Commonly encountered system voltages are 120 VAC, 24 VDC, 12–48 V AC/DC, 5 VDC (transistor–transistor logic [TTL]), or 230 VAC. Hence, among output modules there are analog output modules requiring a digital-to-analog converter to transform digital values from the control processor module into an analog signal for the process devices. There are also high-speed special routines or functions having high timing requirements (e.g., PID for motor speed control). In order to meet the requirements for fast data processing, a processor with sufficient baud rate (above 10 MHz) and up to 64-bit length should be required.

3.9 Fail-Safe Design and Interlocks and Validation Issues

The *fail-safe* design principle is to ensure continuous execution of system operations in the event of a predefined automation system component failure. This makes its control system tolerant to potential hardware failure, such as wiring or component failures. Applications of fail-safe design are critical in the electrical power management (generation and distribution) supplied to hospital facilities, telecommunication systems, water treatment systems, inner-city traffic lights, highway automated gates, and other important social infrastructure. Here, large circuit breakers are switched ON/OFF by control signals from protective relays. In the event of excessive current, the relay should be chosen such that its switch contact could be open to interrupt a signal while initiating a redundant alternative power source in nearly simultaneous time. Hence, this would be done with no disruption of service supply. A typical illustration of fail-safe design is given in Example 3.1. The *interlock* design principle consists of setting preventive action and technology to ensure no harm to the shop floor operator and no destruction to involved system equipment. This is done through a set of detectors and sensors that activate system interruption in case of a predicted failure. Thus, the interlock design consists of implementing some interlocking switches at key locations in the system plan layout with defined activation rules. Figure 3.37 illustrates the interlock principle within generic process automation architecture.

Usually, some system operating safety measures are implemented through a three-hierarchy SFC level that could be described as follows: (1) a safety SFC (GS) managing process safety modes, such as emergency start, general process faults, process initialization, and after process defaults start; (2) a traded SFC (GT) defining the type of operational modes, such as overall

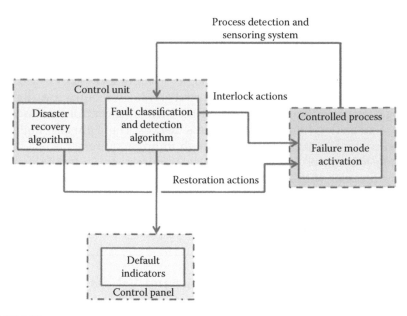

FIGURE 3.37
System interlock design principle.

process start and selection of type of modes (semiautomatic, manual, and automatic); and (3) SFC operational modes, such as the manual operation SFC (GM) or the production SFC (GPNA). This is further covered in Chapter 4.

3.9.1 Logic Control Validation (Commissioning)

General specifications of automation system commissioning can be summarized through a set of test protocols regarding (1) each hardware and software automation system component; (2) the integrated software and hardware automation system (especially for data processing), for normal production operation conditions and for compliance in control and command execution of automated system action in front of changing operating conditions, such as the speed variation control of a motor; (3) the assessment of system operating conditions (e.g., check detector activation condition); (4) the interlock assessment in front of system failure, such as operator security and detector activation; (5) the whole automation system commissioning and compliance test of electrical wiring; and (6) the final electrical wiring diagram and power flow diagram commissioning. In the case of hardwired circuit design, the process simulation for the designed logic controller could be validated through computer tools, such as Verilog, ABEC, VHDL, and PALASM.

3.10 Illustrative Case Studies

In this section, illustrative design steps of logic controllers for several industrial processes are presented, from the system functional analysis to formal modeling based on the methodology list below.

1. Process schematics
2. I/O-involved equipment listing
3. Table of sequence or truth table
4. Switching theory or state diagram
5. State transition table and Karnaugh maps
6. I/O Boolean function
7. Timing diagram
8. Logic control circuitry

3.10.1 Fruit Packaging System Using Sequence Table

Consider an automated natural fruit packaging process, as illustrated in Figure 3.38. The equipment involved is listed in Table 3.20A.

This is a two-stage process: box positioning and box filling with fruit. Initially, the package conveyor driven by motor MTR1 carrying boxes is activated by the START PB push button, and it stops at the package limit switch.

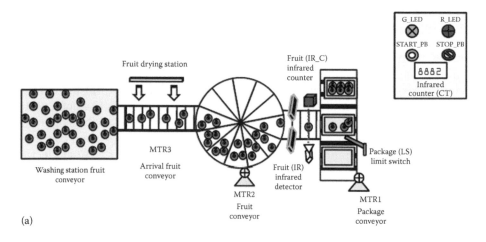

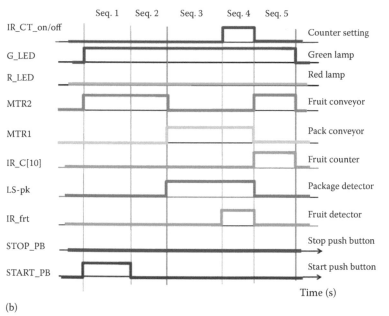

FIGURE 3.38

(a) Fruit conditioning (packaging) unit. (b) One-cycle timing diagram of fruit packaging process.

(Continued)

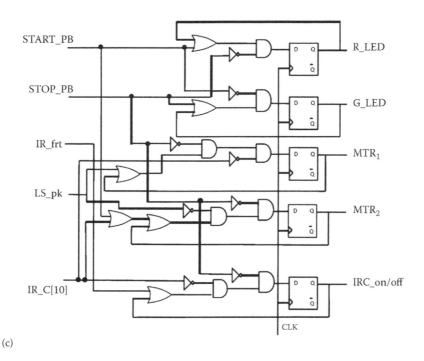

(c)

FIGURE 3.38 (CONTINUED)
(c) Fruit packaging process logic controller circuit using D flip-flop gates.

TABLE 3.20A

Equipment Involved in Fruit Packaging Process

Input Equipment	Symbol	Output Equipment	Symbol
Start push button	START_PB	Green_LED for boxes	G_LED
Stop push button	STOP_PB	Red_LED for fruits	R_LED
Full package limit switch	LS_pk	Counter up [ON/OFF]	IRC_on/off
Fruit infrared detector	IR_frt	Package conveyor	MTR1
Counter []	IR_C[10]	Fruit conveyor	MTR2

Then, fruit arriving on a conveyor driven by motor MTR2 is dropped in boxes of a lot size of 10. After each of the 10 fruits dropped is counted by the infrared counter IR_C positioned at the end of the fruit conveyor course, the package conveyor is activated to release a new box at the limit switch position. A control panel allows the operator to activate or deactivate conveyors, as well as to monitor through light indicators the process operations execution, as depicted in Figure 3.38a.

From the analysis of the sequence table and using switching theory as summarized in Table 3.20B, note that the limit switch *LS_pk* activates *MTR1*, while the system input conditions *IR_C* [10] or *STOP_PB* deactivate MTR1. Hence,

TABLE 3.20B

Sequence Table Analysis Using Switching Theory for Fruit Packaging Process

Sequence	Input Devices						Output Devices				
	StartPB	StopPB	IR_frt	LS_pk	IR_C[10]	G_LED	R_LED	MTR2	MTR1	IRC_on/off	C
1. Start process	1	0	0	0	0	1	0	0	1	0	
2. Moving boxes	0	0	0	0	0	1	0	0	1	0	
3. Positioning boxes	0	0	0	1	0	1	0	1	0	0	MTR2, MTR1
4. Filling boxes	0	0	1	1	0	1	0	1	0	1	IRC_on/off
5. Boxes filled	0	0	0	0	1	1	0	0	1	0	IRC_on/off, MTR2, MTR1

MTR1 [1] = *IR_C* [10] while MTR1 [0] = *LS_pk+STOP*, and so on. Therefore, from analysis of the activation and deactivation conditions of *MTR1*, *MTR2*, *IRC$_{on}$/off*, *G_LED*, and *R_LED* system outputs, the I/O Boolean functions yield

$$MTR1 = (MTR1 + LS\_pk) \cdot \overline{IRC\_C[10]} \cdot \overline{STOP\_PB}$$
$$MTR2 = (MTR2 + IR\_C[10]) \cdot LS\_pk \cdot \overline{STOP\_PB}$$
$$IR\_CT\_on/off = (IR\_CT\_on/off + IR\_frt) \cdot (\overline{IR\_C[10] \cdot STOP\_PB})$$
$$G\_LED = (G\_RED + START\_PB) \cdot \overline{STOP\_PB}$$
$$R\_LED = (R\_LED + STOP) \cdot \overline{START}$$

The equivalent one operating cycle timing diagram and logic circuit are depicted in Figure 3.38b and c.

3.10.2 Fruit Picker Process Design Using Sequence Table

Figure 3.39 depicts a typical automated agricultural fruit picker system. The equipment involved is listed in Table 3.21. This is a four-stage robot-based picking process. Once, the StartPB is activated from the control panel, the empty box is carried over the conveyor belt (MTC) up to the limit switch

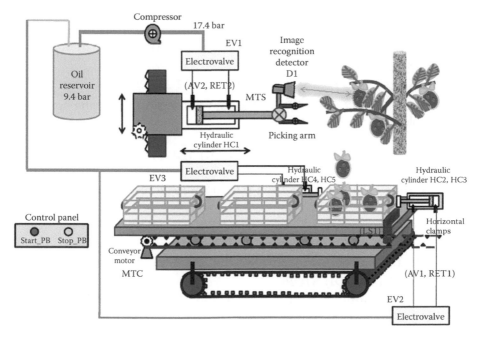

FIGURE 3.39
Fruit picker machining process.

TABLE 3.21

Equipment Involved in Fruit Picker Machining

Input Equipment	Symbol	Output Equipment	Symbol
Start push button	StartPB	Bidirectional clamps	AV1 and RET1
Limit switch sensors	LS1	Bidirectional motor advance	AV2 and RET2
Timer delay of 5 s	T(5 s)	Timer [ON/OFF]	T1
Recognition detector	D1	Motor conveyor	MTC
Stop push button	StopPB	Motor fruit picker	MTS

(LS1). Then, box clamping is horizontally performed by hydraulic cylinders HC2, HC3, HC4, and HC5 by releasing pressure over electrovalves EV3 and EV2. Then, electrovalve EV1 (through AV2) releases enough hydraulic pressure on the cylinder (HC1) to advance toward the detected fruit on the tree. This cylinder is carrying on the picker robot arm. Once the robot arm is positioned over the vision-based fruit-detected position (D1), the robot picker motor is activated (MTS). For a timer of T1 (5 s), the robot arm advances while cutting off the fruit from its tree. Then the pressure is released by EV1 (RET2), forcing the retreat of the robot arm until the double switching of the limit switch D1 activates the stopping of the robot arm retreat. At this stage, it is completely retracted to its initial position. Then, the valves retract the fruit boxes clamping through hydraulic cylinders HC2, HC3, HC4, and HC5. From the functional analysis, the table of sequences can be established as in Table 3.22.

From the sequence table analysis using switching theory, I/O Boolean functions yield $AV1\ [0] = D1$ while $AV1\ [1] = LS1$ and then $AV1 = (AV1 + LS1) \cdot \overline{D1} \cdot \overline{STOP}$.

Similarly, with $RET1\ [0] = \overline{LS1} \cdot \overline{D1}$ and $RET1\ [1] = D1$, $RET1 = (RET1 + D1) \cdot (LS1 + D1) \cdot \overline{STOP}$. For the conveyor belt carrying the boxes, the I/O Boolean equation yields

$$MTC = (MTC + \overline{LS1} \cdot \overline{D1}) \cdot \overline{LS1} \cdot \overline{STOP}$$

For the motor for the robot arm (MTS) advance, it results in

$$AV2 = (AV2 + \overline{T[5s]} \cdot V1) \cdot \overline{D1} \cdot \overline{STOP}$$

While the robot arm returns, the Boolean function is $RET2 = (RET2 + D1) \cdot (LS1 + D1) \cdot \overline{STOP}$.

For the robot arm, $MTS\ [0] = \overline{LS1} \cdot \overline{D1}$ and $MTS\ [1] = LS1$; thus, the motor for the picking arm is given by

$$MTS = (MTS + LS1) \cdot (LS1 + D1) \cdot \overline{STOP}$$

TABLE 3.22

Sequence Table Analysis and Switching Theory for Fruit Picker Machining

Sequence	Input Devices									Output Devices								
	StartPB	LS1	D1	T[5 s]	AV1	RET1	AV2	RET2	MTC	MTS	T1	S	V1	V2
1. Fruit box advance to LS1	1	0	0	0	0	0	0	0	1	0	0		1	0
2. Robot arm activation and box clamping	1	1	0	0	1	0	0	0	0	1	1	MTC, AV1, MTS, T1	0	1
3. Fruit picking operating timer activated	1	1	0	1	1	0	0	0	0	1	1		0	1
4. End of fruit picking	1	1	0	0	1	0	1	0	0	1	0	AV2, T1	1	0
5. Box unclamping	1	1	1	0	0	1	0	1	0	1	0	MTC, AV1, AV2, RET2	1	0
6. Next fruit box arrival	1	0	0	0	0	0	0	0	1	0	0	MTC, RET1, RET2, MTS	1	0

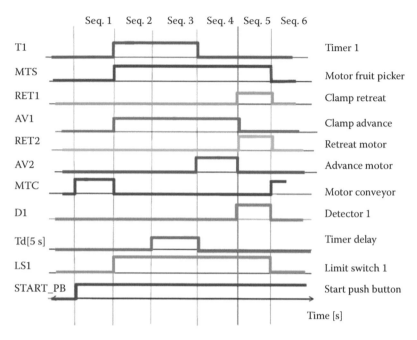

FIGURE 3.40
Timing diagram of the Fruit picking process.

For the timer, $T1\,[0] = \overline{T[5s]} \cdot \overline{V2}$ and $T1\,[1] = LS1 \cdot V2$. Then,

$$T1 = (T1 + LS1 \cdot V2) \cdot (T[5s] + V2) \cdot \overline{STOP}$$

For virtual output variables, $V2 = (V2 + LS1) \cdot T[5s]$

$$V1 = (V1 + \overline{T[5s]}) \cdot \overline{LS1}$$

The equivalent timing diagram of those I/O Boolean functions derived is illustrated in Figure 3.40.

3.10.3 Elevator-Based Motion Process

Consider a three-floor elevator system as depicted in Figure 3.41, while the I/O variables are listed in Table 3.23. This elevator's up or down motion starts when a direction (Up or Dn) is selected. When the elevator reaches the requesting floor, the cabin gate opens (OG) for 30 s and then closes (CG).

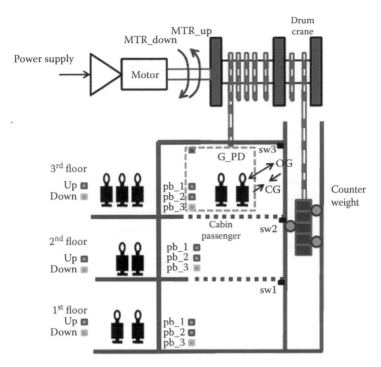

FIGURE 3.41
Three-floor elevator system.

TABLE 3.23

Equipment Involved in Three-Floor Elevator System

Input Equipment	Symbol	Output Equipment	Symbol
Selecting up motion	Up	Bidirectional open cabin motor	OG
Selection down motion	Dn	Bidirectional closed cabin motor	CG
Level 1 sensor	sw1	Bidirectional up motor	M1
Level 2 sensor	sw2	Bidirectional down motor	M2
Level 3 sensor	sw3	Timer [ON/OFF]	T1
Calling for floor 1	Pb1		
Calling for floor 2	Pb2		
Calling for floor 3	Pb3		
Timer delay 1	T[30]		
Emergency stop	AU		
Cabin gate photodetector	GPD		

Then, based on the selection of the *i*th floor by the Pb*i* (Pb1 or Pb2 or Pb3) activated, the motor contactor for the selection direction (M1 or M2) turns on until the corresponding floor limit switch SW*i* (SW1 or SW2 or SW3) is activated. Then, the motor stops (M1 or M2 deactivated) and the cabin gate opens for another 30 s, after which it closes (CG). A gate detector (GPD) is used to activate the gate opening contactor by detecting a person's presence over the gate ramp. Also, when the elevator is waiting at a specific floor level, pressing the same floor level push button keeps the door open. Otherwise, it will request the elevator motion to this specific level. In order to model this elevator system, I/O Boolean functions can be derived from the combinational logic either based on a truth table covering all possible floor selections or by using advanced floor counting strategies and computing devices.

For space reasons, only a two-floor elevator motion cycle (floor 1 – floor 2; floor 2 – floor1) is considered and analyzed using a sequence table (Table 3.24) and a state diagram (Figure 3.42a). In order to reduce the number of variables, it is considered that the floor selection combines two variables: the direction and the selected floor level. For example, pushing the Pb1 button activates Pb1 and Dn, while pushing the Pb2 button activates Pb2 and Up.

I/O Boolean equations derived from the sequence table established in Table 3.25 are

$$OG = (OG + Pb1 + sw2 + Pb2 \cdot T[30s] + sw1 \cdot D1) \cdot \overline{AU} \cdot \overline{T[30s]}$$

$$CG = (CG + T[30s]) \cdot \overline{T[30s]} \cdot \overline{AU}$$

The I/O equations with respect to the time diagram are

$$MTR\_up = M1 = (M1 + sw1 \cdot \overline{T[30s]} \cdot V3) \cdot \overline{sw2} \cdot \overline{AU}$$

$$MTR\_down = M2 = (M2 + sw2 \cdot \overline{T[30s]} \cdot V4) \cdot \overline{sw1} \cdot \overline{AU}$$

$$T1 = (T1 + Pb1 + sw2 + Pb2 \cdot T[30s] + sw1 \cdot V5) \cdot \overline{T[30s]}$$

Using the following virtual output variables, their Boolean equations yield

$$V1 = (V1 + sw1) \cdot \overline{T[30s]}$$

$$V3 = (V3 + sw1 \cdot \overline{T[30s]}) \cdot (sw2 + \overline{T[30s]})$$

$$V4 = (V4 + sw1 \cdot \overline{T[30s]}) \cdot (sw2 + \overline{T[30s]})$$

$$V5 = (V5 + sw1) \cdot \overline{T[30s]}$$

An equivalent state diagram is sketched in Figure 3.42a. Notice that the state diagram offers wider and easier modeling options, as more than

TABLE 3.24

Sequence Table of Two-Floor Elevator System Analysis Using Switching Theory

Sequence	Input Devices						Output Devices					Changing					Virtual			
	GPD	sw1	sw2	Pb1	Pb2	T[30]	OG	CG	M1	M2	T1	OG	CG	M1	M2	T1	V1	V3	V4	V5
Rest	0	1	0	0	0	0	0	1	0	0	0						0	0	0	0
Selecting and door open	1	1	0	0	1	0	1	0	0	0	1	X^1	X^0			X^1	0	0	0	0
Door closed	1	0	0	0	0/1	1	0	1	0	0	0	X^0	X^1			X^0	0	0	0	0
Move up 1 → 2	0	0	0	0	0/1	0	0	1	1	0	0			X^1			0	0	0	0
Door open	0	0	1	0	0	0	1	0	0	0	1	X^1	X^0	X^0		X^1	0	1	1	0
Door closed	0	0	1	0	0	1	0	1	0	0	0	X^0	X^1			X^0	0	1	1	0
Selecting and door open	1	0	1	1	0	0	1	0	0	0	1	X^1	X^0			X^1	0	1	1	0
Door closed	0	0	1	0/1	0	1	0	1	0	0	0	X^0	X^1			X^0	0	1	1	0
Move down 2 → 1	0	0	0	0/1	0	0	0	0	0	1	0		X^0		X^1		0	0	0	0
Door open	0	1	0	0	0	0	1	0	0	0	1	X^1			X^0	X^1	1	0	0	1
Door closed	0	1	0	0	0	1	0	1	0	0	0	X^0	X^1			X^0	1	0	0	1
Rest	0	1	0	0	0	0	0	1	0	0	0		X^0				0	0	0	0

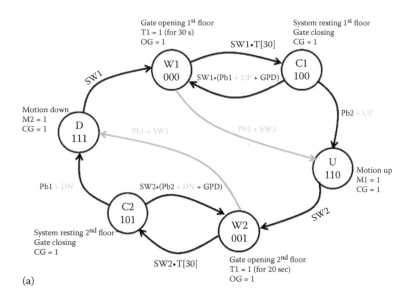

(a)

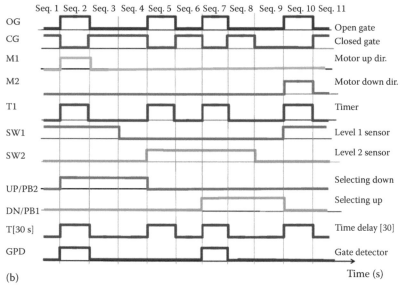

(b)

FIGURE 3.42

(a) State diagram of two-floor elevator system (Up/Dn is optional). (b) Elevator timing diagram.   *(Continued)*

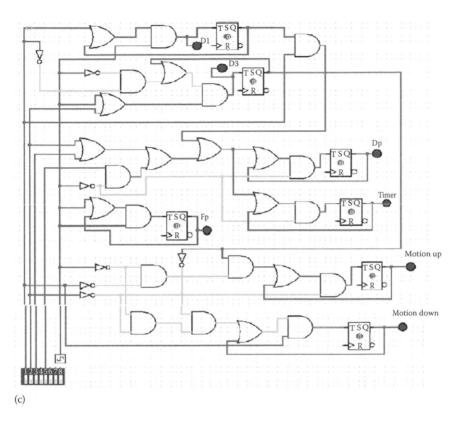

(c)

FIGURE 3.42 (CONTINUED)
(c) Circuit logic of two-floor elevator system.

TABLE 3.25

Equipment Involved in the Automatic Ticket Gate Entrance

Input Equipment	Symbol	Output Equipment	Symbol
Push-button gate entrance ticket request	B	Solenoid supplying parking ticket	S
Limit switch ticket taken	T	Red LED	L2
Load sensors client moving in	LS1	Green LED	L1
Load sensors client moving out	LS2	Alarm	H
Gate obstructed sensor	STD	Motor raise gate	M1U
Photodetector gate up position	CP1	Motor lower gate	M1D
Photodetector gate down position	CP2	Client counter	CC
Timer delay 1 []	T1[30]	Timer 1	T1
Timer delay 2 []	T2[60]	Timer 2	T2
Counter up []	CM		

one operating cycle can be represented. The corresponding system states' Boolean equations yield

$$W1 = (W1 + D \cdot SW1 + C1 \cdot SW1 \cdot (Pb1 + GPD)) \cdot \overline{SW1 \cdot T[30s]}$$
$$C1 = (C1 + W1 \cdot SW1 \cdot T[30]) \cdot \overline{Pb2 \cdot SW1 \cdot (Pb1 + GPD)}$$
$$U = (U + C1 \cdot Pb2) \cdot \overline{sw2}$$
$$W2 = (W2 + U \cdot SW2 + C2 \cdot SW1 \cdot (Pb1 + GPD)) \cdot \overline{SW2 \cdot T[30s]}$$
$$C2 = (C2 + W2 \cdot SW2 \cdot T[30]) \cdot \overline{Pb1 \cdot SW2 \cdot (Pb2 + GPD)}$$
$$D = (D + C2 \cdot Pb1) \cdot \overline{sw1}$$

with the system outputs' Boolean equations given by

$$CG = C2 + D + C1 + U$$
$$OG = W1 + W2$$
$$M1 = U$$
$$M2 = D$$
$$T1 = W2 + W1$$

The logic functions above should be equivalent to the I/O Boolean functions derived from the sequence table analysis using switching theory. The resulting timing diagram and logic controller circuit are depicted in Figure 3.42b and c.

3.10.4 Automatic Ticket Gate

An automatic ticket gate is depicted in Figure 3.43, with the equipment listed in Table 3.25. When a client arrives at the entrance gate, whose presence is detected by load sensors (CP2), a request is made (B). After the ticket is redrawn, the green light (L1) turns ON and the gate moves up (M1U contactor activates) to the level given by the detector (CP1). Then, the client can come in. The load sensors for the client's presence turn off (LS1), allowing the gate to come back down; the motor (use M1D) contactor becomes active until the gate reaches the level given by the detector (CP2). Figure 3.44 depicts the corresponding timing diagram.

This is a two-stage process with three cases (stadium not full, stadium full, and entrance gate stuck). Due to spacing from the page layout, the correspondence between the sequence table, variable symbols, and entrance gate equipment names is given in brackets. Also, note that the following system inputs have been discarded: gate obstructed (STD), client counter (CM), client moving out of the stadium as detected by the load sensor (LS2), and output system counter (CC), as illustrated in the sequence table depicted in Table 3.26.

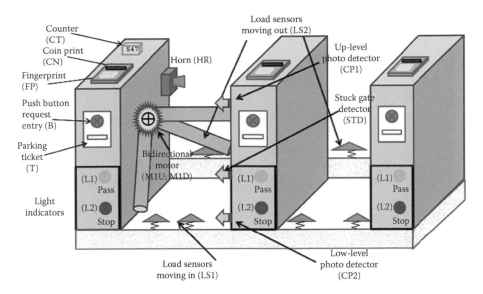

FIGURE 3.43
Automatic stadium access gate entrance.

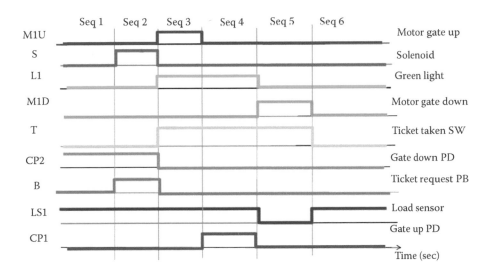

FIGURE 3.44
Timing Diagram of Operating Stadium Access Gate for Cycle 1 (when stadium is not full and gate not stuck).

TABLE 3.26

Table of Sequence Analysis for Operating the Stadium Access Gate Using Switching Theory

	Input Devices								Output Devices								Changing
	B	T	LS1	CP1	CP2	CM	T1[30]	T2[60]	S	M1D	M1D	L1	L2	T1	T2	H	
Cycle 1																	
Gate down	0	0	1	0	1	0	0	0	0	0	0	0	0	0	0	0	
Ticket called, stadium not full	1	0	1	0	1	0	0	0	1	0	0	0	0	0	0	0	S
Gate moving	0	1	1	0	0	0	0	0	0	1	0	1	0	0	0	0	S, M1, L1
Gate up	0	1	1	1	0	0	0	0	0	0	0	1	0	0	0	0	M1
Car entering / Gate moving	0	1	0	0	0	0	0	0	0	0	1	0	0	0	0	0	L1, M2
Gate down	0	0	1	0	0	0	0	0	0	0	0	0	0	0	0	0	M2
Cycle 2																	
Gate down	0	0	1	0	1	0	0	0	0	0	0	0	0	0	0	0	
Ticket called, stadium full	1	0	1	0	1	0	0	0	0	0	0	0	1	1	0	0	L2, T1
Delay 30 s	0	0	1	0	1	0	1	0	0	0	0	0	1	1	0	0	L2, T1
Gate down	0	0	1	0	1	0	0	0	0	0	0	0	0	0	0	0	
Cycle 3																	
Gate down	0	0	1	0	1	0	0	0	0	0	0	0	0	0	0	0	
Gate stuck	0	0	0		0	1	0	0	0	0	0	0	0	0	1	1	T2, H
Delay 60 s	0	0	0	0	0	0	0	1	0	0	0	0	0	0	0	0	T2, H
Gate down	0	0	1	0	1	0	0	0	0	0	0	0	0	0	0	0	

Then, using the switching theory to analyze the sequence table above, I/O Boolean functions can be derived, such as with S, $S(1 \rightarrow 0) = S[0] = B \cdot LS \cdot CP2$ while $S(0 \rightarrow 1) = S[1] = B$. Thus, it yields

$$S = (S+B) \cdot (B + \overline{LS} + CP2)$$

Similarly for M1U, M1D, L_1, L_2, T_1, T_2, H, I/O Boolean functions result in

$$M1U = (M1U + \overline{B} \cdot T \cdot \overline{CP2}) \cdot \overline{CP1}$$

$$M1D = (M1D + LS1 \cdot \overline{CP1}) \cdot (\overline{C1} + CP2)$$

$$L_1 = (L_1 + \overline{B} \cdot LS \cdot \overline{CP2}) \cdot (CP1 + LS1)$$

$$L_2 = (L_2 + B) \cdot (B + \overline{T1[30]})$$

$$T_1 = (T_1 + B) \cdot (B + \overline{T1[30]})$$

$$T_2 = (T_2 + SD) \cdot (SD + \overline{T2[60]})$$

$$H = (H + SD) \cdot (SD + \overline{T2[60]})$$

3.10.5 Mixing Juice Tank

Consider a two-stage mixing juice process consisting of filling a tank with different beverages and homogenizing the mixed liquid before conditioning it into bottles, as illustrated in Figure 3.45. When the start PB is selected, both

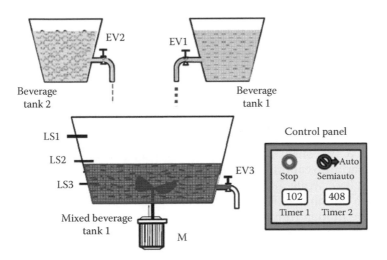

FIGURE 3.45
Mixed juice processing unit.

TABLE 3.27

Equipment Involved for Mixed Juice Processing Unit

Input Equipment	Symbol	Output Equipment	Symbol
Start push button	Start	Electrovalve 1	EV1
Stop push button	Stop	Electrovalve 2	EV2
Timer delay 1	TD1 (3 s)	Mixing motor	M
Timer delay 2	TD2 (3 s)	Timer 1	T1
Timer delay 3	TD3 (3 s)	Timer 2	T2
		Timer 3	T3

electrovalves EV1 and EV2 open to proportionally fill the tank for time durations TD1 and TD2, respectively. Once both time durations have expired, the homogenizing motor contactor is activated for the time duration TD3. Notice that the level sensor can replace the time delays (TD1 and TD2) for synchronization of the filling process. The involved process equipment and the resulting sequence table are presented in Tables 3.27 and 3.28. Figure 3.46 depicts the corresponding timing diagram.

The resulting I/O Boolean equations yield

$$EV1 = \left[(EV1 + START) \cdot \overline{TD1 \cdot \overline{START}}\right] \cdot \overline{STOP}$$

$$EV2 = \left[(EV2 + START) \cdot \overline{TD2 \cdot \overline{TD1}}\right] \cdot \overline{STOP}$$

$$M = \left[(M + \overline{TD1} \cdot TD2) \cdot \overline{\overline{TD2} \cdot TD3}\right] \cdot \overline{STOP}$$

$$T1 = \left[(T1 + START) \cdot \overline{TD1 \cdot \overline{START}}\right] \cdot \overline{STOP}$$

$$T2 = \left[(T2 + START) \cdot \overline{TD2 \cdot \overline{TD1}}\right] \cdot \overline{STOP}$$

$$T3 = \left[(T3 + \overline{TD1} \cdot TD2) \cdot \overline{\overline{TD2} \cdot TD3}\right] \cdot \overline{STOP}$$

3.10.6 Seaport Gantry Crane

Consider a two-stage gantry crane handling process capable of loading and unloading seaport containers from specified locations, as illustrated in Figure 3.47, while Table 3.29A lists all I/O variables involved. Once the start push button (START) is activated, the crane is expected to perform yz-coordinate motion from the (LS3, LS1) location to the (LS4, LS2) location to pick up the container and drop it into the (LS4, LS1) location. During this motion, the motor contactor (M1B) is activated, causing the trolley to move down horizontally from position level switch (LS1) to position level switch (LS2). Simultaneously, the motor contactor (M2B) is activated, causing the

TABLE 3.28

Analysis of the Sequence Table for Mixed Juice Process Using Switching Theory

Sequence	Input Devices					Output Devices						Switching				Virtual		
	Start	TD1	TD2	TD3	Stop	EV1	EV2	M	T1	T2	T3	OP	M	D	T1	D1	D4	D5
Rest	0	0	0	0	0	0	0	0	0	0	0					0	0	0
Valves opening and tank filling	1	0	1	0	0	1	1	0	1	1	0	X^1			X^1	0	0	0
EV1 closed	0	1	0	0	0	0	1	0	0	1	0	X^0 X^1			X^0	0	0	0
EV2 closed and start motor	0	0	1	0	0	0	0	1	0	0	1	X^0	X^1			0	0	0
Stop motor	0	0	0	1	1	0	0	0	0	0	0	X^1	X^0		X^1	0	1	0

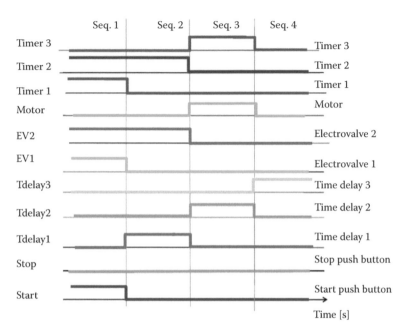

FIGURE 3.46
Timing diagram for mixed process.

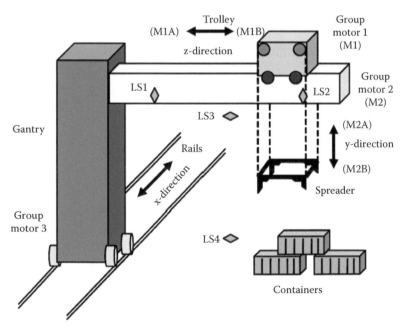

FIGURE 3.47
Gantry crane handling system of container seaport.

TABLE 3.29A

Equipment Involved in Container Handling Process Using Gantry Crane

Input Device	Symbol	Output Device	Symbol
Start push button	Start	Trolley motor contactor (right direction)	M1B
Stop push button	Stop	Trolley motor contactor (left direction)	M1A
Trolley level limit switch 1	LS1	Spreader motor contactor (up direction)	M2A
Trolley level limit switch 2	LS2	Spreader motor contactor (down direction)	M2B
Spreader level limit switch 3	LS3		
Spreader level limit switch 4	LS4		

spreader to vertically move from position level switch (LS4) to position level switch (LS2). When the container is clamped into the spreader, it moves back to the final position (LS4, LS1) using activation for both motor contactors (M1A and M2A) for simultaneous motion of the trolley and spreader. Those motor contactors are deactivated when they reach the locations LS4 and LS1, respectively. The start and stop push buttons from the control panel are not shown in Figure 3.47.

From the system function analysis and applying switching theory in the sequence table (Table 3.29B), the I/O Boolean functions are

$$M1A = (M1A + \overline{LS2} \cdot LS4 \cdot \overline{LS3}) \cdot (LS4 + \overline{LS1})$$
$$M1B = (M1B + \overline{LS2} \cdot LS4 \cdot \overline{LS3}) \cdot (LS4 + \overline{LS1})$$
$$M2A = (M2A + \overline{LS3}) \cdot (LS4 \cdot \overline{LS1})$$
$$M2B = (M2B + \overline{LS2}) \cdot (LS4 \cdot \overline{LS1})$$

Exercises and Problems

3.1. a. Derive a logic circuitry using an R-S logic gate from the Boolean function below.

$$L = (L + START) \cdot \overline{STOP}$$
$$M = (M + Td[5\,sec] \cdot SW1) \cdot \overline{\overline{SW1} \cdot SW2 \cdot STOP}$$

b. Derive the I/O Boolean function corresponding to the logic circuitry to control a delayed motor starter (M), illustrated in Figure 3.48.

TABLE 3.29B

Sequence Table Analysis and Switching Theory for Container Handling Process Using Gantry Crane

Sequence	Input Devices					Output Devices				Changing			
	Start	LS1	LS2	LS3	LS4	M1A	M1B	M2A	M2B	M1A	M1B	M2A	M2B
Start	1	0	0	0	0	0	0	0	0				
Moving trolley	0	1	0	0	1	0	1	0	0	X^0		X^0	
Moving spreader	0	0	1	0	0	0	0	0	1		X^1	X^1	X^1 X^1
Hosting spreader with container	0	0	1	1	0	0	0	1	0	X^0		X^0	
Pulling back trolley	0	0	0	0	1	1	0	0	0	X^1	X^1	X^1 X^1	X^1
Moving spreader with container	0	1	0	0	0	0	0	0	1	X^0		X^0	
Hosting of spreader	0	1	0	1	0	0	0	1	0	X^0		X^0	
Moving trolley (new cycle)	0	1	0	0	1	0	1	0	0				

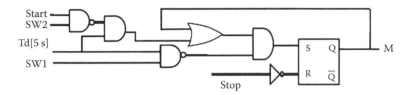

FIGURE 3.48
Logic circuit of a controller for a delayed motor starter.

   c. Consider the process input variables START and STOP push buttons interfacing the logic controller and the process output variables MOTOR and MOTOR_STARTER such that for any change input combination, their values are captured by the truth table in Table 3.30.

      i. Derive the simplified I/O Boolean functions from the truth table and K-maps if necessary.

     ii. Build the equivalent logic circuitry.

   d. Considering the Karnaugh table of a motor with two inputs (stop and start push buttons) given in Table 3.31. Derive the resulting I/O Boolean functions.

TABLE 3.30

Truth Table of Motor Starting Process

START	STOP	MOTOR	MOTOR_STARTER
0	0	0	0
0	1	0	0
1	0	0	1
1	1	0	0
0	0	1	1
0	1	1	0
1	0	1	1
1	1	1	0

TABLE 3.31

K-Map of a Motor Starter System

		START		STOP	
		00	01	11	10
MOTOR	0	0	0	X	1
	1	1	0	0	1

3.2. Consider a car alarm system consisting of a sound alarm and four direction lights that are activated if an intrusion is detected by the car window broken sensor (WS) and the door locker broken sensor (DS). The window sensor is a loop of wire encircling the window, while the door locker sensor activation is done by the key. Furthermore, the alarm key switch (KS) is used to activate or deactivate the alarm once the sensor's safe conditions are satisfied. The truth table in Table 3.32 summarizes the three inputs (four window sensors, a door sensor, and an alarm key switch) and two car alarm outputs (the alarm horn and car four-direction lights).

    a. By examining the truth table, derive the I/O Boolean functions.

    b. Derive the equivalent logic circuit for the car alarm.

3.3. Consider the two 3-state state diagrams depicted in Figure 3.49a and b. Write the feasible I/O Boolean states and the output equations describing them.

3.4. A logic circuit for the activation and deactivation of a remotely controlled electric motor–driven pump is illustrated in Figure 3.50.

TABLE 3.32

Truth Table of a Car Alarm

Door Sensors (DS)	Windows Switch (WS)	Key Switch (KS)	Car Horn (CH)	Car Lights (CL)	Comment
0	0	0	0	0	No alarm
0	0	1	0	0	No alarm
0	1	0	0	0	No alarm
1	0	0	0	0	No alarm
1	1	0	0	0	No alarm
1	0	1	1	1	Intrusion
0	1	1	1	1	Intrusion
0	0	1	0	0	Alarm active
1	1	1	1	1	Intrusion

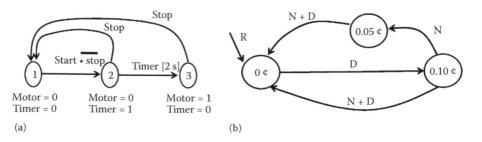

FIGURE 3.49

(a) Motor starter system state diagram. (b) Three-state state diagram.

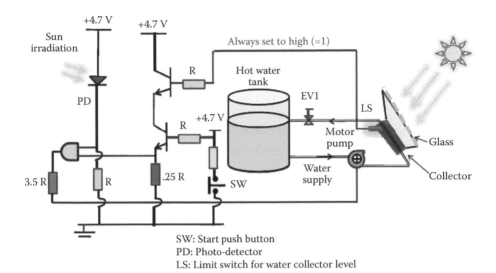

FIGURE 3.50
Motor-driven pump control circuit.

Here, the motor is energized by three digital sensors: an infrared red remote control source toward a photodiode (PD), a switch (SW) turned ON, and a limit switch LS always set to 1. Those signals are connected into a two-input AND gate, all corresponding to the I/O logic Boolean function that can be summarized by

$$Motor = SW \cdot PD$$

It is desired to upgrade the circuit of the logic control for the motor pump switching operations according to the truth table below, including deactivation operations (e.g., PD being OFF and SW set to 1).

a. Derive the upgraded I/O Boolean function of this system corresponding to the truth as summarized in Table 3.33.

b. Build the circuit to implement the logic controller derived in (a).

TABLE 3.33

Truth Table

PD	Switch	Motor
1	1	1
1	0	1
0	1	0
0	0	0

3.5. Consider a walker road crossing with individual lamp adjustment of three unknown variables: LAMP1 (green), LAMP2 (amber), and LAMP3 (yellow). The controller logic is functioning based on possible values of the input variables PB1 and PB2 (push buttons). For example, when the LAMP1 is set to 1, neither button is pushed. LAMP3 is set to 1 when either button is pushed. LAMP2 is set to 1 when both buttons are pushed, as shown in the Table 3.34. Using this truth table, derive the I/O Boolean equations.

3.6. Consider a driverless car with an automatic speed and direction and a crash avoidance system (Figure 3.51). The input and output variables used to model this system are depicted in Table 3.35. Key vehicle

TABLE 3.34

Truth Table of Outputs LAMP1, LAMP2, and LAMP3

PB1	PB2	LAMP1	LAMP2	LAMP3
0	0	1	0	0
1	0	0	0	1
0	1	0	0	1
1	1	0	1	0

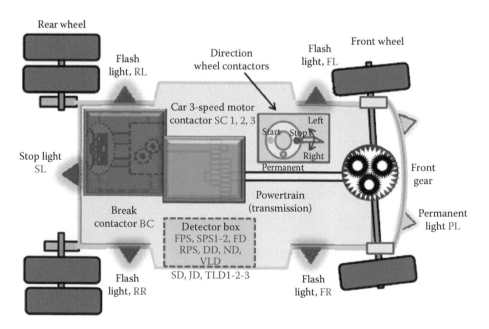

FIGURE 3.51
Driverless car system.

TABLE 3.35

Equipment Involved in Electric Driverless Car Motion Control System

Input Devices	Symbol	Output Devices	Symbol
Start push button	Start	Left flash light indicators (rear and front)	RL, FL
Stop push button	Stop	Right flash light indicators (rear and front)	RR, FR
Front position sensors	FPS	Car direction motor contactor	MC
Side position sensors 1 and 2	SPS1, SPS2	Car three-speed motor contactor	SC1, SC2, SC3
Relative position sensor	RPS	Break contactor	BC
Front car presence detector	FD	Permanent light	PL
Side car presence detector	SD		
Traffic light detector (red, yellow, green)	TLD (1, 2, 3)		
Junction detector	JD		
Day and night detectors	DD, ND		
Visibility length detector	VLD		

maintenance fault detectors (for the engine, flat tire, etc.) have been discarded for simplification. It is assumed that the car trajectory is known in advance.

a. Based on the GPS-given trajectory, derive a sequence table.

b. Develop a state diagram for such a system.

c. Derive an equivalent I/O Boolean function.

3.7. Consider the Lathe machining process using four photodetectors to control the advance of motor 1. Based on the sequence table shown in Table 3.36, derive the equivalent I/O Boolean functions.

3.8. a. Write an assembly language program that computes the following control equation:

$$A1 \leftarrow (A2 + B \cdot C) \cdot \overline{D} + E$$

with data memory locations of A1, A2, B, C, D, and E being, respectively, 0001, 0010, 0011, 0100, 0101, and 0110.

b. Write an assembly language program that computes $A \times B$ using addition operations within a register.

c. Consider the screw table motion system illustrated in Figure 3.52. When the START button is pushed in the system, the HALT lamp is turned OFF while the RUN lamp is turned ON, and the motor contactor (MTR) is energized to provide an output voltage $m(n)$ to the stepper motor. The motor contactor (MTR) is

TABLE 3.36

Sequence Table of the Lathe Machining Process

	Input Devices							Output Devices						
Sequence	Start	Stop	T (2 s)	PD1	PD2	PD3	PD4	KMP	KM1	KM2	KMC	KMCU	T1	CT
1. Rest	0	0	0	1	0	1	0	0	0	0	0	1	0	0
2. Start	1	0	0	1	0	1	0	0	1	0	1	0	0	1
3. Advance to PD2	0	0	0	0	0	1	0	0	1	0	1	0	0	1
4. PD2 activated	0	0	0	0	1	1	0	1	1	0	1	0	0	1
5. Retract to PD1	0	0	0	0	0	1	0	1	1	0	1	0	0	1
6. PD1 activated	0	0	0	1	0	1	0	0	1	0	1	0	1	0
7. Timer ON	0	0	0	1	0	1	0	0	1	1	1	0	1	1
8. Timer delay OFF	0	0	1	1	0	1	0	0	1	1	1	0	1	1
9. Advance to PD4	0	0	0	1	0	0	0	0	1	1	1	0	0	1
10. PD4 activated	0	0	0	1	0	0	1	1	1	1	1	0	0	1
11. Retract to PD3	0	0	0	1	0	0	0	1	1	1	1	0	0	1
12. PD3 activated	0	0	0	1	0	1	0	0	0	0	1	0	0	0

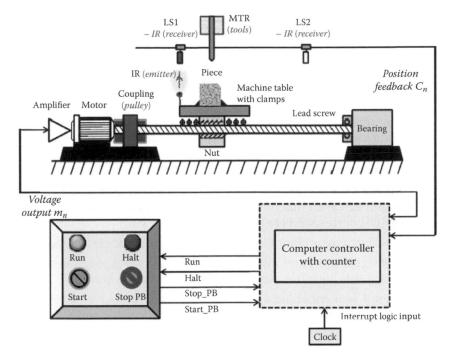

FIGURE 3.52
Logic control for screw table motion system.

deactivated when either limit switch LS1, limit switch LS2, or the STOP button is activated. This also sets the RUN lamp OFF, while the HALT lamp turns ON. In order to design a logic controller, for this screw table motion, write two assembly language programs: (i) using the asynchronous mode, where the system is synchronized with external event occurrence using an interrupt program, and (ii) for the synchronous mode, where the system uses a clock signal to ensure synchronization with event occurrence.

3.9. Recall the system from Figure 3.52, where a motor rotates by one angular increment in position whenever a pulse is received at its Step_motion () input signal. The Dir_motion() input signal determines the rotational direction (logic level 0 for counterclockwise motion, logic level 1 otherwise).

a. For a motor resolution of 250 steps per revolution, design a logic controller to perform a three-quarter revolution in a clockwise direction and two revolutions in another direction. Edit an interrupt-based logic controller program for the system shown in Figure 3.52 to output signals STEP and DIR to the motor.

A 1 GHz clock is used to construct interrupts setting the STEP puls output rate at 50,000 Hz.

b. Consider a 12 cm motion (from LS1 to LS2) with a pitch of 6 revolutions/cm. Design a logic controller program where LS1 and LS2 are used to reverse the rotational direction of the motor-driven table.

c. By defining input and output variables and setting their logic addresses, design an assembly language program to continuously move the table from one limit switch to another at a constant velocity of 1.8 cm/s.

3.10. Consider a system made of a LED and four switch elements LS1, LS2, LS3, and LS4 such that the LED turns when limit switches are set to 0101. (LS1 corresponds to the least significant bit.)

a. Design a logic circuit to perform the required logic function.

b. Design a computer interface and a logic control program to perform this logic function. Compare this solution with (a).

3.11. Design a sequential logic circuit for the secure door of a building main entrance using a four-digit electronic identifier code. In order to open the door, the user slides his or her card along a photodiode-based barcode reader to identify the cardholder and then enters his or her personal code. Once the identifier code 1110, for example, is validated, the door opens; otherwise, the computer keeps the door closed. *Hint*: Draw the state transition diagram.

3.12. Consider the automated pedestrian crossroad illustrated in Figure 3.53. The operating sequences for the road junction traffic management consist of a green traffic light in one direction for 15 s, followed by a yellow traffic light for 5 s. Simultaneously, the traffic lights are

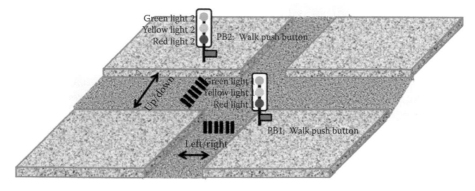

FIGURE 3.53
Traffic light junction.

turned ON in the other direction (up-down/left-right) such that a green or yellow light in one direction immediately induces the red light to turn ON in the other direction. When a pedestrian activates the walk push button, a crosswalk light is turned ON, the green light time is reduced by 5 s, and the crossroad red light time is also reduced by 5 s.

System inputs are timer (15 s), timer (5 s), and walk push buttons PB1 and PB2. System outputs are green light 1 (GL1), yellow light 1 (YL1) and red light 1 (RL1); green light 2 (GL2), yellow light 2 (YL2), and red light 2 (RL2); and timer. The state is represented as the bit pattern of the six lights. Thus, from the operating sequence analysis, activation of a crosswalk button PB1 or PB2 causes a sudden transition from a green light to yellow only when the remaining green light time is less than 5 s.

If those crosswalk buttons are pushed during the red light period, no transition is activated. Operating sequences are summarized in Table 3.37, with the time varying for the green lights. Notice that there are only four states given by the 6-binary-bit sequence derived.

a. Derive the state transition diagram.

b. Derive I/O Boolean equations for each process output.

c. Derive the ladder logic program.

3.13. The electrical wiring diagram of an automatic access gate consists of a logic control unit interfacing with input devices (infrared detector IR, switches SW1 and SW2) and output devices (gate MOTOR connected to a relay coil CR), as shown in Figure 3.54. The gate (MOTOR) could be activated either by switches located on each side or remotely by the IR detector. When MOTOR is activated by either the switches or IR detector, it opens the gate and remains activated for 20 s before being deactivated. This causes the immediate gate closure. If a car is stuck at the gate, as this detected by IR, the control relay turns OFF to allow MOTOR to be activated for an additional 20 s.

a. List all I/O process variables.

b. Develop a state diagram of this automatic gate.

c. Derive the corresponding I/O Boolean expression.

d. Insert an interlocking design to prevent the door from closing when a person moves in. Resketch the state diagram accordingly.

TABLE 3.37

Sequence for Traffic Light

GL1	YL1	RL1	GL2	YL2	RL2

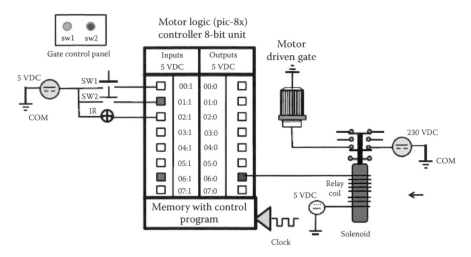

FIGURE 3.54
Power supply circuit of a motor-driven garage access gate opener interfacing with a logic controller.

3.14. Consider a motor-driven garage door that can be activated by either pushing a push button (PB) or a remote control infrared-based device (PD). There are top and bottom limit switches (LS_U, LS_T) to stop the garage door motion. When either button is pushed, the garage door should move up or down. If the buttons are pushed once while the door is moving, the door stops, and a second push starts the motion again, in the opposite direction. A light presence detector (LD) is positioned at the bottom of the garage door. When the door moves down (i.e., the door is closing) while this LD beam is cut, it stops and moves up to keep the door open. A light (LAMP) turns ON for a time delay of 3 min anytime the door opens. It can be reactivated for another time delay by pressing the pulse operating mode switch (SW) inside the garage door.

   a. Derive the I/O Boolean function for the controller using a sequence table and the switching theory method, as well as using the state diagram method. Compare your results from both methods.

   b. Design the corresponding logic circuit.

3.15. For compliance with car driving safety and security regulations, the engine (ENGINE) must validate some checks before it can start. The I/O Boolean function for ENGINE could be described as

$$ENGINE = (ENGINE + PD\_KEY \cdot SAFE) \cdot \overline{ESTOP} \cdot SAFE$$

   a. Develop an assembly language program (using AND, OR, LOAD, etc.) to implement this logic using bit 0 as the least significant bit and the I/O interface addresses given in Table 3.38A and B.

TABLE 3.38A

Mnemonic for Process Control

Process Variable	Variable Type: Address
ENGINE	Output: 0 (O/0)
PD_KEY	Input: 0 (I/0)
STOP	Input: 2 (I/2)
SAFE	Input: 3 (I/3)
TEST	Input: 4 (I/4)

TABLE 3.38B

Mnemonic for Process Control and Bit Adressing

Signal	Variable Type: Address	Bit
ENGINE	Output: 0 (O/0.0)	0
PD1	Output: 0 (O/0.1)	1
PD_KEY	Input: 1 (I/1.0)	0
ESTOP	Input: 1 (I/1.1)	1
SAFE	Input: 1 (I/1.2)	2

b. Develop a logic control program for the case above, where all I/O process variable signals are exchanged according to the interface assignments given in the Table 3.38B.

3.16. Consider a solar-powered electric car as depicted in Figure 3.55. In order to maximize the battery-charging process, it is suitable to ensure maximum sun irradiation exposure despite any possible

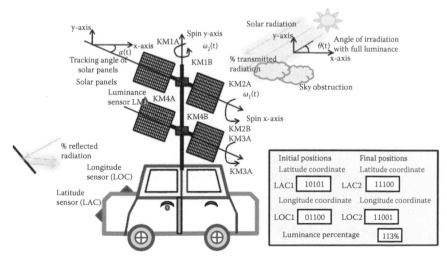

FIGURE 3.55
Solar car with motorized sun tracking system.

obstruction. This is done by establishing and maintaining an approximately 90° orientation of the attached solar panel with sun irradiation. Hence, intermittent panel orientation can be done through two bidirectional motorized supports whose direction are activated by respective contactors (KM1A, KM1B, KM2A, KM2B, KM3A, KM3B, KM4A, and KM4B). Those supports allow the two panel pairs to position along the x- and y-directions based on sunlight detection. Assume that each 2 s activation of the step motor of each panel pair produces a force capable of moving the panel orientation by 2.5° in any specific direction.

It is desired to derive a logic control algorithm panel positioning process by performing time-based ON/OFF step motor operations.

a. List all involved input and output devices, including those required for the tracking system of the panel orientation trajectory. (Hint: List any timers and counters involved.)

b. For a specific given trajectory, develop the sequence table describing the panel positioning process and the equivalent Mealy state diagram.

c. Write state and process output Boolean expressions from the state diagram.

d. Accordingly to (c), write an assembly language software to implement the logic control algorithm.

e. Write the equivalent LD.

f. Design a logic circuit to perform the required I/O Boolean functions derived above.

g. Sketch the corresponding timing diagram and discuss your logic control algorithm.

3.17. The relay logic circuit for a vehicle engine starter is drawn in Figure 3.56, while the operating sequence is summarized in the sequence table depicted in Table 3.39. It is desired to migrate the control system of this electromechanical relay-based motor starter into a logic control algorithm. Here, the activation and deactivation of the start and stop push buttons are considered inputs in order to energize and de-energize a 12 VDC motor starter system.

a. Based on the system schematic and table of sequence, derive I/O Boolean equations capturing the motor starter operating sequences.

b. Build the equivalent logic control circuit.

c. Write the assembly language program to implement this logic control algorithm.

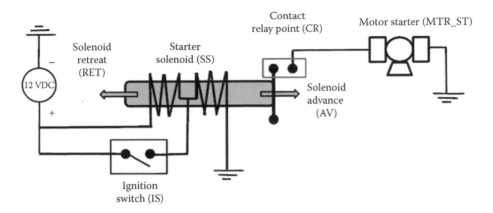

FIGURE 3.56
Relay-based control logic of a car engine starter.

TABLE 3.39

Sequence State Table of a Car Engine Starter

	System Inputs		System Outputs		
Sequence	IS	CR	AV	RET	MTR_ST
1. Solenoid advance	1	0	1	0	0
2. Motor running	1	1	1	0	1
3. Solenoid retreat	0	1	0	0	1
4. Motor driving end	0	0	0	1	0

3.18. Recall the elevator for the three-story building illustrated in Figure 3.42a. It is desired to design a logic controller using, as inputs, three-floor selection push buttons (Pb1, Pb2, Pb3) located outside of the elevator at each floor (instead of Up and Down push buttons) and three-floor limit switches located on the floor strip (sw1, sw2, sw3), and as output, the elevator bidirectional motor (M1, M2).

   a. Derive a state diagram for this elevator ensuring a suitable floor selection.

   b. Derive the corresponding I/O Boolean equations.

   c. Develop an equivalent ladder logic program that implements the resulting I/O Boolean equations.

3.19. A speedway crossing gate motion is dictated by a bidirectional motor for a raising and lowering motion, as depicted in Figure 3.57.

   Consider the following operating constraints: when the opening gate is stuck, a current detector (DS) sets a logic controller input ON,

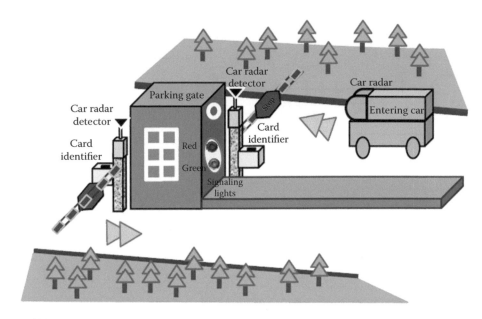

FIGURE 3.57
Automated speedway crossing gate.

causing the gate to reverse its motion and set the red light ON indefinitely; if a valid keycard is entered, a logic controller input is set ON, causing the gate to rise and remain open for 10 s.

When a registered car moves across the car radar detector (valid for car within less than 1 meter of distance), a logic controller input is set ON. The gate is maintained open while this detector is active. In the case where the speedway entrance gate is obstructed, a car would wait longer, such that this detector would be active for more than 30 s, causing the gate signaling red light to turn ON until the gate closes. Otherwise, the gate signaling green light is ON.

a. Based on the system schematic, build a table of sequence.

b. Draw the state diagram reflecting the gate operating sequences.

c. Derive the I/O Boolean functions corresponding to the gate operation.

3.20. It is desired to design a logic controller for a medical tablet press machine. This machine is capable of producing a tablet batch size of 100. The machine receives granules of a uniform size from a powder processing granulator substation. Then, this machine press up to 20 granules for 30 s. Afterward, it applies a sweetened paste to the bitter-tasting tablets in batches of 100, before drying them for 5 min. Finally, those tablets are manually removed from the machine for weight-oriented quality control manipulation, while a new batch of granules is supplied. The start button is pressed in order to start loading the new batch of granules onto the recipients encapsulated by a motor-driven wheel, and a light green indicator turns ON. Then the motor moves until a switch (Switch 1) is closed. Consequently, for 10 s, the machine press of the granules is activated. Subsequently, the motor moves farther until another switch (Switch 2) is closed, causing the release of the resulting tablet from the recipient. When the motor moves the wheel for more than 12 s without activating the switch (Switch 2), there is a machine failure, causing the motor to stop and a red light indicator to turn ON. The remedy operations consist of switching to manual operating mode, bringing the process back to its initial state, and pushing the start button twice. At the end of each successful processing batch of granules, a counter should update the count of drug condition and an amber light indicator should be set ON.

a. Draw a state diagram for this drug conditioning process from the injection of a granule batch into the recipient up to the pressing operation occurring before the sweet coating operation.

b. List the variables needed to indicate when each state is ON, as well as any timers and counters used.

c. Write an I/O Boolean expression for each transition in the state diagram.

d. Sketch a wiring diagram for the PLC.

e. Write the corresponding ladder logic diagram for the state using the mnemonic in Table 3.40.

3.21. Consider the automatic tablets press machine depicted in Figure 3.58. When Power (PB_power) is turned ON, the bottle valve outlet (EV1) opens for pressured air to flow through SOL1 and SOL2 up to 20 psi. The push-button STOP or limit switches (LS1, LS2) protect the system from overtravel by turning off the main bottle valve. PB_Step allows adjustment of the press speed to be either low or high by restricting the airflow through the electrovalve (EV2).

When the PB_Resume push button is activated, the press quickly retracts to its initial position due to the electrovalve (EV3) changing

TABLE 3.40

I/O Process Variable Addressing

Process Inputs	Process Outputs
I:1—Start push button	O:1—Green light indicator
I:2—Stop push button	O:2—Machine ON
I:3—Switch 1: recipient of batch of granules out	O:3—Red light indicator
I:6—Switch 2: recipient of batch of granules out	O:4—Amber light indicator
	O:5—Motor ON

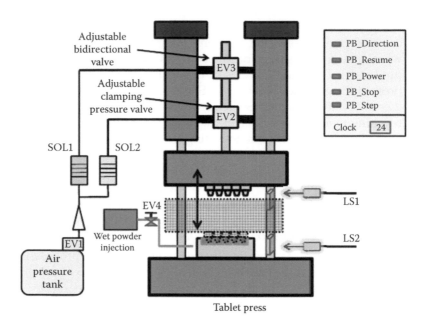

Adjustable bidirectional valve

EV3

Adjustable clamping pressure valve

EV2

SOL1   SOL2

EV4

Wet powder injection

EV1

Air pressure tank

PB_Direction
PB_Resume
PB_Power
PB_Stop
PB_Step

Clock   24

LS1

LS2

Tablet press

FIGURE 3.58
Automatic tablet press.

the direction of the pressured airflow (1 for up or 0 for down position, respectively).

a. Using Figure 3.58, build the sequence table or state diagram for this process.

b. Derive the I/O Boolean functions capturing the press machine operating sequences.

c. Sketch the corresponding logic circuit to implement this logic.

3.22. An operator in charge of power flow monitoring typically traces it from the generator plant to the load substations. The primary components of an electric power grid network are the power generator, motors, transformers, breakers, fuses, switchgears, starters, and switches. Here, power supply lines link network components, such as power dispatching switches, ammeters, power converter substations, and power flow monitoring rooms. Conversion from high-voltage signal to low-voltage signal is ensured at the substation by power transformers. Power breakers regulate the current flow by interrupting it only if the operating conditions are abnormal. The ammeters and voltmeters are electric devices used to monitor the current flow and the voltage in an electric system. Consider a small electric power network connecting through 66/33 kV power lines the boiler plant to lower voltage 240/430 V power lines ongoing into the meter associated to the machine processing unit. It is desired to dispatch the power flow over the load. Considering the grid depicted in Figure 3.59, answer the following questions:

a. List all I/O devices involved and perform a functional analysis of the power flow process.

b. Based on the power dispatching strategy chosen, write a logic control algorithm that enables us to dispatch power based on the process energy load required over those available, as well as to protect the operating devices.

3.23. Consider an electric vehicle (MOTOR) that is energized when simultaneously a recorded fingerprint (FGP) presses the start push button (START), the safety belt detector (BELT) of the driver is closed, the in-board engine failure detector (ENGINE) is at level 0, and the driver passes the in-vehicle alcohol-level breath detector (ATL). The electric vehicle will turn immediately OFF if the STOP push button is activated or an engine problem is detected.

a. Develop an I/O Boolean equation of this system.

b. Build the corresponding logic circuit of this system using the D flip-flop device.

c. Construct the equivalent LD program.

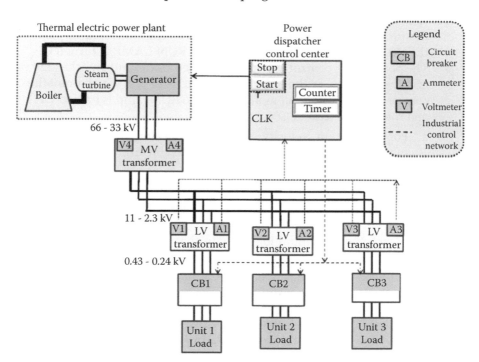

FIGURE 3.59
Plant electrical power supply network.

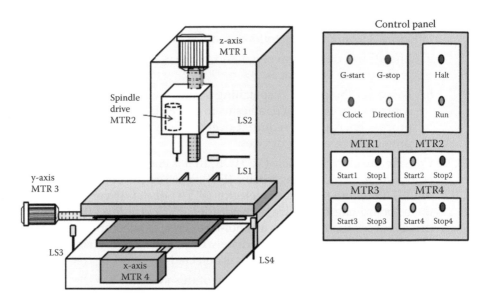

FIGURE 3.60
Automatic machine tools.

3.24. Consider the automated machine tools depicted in Figure 3.60. When hitting the pulse-type G-START push button on the control panel, the HALT lamp is turned OFF while the RUN LAMP is turned ON. The positioning control system is activated until limit switches LS1, LS2, LS3, and LS4 or the STOP button is activated, at which point the control is deactivated, all motors stop, the RUN lamp is turned OFF, and the lamp is turned ON.

    a. For a constant velocity of motion, develop a time-triggered machine tool positioning subroutine to synchronize the motion of all three motors.

    b. Based on the I/O devices shown in Figure 3.60 and the positioning subroutine in (a), write a logic control algorithm for the machine tool positioning system.

    c. Sketch the sequential logic circuit that could be used to implement this logic.

3.25. As depicted in Figure 3.61, consider a multicooker device with two mutually exclusive operating cycles: baking and cooking operating cycles. If the baking cycle selection (PB_bake) and START push buttons are activated, an electric power of 1100 W (ENRG1), light indicator (L2_bake), and motor-driven rotating table (MTR) are energized for an elapsed time of 540 s/200 g of baking raw product. Similarly, if the cook cycle selection (PB_cook) and START push buttons

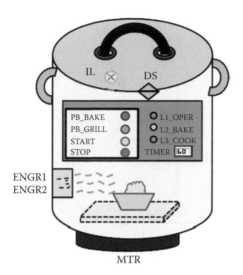

FIGURE 3.61
Multicooker device.

are pressed, 550 W of electric power is supplied toward the cooking product (ENGR2), and the light indicator (L3_cook) and motor (MTR) driving the rotating table for 60 s are energized, and then de-energized for 30 s. This is repeated five times. Note that the cycle selection must be done before activation of the START push button; the activation of either PB_bake and PB_cook push buttons during the heating process should have no effect on the cycle execution; the last operating cycle selected is maintained from cycle to cycle, except if another selection occurred; no cycle can start if the multicooker is not properly closed, meaning the door is unlatched (DS = 0); and when the door is unlatched (DS = 0) and power is supplied, an inside multicooker light (IL) is energized.

Any activation of the STOP push button or opening of the door (DS) causes an immediate termination of the operating cycle. The light indicator (L1_OPER) is ON when the multicooker operating cycle is in progress.

a. List in a table all I/O devices involved for both cycles.

b. Establish the sequence table for both cycles.

c. Derive the state diagram and subsequent I/O Boolean functions.

d. Derive the corresponding logic controller program in an LD.

e. Build the circuitry for the logic controller using the D flip-flop gate.

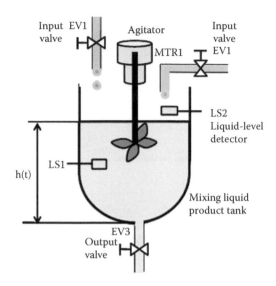

FIGURE 3.62
Mixing tank system.

3.27. Consider an automatic mixing station, as illustrated in Figure 3.62.

The operating sequences of events for this mixing process are given as

a. Open valve 1 (EV1) until level 1 (LS1) is reached for the first liquid.

b. Close valve 1 (EV1).

c. Open valve 2 (EV2) until the level 2 limit switch (LS2) is reached for the second liquid.

d. Close valve 2 (EV2).

e. Start the motor (MTR1) and agitate two liquids for up to 20 min.

f. Stop the motor (MTR1).

g. Open valve 3 (EV3) for up to 5 min.

h. Close valve 3 (EV3).

i. Repeat the mixing process.

   i. List all I/O devices involved in the mixing process and sketch the state diagram corresponding to the operating sequences described above.

   ii. Draw the wiring diagram and the corresponding ladder logic program for this mixing process.

3.28. Consider a packaging process of a bottle milk filling process, as illustrated in Figure 3.63. Here, a pump removes the milk from a tank at a constant flow rate to inject it into a bottle positioned by a motorized conveyor. When the bottle is filled as expected from the

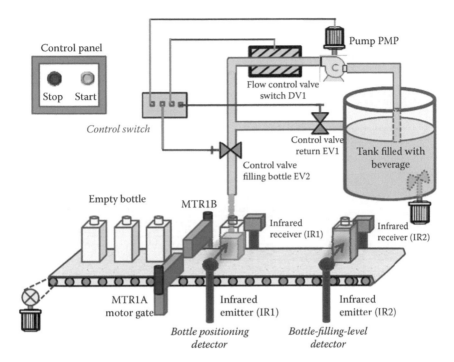

FIGURE 3.63
Bottle milk filling process.

scale platform within the conveyor, the filling valve is closed, the gate moves down, and the conveyor brings in another empty bottle to replace the full one. Then, the filling valve is reopened and the filling cycle restarts. If the fluid level sensor or the bottle position switch is OFF, then the pump and filling valve are OFF.

The digital flow switch is used to regulate a constant flow rate of milk. There is a recycling valve that is opened to redirect the milk when the filling valve is closed. It is closed when the filling valve is opened. The recycling valve and filling valve use the same solenoid, which is normally closed. One pack contains up to 10 milk bottles. It is expected to be uploaded manually in 45 s. This is done when the production line is stopped. A counter is used to stop the production line, and a timer is used to adjust the uploading of the bottles in a pack.

a. Using the operating conditions presented above and the schematic in Figure 3.64, draw the state diagram and equivalent sequence table.

b. List all I/O process devices involved, along with their assigned address, and then draw the wiring diagram schematic.

c. Sketch the corresponding sequential function chart for this logic based on the corresponding mnemonics of process devices.

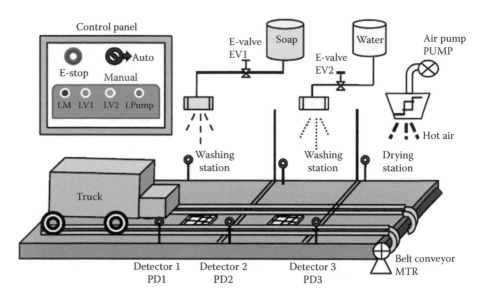

FIGURE 3.64
Car washing process with three substations.

3.29. A bank branch has a two-door security system such that door 1 requires a two-digit code, while door 2 requires a two-digit code and door 1 being closed after being opened. The operating sequence could be a combination of unlock–delay–lock.

   a. List all I/O devices involved in this security door opening process, and then the table of sequences from prior functional analysis being developed.

   b. Derive the SFC logic control program for the two-door security system.

3.30. Consider an automatic speed control of a motor through variable speed (VS) drives commutating its output voltage between 0 and 10 V. Its control panel consists of an operating speed adjustment through the potentiometer device, the toggle switch to select the direction of rotation, and start and stop push buttons.

   a. Using all these inputs from the panel, develop an LD program to operate the AC motor at various speeds integrating any interlock functions. This should contain the logic to start and stop the commands for the direction of rotation, as well as be able to either automatically or manually select operating options.

   b. Sketch the wiring diagram for this system if it is connected to a PLC (for each device involved, specify the corresponding address and mnemonics).

3.31. Consider an *n*-place parking garage management system associated with a parking gate controller. Trucks and vehicles can be parked in this garage, but a truck requires the double of vehicle place. The parking place management system should indicate by colored light signals when any of the following scenarios are occurring: places available for vehicles and trucks (FP), places available only for vehicles (FPC), and no place available (FF).

   a. List all I/O devices involved in this process.

   b. Establish the truth table of possible combinations between input variables and output variables defined by the parking indication status (FP, FPC, or FF) to control the entrance of a five-place parking.

   c. Using POS, POS methods, and K-maps if necessary, derive the corresponding I/O Boolean functions from the truth table above.

   d. Determine the Mealy state diagram of a process to control the entrance of a 20-place parking. (Hint: This is similar to the logic control of an automatic vending machine).

3.32. Consider an automatic vending machine distributing sweets at 75 cents/unit. This machine is capable of receiving coins of 10, 25, and 50 cents. For any sum of inserted coins being equal to or higher than 70 cents, a sweet is delivered and the round-up change is ejected in the return box when necessary. When the machine fails to deliver, a failure detector (FD) is activated and energizes the red light indicator; otherwise, a green light indicator is activated.

   a. List all I/O devices involved in the automatic vending process.

   b. Derive the sequence table and fill the equivalent state transition table for this automatic vending machine.

   c. Draw the equivalent state diagram.

   d. Derive the subsequent state Boolean functions.

   e. Sketch the logic circuit and implement it using the D flip-flop gate.

3.33. Consider an automatic car wash process with a truck parked on a driving conveyor belt, as illustrated in Figure 3.64. When the START push button is activated, the conveyor belt moves the truck up to photodetector 1 (PD1). Once PD1 is activated by the truck presence, the soapy water shower valve (EV1) of the washing station is opened for timer T1 for 60 s. When the detector (PD2) is activated by the presence of the car in the second station, the clean water shower valve (EV2) activates timer T1 for 60 s. Similarly, when the detector (PD3) is activated by the presence of the truck in the third station, the air pump (PUMP) is activated. PD3 is deactivated when the truck leaves this washing process (due to similarity, this last station is discarded). The control panel eases the monitoring of active system outputs through light indicators, conveyor (LM), EV1, LV1, EV2, LV2,

PUMP, and LPUMP, as well as the selection of the process operation type (manual or automatic).

a. List all I/O devices involved in the car washing process.

b. Derive the sequence table and fill the equivalent state transition table for this process.

c. Draw the equivalent state diagram.

d. Derive the subsequent state Boolean functions.

3.34. Consider a four-stage pozzolana removal process, as depicted in Figure 3.65. First, by pushing StartPB, the scratcher driven by motors M5 and M6 positions transversally just above the pozzolana stockpile. Then, while secondary and primary arms, driven by the activation of motors (M2A and M3A), are moving down over the pozzolana stockpile, the motors (M1 and M4) are scratching the pozzolana from one side to another, up to the conveyor, for 139 s. After a short stop, the motor contactors M2A and M3A reposition for 2 s farther down the pozzolana stockpile. Then, for another 139 s, M1 and M4 scratch the pozzolana. This scratching stage lasts until the primary arm reaches the lower-level limit switch LS1. Here, both motor contactors M2B and M3B are activated to retract simultaneously until the secondary arm reaches the upper position given by LS2. At this position, motors M5 and M6 move to a new transversal position above the pozzolana and the scratching stage of the process can restart to the lower-level limit switch LS1. The conveyor belt collects the pozzolana scratched continuously toward the next cement drying station. Note that motors (M3 and M2) or (M1 and M4) or (M5 and M6) run simultaneously,

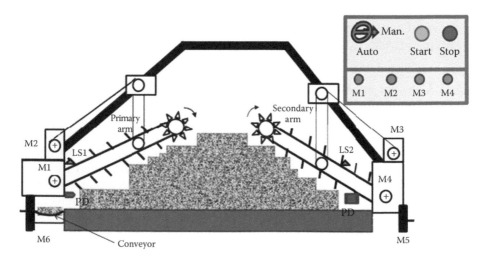

FIGURE 3.65
Schematic of the cement pozzolana scratching process.

usually in the same direction, except for M2 and M3, which run in opposite directions. Only M2 and M3 are bidirectional motors; as M2 has two motor contactors (M2A, M2B) and M3 has two motor contactors (M3A and M3B). Therefore, the number of outputs can be reduced from eight to six. The transversal positioning is discarded. Activation of the photodetector (PD = 1) indicates that the scratcher is over the pozzolana and needs a transversal repositioning.

a. List all I/O devices involved in the cement pozzolana scratching process.

b. Derive the sequence table for this cement pozzolana scratching process.

c. Draw the equivalent state diagram.

d. Derive the subsequent I/O Boolean functions and draw the resulting timing diagram.

3.35. Consider a drilling process as depicted in Figure 3.66. Pushing the start button activates the clamping of the workpiece, which has been detected by the contactor (KMC). And the motor (MTR1) is energized by the contactor (KM1), allowing the piston rod of cylinder 1 to advance downward for a 2 s time delay (T1/TD[2]). When the time delay occurs, the system reaches the position detector (PD3). Then, the drilling motor MTR2 contactor (KM2) is activated, while simultaneously the piston rod of cylinder 2 advances up to PD4. At this position, the motor (MTR2) is de-energized while the activation

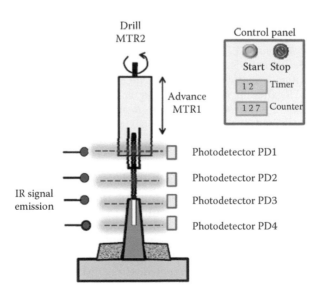

FIGURE 3.66
Drilling lathe machine.

of the contactor (KMP) forces the rod to retract up to its initial position (PD1), where the contactor (KMP) is deactivated. The system remains at rest, awaiting the start push button to be activated again. Notice that the workpiece is clamped from the time the start button is activated until the rod retracts to the initial position (PD1). Otherwise, it is unclamped (KMCU). The counter is used to estimate the amount of holes drilled.

a. List all I/O devices involved in the drilling machine process.
b. Considering the I/O table, derive the sequence table and fill the equivalent state transition table for this drilling machine.
c. Draw the equivalent state diagram.
d. Derive the subsequent I/O Boolean functions.
e. Sketch the resulting timing diagram.

Bibliography

Alciatore D.A., M.B. Histand, *Introduction to Mechatronics and Measurement Systems*, 4th ed., McGraw-Hill, New York, 2012.

Ashar P., S. Devadas, A.R. Newton, *Sequential Logic Synthesis*, Springer, Berlin, 2012.

Bemporad A., M. Morari, *Control of Systems Integrating Logic, Dynamics, and Constraints*, Automatica, Elsevier, Amsterdam, 1999.

Bollinger J.D., N.A. Duffie, *Computer Control Machines and Processes*, Addison-Wesley, Reading, MA, 1989.

Bolton W., *Mechatronics: Electronic Control Systems in Mechanical and Electrical Engineering*, Pearson, 2016.

de Silva W., *Mechatronics: A Foundation Course*, CRC Press, Boca Raton, FL, 2010.

Isermann R., *Mechatronics System Fundamentals*, Springer, Berlin, 2003.

Hugh J., *Automating Manufacturing Systems with Plcs*, Lulu.com, 2010, 644 pages.

John K.-H., M. Tiegelkamp, *IEC 61131-3: Programming Industrial Automation Systems: Concepts and Programming Languages, Requirements for Programming Systems, Decision-Making Aids*, Springer, Berlin, 2001.

Katz R.H., G. Borriello, *Contemporary Logic Design*, 2nd ed., Prentice-Hall, Upper Saddle River, NJ, 2004.

Kuphaldt T.R., *Lessons in Electric Circuits*, vol. IV, digital, 4th ed., 2007.

Lampérière-Couffin S., J.J. Lesage, *Formal Verification of the Sequential Part of PLC Programs, Discrete Event Systems*, Springer, Berlin, 2000.

Mano M.M., M.D. Ciletti, *Digital Design*, 4th ed., Prentice-Hall, Upper Saddle River, NJ, 2006.

Nof S.Y., *Handbook of Automation*, Springer, Berlin, 2009.

Olsson G., *Industrial Automation*, Lund University, 2003.

Petruzella F.D., *Programming Logic Controllers*, 4th ed., McGraw-Hill, New York, 2010.

Roth C.H., *Digital System Design Using VHDL*, Thomson, 2002.

Vahid F., *Digital Design*, Wiley, Hoboken, NJ, 2007.

4

Process Monitoring, Fault Detection, and Diagnosis

4.1 Introduction

Industrial process monitoring platforms usually support a set of decision-making tools that help to safely run the process and make suitable decisions in the presence of faults or undesired disturbances. This can be achieved mainly by maintaining the process variables within predetermined ranges. Furthermore, the process visualization option enables the manipulation of vast amounts of data and the choice of appropriate decisions among alternative process operations in near real time. Process monitoring tools use system identification techniques and fault diagnosis methods to achieve process fault-tolerant management-oriented decisions. Also, depending on the process knowledge (structural, statistical, fuzzy, functional, behavioral, etc.), they elaborate commands based on distributed control-oriented decisions for large-scale plant operations.

However, the design of monitoring and diagnostic systems is quite complex due to constraints related to (1) the combined continuous and discrete-event nature of process operations, which makes adequate process modeling difficult to build, and (2) the dependency of the process data gathering on the sensing methods and technologies used, which can lead to incomplete and uncertain process data. Hence, a high level of process knowledge and redundant data gathering is required to build an efficient monitoring and diagnostic system.

In this chapter, an integrated methodology to design and implement monitoring, diagnosis, and control systems for the supervision and fault management of industrial process operations is presented. First, requirements for the design of the process monitoring system and the gathering the analysis of the process data structure are defined. Then, formal hybrid process models are developed to integrate discrete and continuous process operations. This lays the foundation for the design of the process monitoring and diagnostic system. Hence, a generic methodology is derived for the design of a process monitoring and fault management system capable of fault detection and isolation, and process control in failure mode. Finally, some case studies on fault monitoring, including fault detection, fault diagnosis, and distributed process control, are presented within various industrial processes, such as breweries, cement drying, and power grids.

4.2 Requirements for Process Monitoring, Control, and Fault Diagnosis

Process monitoring is a function that consists of recognizing anomalies in the behavior of a dynamic system and underlying faults. Hence, process monitoring for abnormal behavior poses a challenging problem of process operation observability and diagnosis, which can be approached through full-scale sensor-based process data gathering and model-based process data analysis. Model-based approaches consist of (1) modeling quantitative and qualitative physics-based change of process behavior or characteristics over the time, (2) designing efficient and real-time process data gathering, and (3) analyzing and updating in real time the process operating conditions. In short, the idea is to mimic the condition of a physical system in the model, as illustrated in Figure 4.1.

Commonly encountered process monitoring and control system objectives are

1. Monitoring the execution of process operations in order to take appropriate counteraction in front of abnormal process behavior by

   a. Detecting fault from measurable process variables based on defined tolerances, and generating alarms for the process operator

   b. Diagnosing fault features through symptoms of process anomalies

   c. Selecting remote process maintenance operations and counteractions to fault occurrence

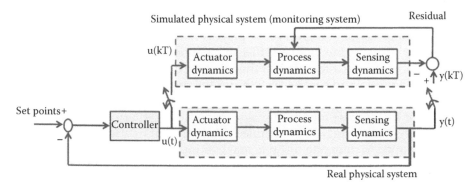

FIGURE 4.1
Generic process monitoring and control system block diagram.

2. Designing a command and control system for the process operations in order to automatically maintain the process values close to the targeted process values

These objectives are translated into software, hardware, and mechanical design specifications to ensure integrability and interoperability in terms of diagnostic, condition monitoring, control, maintenance, and safety applications. The software specifications are defined based on a combination of requirements defined as

- Functional requirements, which are related to the detection of process operation deviation from known patterns of normal operations
- Technical operational requirements, which are related to the type of fault management plan (reactive or predictive), as well as the definition, classification, and prioritization of operation maintenance interventions
- Operational requirements, which are related to performance, reliability, robustness, and fault tolerance and can be listed in terms of (1) process independence maintainability from the scale and from the process sensing and data acquisition technology, (2) process configurability, (3) real-time constraints, and (4) process safety

4.3 Monitoring System Architecture and Components

Monitoring systems gather process data in order to assess whether the processes are operating within normal ranges. For example, information on process operations such as accuracy is used to describe the quality of process operating conditions. Hence, in the occurrence of an abnormal process operating behavior, the monitoring system should detect the origin and type of faults. Thus, it ensures that reliable process operations are maintained within defined ranges. As such, in order to perform process monitoring, those systems gather process data and store them in a *system database*, upon which process behavior analysis is performed through a *fault-tolerant decision support system*.

4.3.1 Process Database Structure Design

A typical process database structure is expected to contain information related to the technical constraints of the process equipment (actuators and sensors or detectors), as well as process operating conditions, and functional characteristics of process or machine failure modes and fault classification. The database structure is a hierarchical group of process data types classified as follows:

1. Data describing classes of process devices (e.g., actuators) and their operating constraints and qualified fault types
2. Data describing subprocesses model parameters of various components of operational units
3. Data describing process operating modes, including start and shutdown operating conditions
4. Data describing various types of controllers and their configuration settings
5. Data describing rule-based logical decision support
6. Data related to process interlocks

Table 4.1 summarizes the database components of a process monitoring system.

The data derived above are either (1) variables associated with the execution of process operations, (2) intermediate variables related to the control panel, (3) threshold values for interlocking, (4) manipulated variables related to process operating modes, or (5) configuration variables associated with process operating conditions. Table 4.1 summarizes typical process equipment operating data, while Table 4.2 lists the nomenclature of information used to characterize the process equipment. For example, data associated with monitoring screens are (1) process variables and parameter configurations (i.e., actuator start-up conditions), (2) start-up and shutdown variables, (3) control variables, and (4) interlock variables and forcing condition variables. These data are used to display on monitoring screens information related to (1) process data acquisition readings, (2) process safety indicators and status, (3) process parameter configurations, and (4) process command and control of the data exchange between the controller unit and

TABLE 4.1

Typical Database Structure of a Process Monitoring System (Table Presentation)

Process Equipment				Operating Condition			Calibration		Failure		Functional Characteristics
Ac	Tr	Hm	Se	Rg	Au	A/D	Ca	Pw	Fm	Fc	Fu

TABLE 4.2

Glossary of Terms Used in Table 4.1

Symbol	Nomenclature
Ac	Actuator
Tr	Transmission media
Hm	HMI panel
Se	Sensor/detector
Rg	Range
Au	Accuracy
A/D	Analog or digital
Ca	Calibration
Pw	Power supply
Fm	Failure mode
Fc	Fault classification
Fu	Functional characteristics

human–machine interface (HMI) station. Thus, the design procedure of the process monitoring and control databases consists of

- Establishing relational database tables for process variables and parameters, operating conditions, and command and control variables
- Adding variables associated with operating modes (normal, maintenance, etc.), as well as start-up and shutdown sequences
- Adding all variables associated with interlocks and those related to process operation safety

4.3.2 Process Data Acquisition and Monitoring Decision Support System

Process fault detection methods are either signal based or model based. They use measured process data or estimated process parameters and variables through a data acquisition system, ensuring data transmission integrity, as well as the real-time status and completeness of process information. Usually, a redundant data network is required to avoid data losses in the transmission channel. Hence, for process operation command and control, as well as process fault management, a procedure for the design of a monitoring decision support system would consist of

- Identifying the critical process variables affecting process stability and normal process operations
- Identifying variable values related to potential process failure modes and their associated remedy actions

- Deriving a logical set of rules based on understanding of the process behavior (signal) or process model
- Deriving real-time decisions related to process monitoring by applying either a combination of remedy actions in response to a process failure occurrence, or fault preemptive actions (preventive maintenance or early diagnosis)

4.4 Operating Model and Fault Management of Processes

A large amount of the process data gathered are used to detect fault occurrences, whether associated with process equipment failure or abnormal process behavior. Due to the continuous and discrete-event nature of process data and the accessibility or estimation of related process variables, the design of an automatic-based fault detection and diagnosis system is a challenging system identification problem. There are two methods for the detection of process operation failures: (1) method based on the signal spectral analysis of the quantitative plant dynamics, called signal-based or model-free method, and (2) method based on the process model. In signal-based methods, several symptoms are generated to determine the difference between nominal and faulty status. Then, fault diagnosis procedures determine the fault by applying symptom classification or inference techniques. Fault detection model-based methods compare input and output process signals with process model response signals. This is usually performed through parameter estimation, parity equations, or state observers. As such, the error detection algorithms for fault diagnosis use the difference between the subprocess response and the response from a subsystem model. Then, this difference value is compared with the assigned process variable threshold, also called the "model rate of change." For example, if the mean squared error (MSE) is less than or equal to the threshold value, the error detected is assigned a 0 value; if the MSE value is more than the threshold value, the error detected is assigned a 1 value, which triggers the fault detection system. The challenge of this method is the validation of the model and the ability of the error detection filter to detect and flag an error despite slight changes in the system parameters.

4.4.1 Fault Classification, Detection, and Diagnosis Methods

A fault can be defined as a nonallowed deviation of a variable characteristic property from its reference threshold. It is usually related to system malfunction or process failure. As such, faults are time dependent and consequently could be qualified as steplike, linear biased, discrete, or intermittent, as illustrated in Figure 4.2.

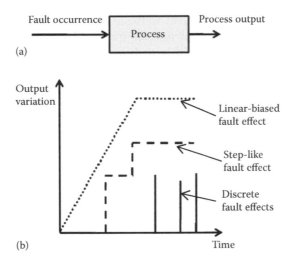

FIGURE 4.2
Various fault effects: (a) schematics of fault event over process output and (b) resulting effects of process output signals.

Discrepancies between process data and process model outputs can be classified according to fault nature and fault occurrence. It is assumed that each fault is related to one or several specific symptoms, and there can simultaneously be multiple fault occurrences. A generic fault detection and classification block diagram is illustrated in Figure 4.3.

Model-based methods for fault detection analyze and classify any deviation between measured process data and estimated process model data,

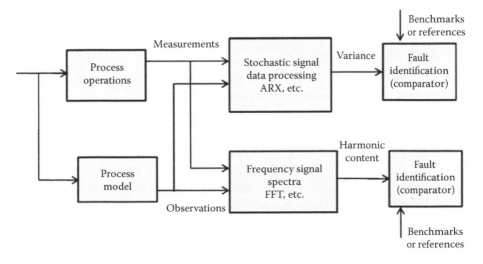

FIGURE 4.3
Example of a fault detection and classification block diagram.

using qualitative or quantitative analysis approaches. Among qualitative approaches there are (1) causal models (fault trees, structural graphs, and qualitative physics) and (2) abstraction hierarchy (functional and structural). Among analysis quantitative approaches there are (1) process and fault modeling, (2) state estimation (observer based and Kalman filter based), (3) parameter estimation (regression analysis, least squares, and recursive least square [RLS]), (4) simultaneous state or parameter estimation (two-stage Kalman filter), and (5) parity space equations (state space and input/output [I/O] observer). Depending on the process model structure and fault nature, strongly isolated or weakly isolated residuals can be separated. Once the occurring fault has been analyzed and classified, a fault diagnosis approach can be applied by determining the type of fault and its location, size, and time of detection. This approach can be either inference based or classification based, which aims to establish the relationship between symptoms and faults, using heuristic or analytical knowledge of the process.

With signal or data-based methods, changes in the I/O behavior of a process are used to simulate changes of the state variables. Then, in order to detect process faults, it is possible to compare them with faulty process-type signal features, by passing the process signals through filters or a bank of observers. Among qualitative signal analysis approaches there are expert systems, fuzzy logic, pattern recognition, and frequency and time–frequency analysis. Among quantitative signal analysis approaches there are neural networks and statistical methods (statistical classifiers). With signal-based fault diagnosis methods, faults are dependent on the amplitude and frequency of input signals, while outputs are considered free of distortion and monitoring irregularities. Usually, there is a limited sensor effect on fault frequency signal. Table 4.3 presents various methods used for process fault detection and diagnostic.

4.4.2 Process Data Reporting

Based on the principle of fault tolerance, the design of a process data reporting system usually contains redundant system modules, in order to maintain a normal data flow in case of system failure. There are four different approaches for *redundant system design*:

1. *Static or hardware redundancy*, when all redundant modules are physically designed to operate simultaneously. This allows us to detect hardware failures among physical components, especially for faulty sensing equipment. Its operational principle is based on m/n redundancy rules such that no fault can occur unless there are multiple defects. Such redundancy is implemented through several (double or triple) computer structures.

2. *Dynamic redundancy*, where redundant modules only operate at the occurrence of a failure. There are two types of dynamic redundancy:

TABLE 4.3

Summary of Methods Used to Design Process Monitoring and Control System

Supervision and Control Objectives	Supervision and Control Functions	Approach	Methods	Tools		
				Techniques	Methods/Tools	
Monitoring	Detection	With process model	Functional and material modeling	Failure modes, effects, and criticality analysis (FMECA)		
			Physical modeling	Physical redundancy	Physically redundant	
				Parameter estimation	Analytically redundant	
		Without process model	Stochastic signal processing	Benchmark test		
				Mean test		
				Variance test		
			Artificial intelligence	Behavior-based model methods	State finite machine	
					Petri network	
				Pattern recognition–based method	Expert systems	
					Statistical tools	
					Case-based learning	
					Neural network	
					Fuzzy logic	
				Methods based on causal analysis of faults	Causal graphs	
					Functional block	
					Petri network	
					Fuzzy logic	
					Neural/fuzzy logic	

(Continued)

TABLE 4.3 (CONTINUED)

Summary of Methods Used to Design Process Monitoring and Control System

Supervision and Control Objectives	Supervision and Control Functions	Approach	Methods	Tools	
				Techniques	Methods/Tools
Diagnostic		With process model	Functional and material modeling	Failure modes, effects, and criticality analysis (FMECA)	
			Physical modeling	Physical redundancy	
				Parameter estimation	
		Without process model	Statistical tools	Benchmark test	
				Variance test	
				Mean test	
			Symbolic tools	Model simulation–based method	State finite machine
					Petri network
				Pattern recognition methods (learning and recognition)	Expert systems
					Statistical tools
					Case-based learning
					Neural network
					Fuzzy logic
					Neural/fuzzy logic
				Event tree analysis (ETA)	Causal graph
				Fault tree analysis (FTA)	Contextual graph
					Petri network
					Fuzzy logic
					Neural/fuzzy logic
Control	Normal production	Control design theory			
	Protection	Interlocks			

(a) blind redundancy, where redundant modules only operate in fault cases, and (b) function-active redundancy, in which redundant modules operate in fault-free cases by running standby functions.

3. *Analytical redundancy* compounds various analytical methods to compare measured and estimated data from process state behavior with data from expected behavior, in order to detect fault occurrence.

4. *Software redundancy* aims to detect errors in software and implement appropriate redundancy measures.

Process data availability requirements are fulfilled through physical redundancy, while analytical redundancy allows estimation of process variables. According to a generally accepted terminology, a fault-tolerant management system covers the following tasks:

- Fault detection in order to make a binary decision (normal or abnormal)
- Fault isolation in order to determine the location of the fault (faulty sensor or actuator)
- Fault evaluation in order to estimate the size and type or nature of the fault

In order to isolate and evaluate process faults, the following process failure modes and effects analysis (FMEA) steps could be used:

1. Break down the process into subprocess components.
2. Determine the system's functional structure and the contribution of each component to system functions.
3. Determine the failure modes of each component.
   - In the case of new components, by using information of similar known components
   - In the case of commonly encountered components, by referring to process knowledge and measurements
   - In the case of complex components, by splitting them into subcomponents and applying FMEA on known subcomponents to determine the failure mode of the components
   - In the case of other components, by deriving possible failures through commonsense reasoning on system functions and operation physical parameters
4. Analyze the failure modes of each component and note all the results in Table 4.4.

All these steps are summarized in process operating and stop mode graphical analysis ("Gemma" in French), as illustrated in Figure 4.4.

TABLE 4.4

Failure-Based Process Component Analysis Data Structure

Device Reference	Failure Analysis Parameters	
	Specifications	
	Failure description	Symptoms
		Category
	Failure analysis	Origin
		Impact
		Diagnosis
	Other constraints	

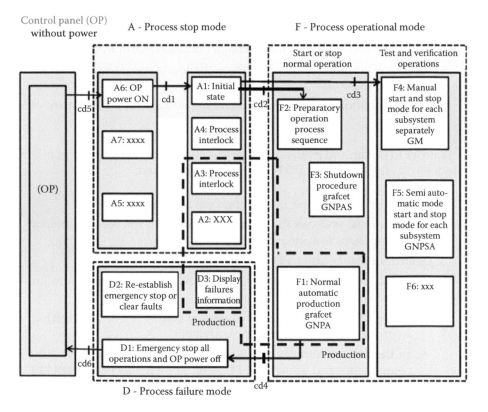

FIGURE 4.4
Graphical analysis of process operating and stop modes. (Adapted from ADEPA, *Guide d'étude des modes de marches et d'arrêts*, GEMMA, Agence pour le development de la productique et l'automatique, Ministère de l'industrie, Paris, France, 1997.)

4.4.2.1 Hybrid Control Strategies

Due to the nature of process operation, process automation usually requires a hybrid control system. A hybrid control system consists of combining sequential logic control and feedback control algorithms. Prior to designing such a system, it is suitable to perform a process functional analysis and decomposition using the FAST method, as illustrated in Figure 5.7. Then, process operating specifications and subprocess functional blocks should be identified, as well as the corresponding control requirements (i.e., open- or closed-loop control feedback, or logic control schemes). Hence, for each process functional block, control algorithms have to be designed. Eventually, those control algorithms will be combined within a hybrid control system. Usually, a logic control unit activates and deactivates process functional block start and stop operations, while a feedback control unit maintains the tracking of the continuous command signal. Process operating modes (such as failure, normal, and stop and start) integrate process functional blocks with process failure recovery operations. Thus, their execution can be structured using sequential function logic. This is usually achieved with sequential function charts (SFCs), also known as "Grafcet" in French, which is a graphical language for the implementation and specification of sequential algorithms. SFC hierarchy is structured in three levels:

1. Safety SFC (GS) to manage process safety modes (emergency start, general process faults, process initialization, after process defaults start, etc.)
2. Traded SFC (GT) to define the type of operating modes (semiautomatic, manual, and automatic), as well as the initial process start mode
3. Execution of any of the available operating modes, such as
   a. The manual operating mode SFC (GM) to manually execute start and stop modes of process operating sequences separately
   b. The normal start automatic mode SFC (GNPA) to automatically execute the start mode of process operating sequences
   c. The normal stop automatic mode SFC (GNPAS) to automatically execute the stop or shutdown mode of process operating sequences
   d. The normal semiautomatic mode SFC (GNPSA) to semiautomatically execute the start and stop or shutdown modes of process operating sequences

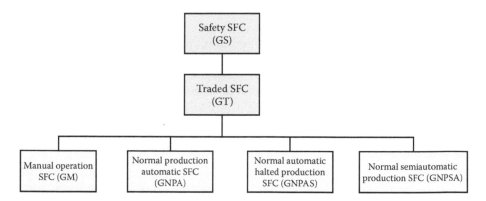

FIGURE 4.5
SFC hierarchy.

This hierarchy of SFC is illustrated in Figure 4.5. In addition, note that a forcing command allows us to move from one SFC step to another SFC step. Such a command is an "internal order" and has priority over process execution sequences. When all forcing commands are coherent between themselves, it is possible to automatically define the SFC hierarchical structure similar to that of GNPAS. As such, the transition condition allows us to automatically proceed on transition between SFCs once the transition condition is fulfilled.

Typical examples of hybrid control systems are the distributed control system (DCS) and the supervisory control and data acquisition (SCADA) system. Such control systems are usually embedded within monitoring and control architecture with a multivariable, multiloop computer control system, and are suitable for large and geographically dispersed plants. Typical DCSs comprise one or more of the following components:

- *Local control unit* (LCU), to handle several individual proportional–integral–derivative (PID) control feedback loops as well as logic control algorithms, with analog input lines and output signals, even with few digital inputs and outputs
- *Data acquisition unit*, which contains several analog I/O channels as the LCU, allowing digital and analog I/O signals to be processed
- *Batch sequencing unit*, which comprises several external events, timers, counters, and function generators
- *Local display*, which provides analog display stations, an analog trend recorder, and sometimes a video display
- *Data archiving unit*, for the storage and recall of process data (e.g., magnetic tapes and storage disks)
- *Computing unit*, to operate specific functions (optimization, advance control, expert systems, etc.) programmed from customer or third-party instructions

- *Central control and monitoring center,* which comprises one or several operator stations for communication and several units for graphics display
- *Data transmission link,* which consists of coaxial cable or other transmission media and enables the serial connection of all other components in the system
- *Local area network,* which connects remote devices to the central control unit

Such hybrid control systems can be structured using FAST and GEMMA methods as specified above, while any subsequent feedback controller, such as a PID-based controller, is designed according to other techniques of discrete controller design.

4.5 Design Methodology for Monitoring and Control Systems

A monitoring and control system requires the integration of a vast amount of data to maintain normal and safe process operating conditions. The design of the software and hardware components of such system should be based on specific process requirements regarding how data should be gathered, analyzed, and reported.

4.5.1 Requirements for Data Gathering and Process Data Analysis

A process data gathering system should continuously provide relevant information on the process execution of all active and inactive system components, from the field process devices to the computing units (e.g., historian server and command and control center). The efficiency of such a system is based on the quality and time-based availability of the process data. Among requirements of process data gathering there are the design and sizing of the data communication field bus, the network protocol, and the data acquisition and measurement technologies with respect to defined process monitoring and control objectives. Network redundancy (analytical or hardware) should also be taken into account, if necessary.

Subsequently, the process data analysis should provide information to the decision support system in order to ensure the normal and safe execution of process operations. This is achieved through fault occurrence detection and diagnosis algorithms for process equipment or software. A typical detection algorithm aims to assess the patterns of correlation between several process variables to derive fault symptoms and the sensitivity of each variable. A basic diagnostic algorithm compares the threshold values with the gathered values of the process variables with regard to tolerances, and triggers alarms to the operator, or activates suitable counteractions in the case of a faulty process state occurrence. Some advanced fault management methods take into account safety requirement specifications, such as low-scale fault process failures, hardware failures, software and human errors, environmental effects (e.g., electromagnetic and temperature), and even power supply disturbances (e.g., loss of supply and reduced voltages). Among some advanced statistical techniques for analyzing process data there are

1. Pattern recognition, which groups data into clusters and then performs similarity analysis between the clusters to identify faulty conditions
2. Statistical analysis of the process data (trend parameters, such as mean and standard deviations)
3. Link and dependency analysis, to understand the process behavior performance and estimate the interactions between variables
4. Process sequence analysis, to generate the time-based sequence patterns of process data and extract deviations, along with trends over time
5. Regression analysis, to develop a predictive process model

In the case of multivariable process analysis, other advanced techniques (e.g., data classification, data mining, and regression analysis) are used to identify patterns and determine the relationships between variables contained in large data sets.

4.5.2 Requirements for Process Operation Data Reporting via Visualization Systems

A typical data reporting operation of a process control and monitoring system should allow us to

- Report the process execution status based on process data analysis
- Monitor the process operations and configure process parameters
- Report any detected process anomalies
- Report any process performance deterioration
- Report preventive maintenance operations, preventing potential system failures
- Implement suitable process control operations or corrective maintenance operations

A key component of the control and monitoring system is the control panel. It displays information on the execution of process operations, the status of process field devices, and the configuration of the process control and command parameters. This information is related to process monitoring, command, and control and that can be classified as follows:

- Reporting variables related to process data analysis, such as process device preventive maintenance (e.g., variables indicating the detection and diagnosis of a tool failure).
- Configuration variables related to process configuration, which are parameters on the operating conditions (e.g., feeding rate, spindle speed, cutting tool geometry, and depth of cut) and controller settings. In addition, some variables and features characterizing the process performance or product quality (e.g., vibration main frequency or surface roughness) may be included.
- Operation monitoring variables related to the execution status of process actuating devices in real time without discontinuity.
- Fault monitoring variables related to safety requirements and fault management, which are associated with (1) safety variables to warn against personnel and equipment harms, (2) performance variables to monitor process performance degradation, and (3) fault (anomaly) variables for the detection and diagnosis of fault occurrence.
- Control variables related to process command and control operations, which are manipulated input variables of actuating devices.
- Intermediated variables related to data exchange between the database and the computing unit (e.g., mean variance in process data analysis routines).

Hence, the control and monitoring software could be divided into the following program modules:

- Process configuration module, containing algorithms related to the definition of process operating conditions (logic and feedback), the controller configuration, and the timing or sequential execution of manipulated variable values for process actuators.
- Fault detection module, containing algorithms related to the detection of process operation faults from the involved equipment (actuators, sensors, etc.) and process performance degradation, using process data analysis techniques (e.g., statistics such as data mining analysis).
- Fault diagnosis and isolation module, containing algorithms related to the development of fault mitigation by deriving suitable interlock actions to be proceeded on manipulated control variables.
- Process monitoring module, containing algorithms related to the reporting in real time of a process device's operating status. Some process variables can be estimated from process data. This should

allow remote supervision of execution operations from top management to the field operator.

- Command and control module, containing hybrid control algorithms (logic and feedback).
- Data availability module, containing algorithms to ensure the availability and reliability of gathered process data, from the data acquisition, transmission, and sensing devices up to archiving in the historian database. Some redundancy strategies can be used.

Table 4.5 summarizes typical process variables used in the design of software monitoring tools.

4.5.3 Design Methodology of a Monitoring and Control System for Industrial Processes

According to the design requirements identified, a generic design methodology of a monitoring and control system consists of

- Performing a technical audit of the process components in order to establish process performance benchmark indexes (e.g., stability, precision, speed of response, robustness, and safety), along with process productivity indexes and even energy consumption indexes. Then, compare those indexes with typical industrial standards.
- Listing the devices and equipment involved in the process operations and their characterizing information, such as
  - Operating conditions (e.g., input voltage, operating range, command profile, sequence execution, and technical specifications) of the actuating equipment (analog and digital such, as motor or solenoid).
  - Operating conditions (e.g., measurand, operating range, and resolution) of the process sensing devices and detectors.
  - Process control input variables representing sensing devices and detectors, as well as visualizing units (man–machine interface [MMI]).
  - Equivalent binary (specified if pulse or level type) or continuous process variables assigned to each input and output from each process sensing device and detector, along with the actuating equipment.
  - Manipulated variables representing outputs of the controller unit.
  - Input and output signals from MMI (e.g., reset push button, stop push button, and limit switch) with their technical specifications (e.g., pulse or level type).

TABLE 4.5

Commonly Used Variable Definitions for Monitoring Software Design

Interlock Variables	Threshold Values	Intermediate Variables	Manipulation Variables	Configuration Variables	Function Variables	Sensors Equipment	Actuator Equipment

- Equivalent binary or continuous process visualizing variables assigned to each input and output from signal conditioning unit.
- Safety input and output variables and interlocks defined based on safety and regulatory requirements. All these variables are expected to be complete after the functional analysis.
- Sketching of piping and instrumentation diagram (P&ID) and process flow diagram (PFD). In addition to the identification of process components and subsystems, P&IDs and PFDs display some key piping and instrument details for process control, activation, and shutdown. Some symbols used to draw several diagrams—PFDs, P&IDs, and electrical wiring—are found in the available literature, and some of them are summarized in Figure 4.6a and b.
- Analyzing the process operating conditions in order to list
  - Process operating cycles
  - Process safety constraints
  - Process output actions in each cycle
  - Transitions between process output action and inputs
- Defining process parameter estimation through
  - Operating threshold values
  - Initial condition values
  - Process modeling and parameter estimation
  - Boolean equation modeling for each state and transition
- Defining process database and process data gathering.
- Proceeding in behavior analysis from the operational (as presented in Figure 4.7), functional, and structural decomposition approaches of the process using the FAST method (Table 4.6) by
  - Identifying the global function for the process control objective
  - Breaking down each primary function into secondary functions or subfunctions, known as functional decomposition
  - Listing all constraints for each key primary function (technical experienced is required) according to the following groups:
    - Operational capacity constraints (maximum level, maximum travel distance, maximum temperature, etc.)
    - Operational time constraints
  - Listing all operational tasks associated with each secondary function, such as
    - Active actions defined for each process operating task
    - Conditions of activation
    - Conditions of termination

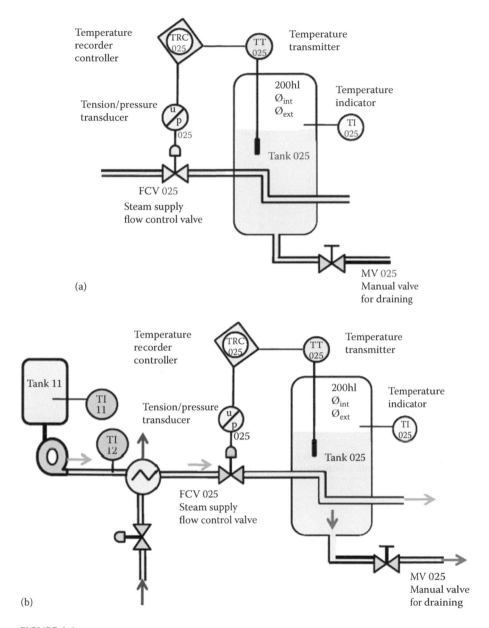

FIGURE 4.6
(a) Example of P&ID process temperature control. (b) Example of PFD for process temperature control.

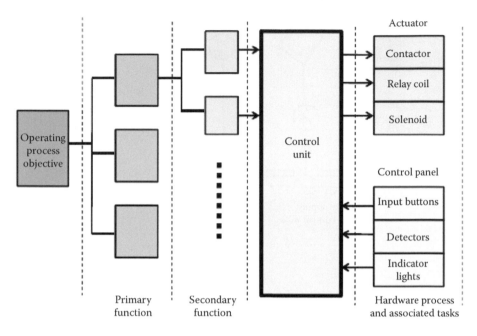

FIGURE 4.7
FAST decomposition method.

TABLE 4.6

Compounding of Expected FAST Results

Primary Function	Constraints	Secondary Function	Associated Operational Tasks	Operational Tasks for Equipment Protection	Operational Tasks for Operator Security	Operational Tasks for Setting Up (Preparation)

- Establishing a coordination between the process operating tasks through
  - Classification between manual, semiautomatic, and automatic operation
  - Definition of the activation or termination condition of each task
- Adding tasks associated with protection of the device for each secondary function, especially interlocks for the process stop condition
- Adding tasks associated with operations and operator security for each secondary function, especially emergency stop in the case of an accident

- Adding all necessary tasks for setting up (preparatory) activities, such as power supply activation for operation, checkup of the start-up operational condition, loading of workpieces to be processed, or even initialization and referencing of the associated tools
- Establishing a descriptive list of action and event sequences under the various process operational modes, including normal production, emergency, start-up, and shutdown modes, and also special operation modes that may not occur very often, and relating those process actions and events to each mode (e.g., the "auto" PB turns the alternating current [AC] motor on)
- Establishing a list of effects from the potential failures within the sensors, the actuators, or even the overall process execution, such as drops in hydraulic pump pressure and failures in top and bottom limit switches
- Designing process interlocks and forcing action
- Modeling the execution of process operations through
  - Sequential logic process formal modeling using Boolean algebra or finite state machine
  - Differential and difference-based equations
- Selecting and designing control systems through
  - Control logic design using techniques such as flowcharts, state diagrams, and petri nets
  - Feedback loop controller design (e.g. using PID controller family design methods)
- Filling the SFC hierarchy based on results from the dependency chart, safety requirements, and FAST decomposition.
- Performing the integration of modules related to the operating modes, control settings, and interlocks using the method (dependency chart) and a process operating and stop mode graphical analysis. This is similar to French graphical representation (GEMMA).
- Sizing and designing hybrid control algorithms by
  - Defining the computing hardware of the control unit and sizing the detection equipment
  - Implementing logic control algorithms through programming languages, such as state function charts, ladder and function blocks, timers and counters, Boolean mnemonics; structured text (ST), and instruction lists (ILs)
  - Implementing feedback control algorithms

- Designing logic programming that fulfills safety requirements by
  - Checking that cycles of operation and logic meet safety specifications
  - Designing algorithm verification and validation
- Developing process data analysis algorithms and implementing a fault-tolerant decision support system.
- Selecting and designing the process data reporting module.
- Editing process automation project documentation by connection diagrams, such as
  - Electrical power supply diagram (power supply of current or voltage required) and command diagram
  - Electronic circuit (if used)
  - I/O wiring connection diagram (in/out of PLC)
  - Detailed system schematic with I/O address assignments over the wiring diagram
  - Mnemonic and data storage address assignments related to process variables

4.6 Fault-Tolerant Process Design Requirements

A safe process operation involves two principles: functional safety and the reliability of the operation. The functional safety of process operations is the ability of a system or device to keep the risk incurred by operators within acceptable safety (nonharmful) limits, while the reliability of process operations is the ability of a system or device to perform its function at any moment in time and for a specified process duration. Hence, a process that can operate safely, even in the case of a single component failure, is called a fault-tolerant process. Levels of fault tolerance in system design strategies include (1) complete fault tolerance, such as a failed operation; (2) constraint operational capacity, such as smooth degradation; and (3) migration into a safe state, such as fail-safe.

The *IEC 61508 safety standard* considers functional safety as various measures to avoid or prevent failures from occurring, as well as those used to detect failures and isolate their negative impact by transitioning the process into a safe state. This standard provides the guidelines to determine the safety integrity level that is required and to list process faults to be detected and process errors to be assessed. Commonly encountered fault types are random component failures (hardware) and systematic faults in design (software and hardware).

Usually, the safety measures are implemented through the monitoring and control system. Those measures prevent hazardous process conditions from occurring. ISO 138491, "Safety Related Parts of Control Systems," outlines the criteria for a benchmarking system categorization with respect to prevention and detection measures. The categories can be summarized as follows:

- *Category B* is related to the safety of controlled system components and/or their protective equipment, which should be selected, designed, and integrated based on relevant standards, in order to withstand the expected influence.

- *Categories 1 and B* are applied to detect faulty components within the controlled system and to ensure its compliance with safety principles. This leads to a controlled system that has higher safety-related reliability and less likelihood of a fault.

- *Categories 2 and B* are applied to periodically check if there are any faulty components within the controlled system. A faulty component detection initiates the migration to a safe state or a warning activation. However, the occurrences of a fault between the checking intervals cause safety compliance loss.

- *Categories 3 and B* are applied such that a single fault in any of the components within the controlled system would not cause a loss of safety compliance. If the fault detection is not exhaustive, especially in the case of undetected faults, the safety function can be degraded.

- *Categories 4 and B* are applied such that in front of undetected fault accumulation, any single fault in any of the components within the controlled system is promptly detected to avoid safety compliance loss.

The design of such fault-tolerant and safety-related systems requires the use of analytical methods, such as fault tree analysis, event tree analysis, or FMEA. Furthermore, such systems are embedded into the design monitoring system, allowing overall process protective measures or interlocks. Typical implementation of safety measures is achieved by stopping process operations, also called process interlocks. Their design principle is presented in the following section.

4.6.1 Process Interlock Design Principles

Based on identified operating and failure modes, process stop modes are designed as protective measures to maintain the process equipment in normal physical and operating condition. Those protective measures can be performed by devices or systems providing the maximum protection with minimum hindrance to the normal execution of process operations. This can be implemented by placing some interlocking switches at key locations in the process plan layout with defined activation rules, which is achieved either by preventing access in the case of destructive or harmful motion or preventing destructive or harmful motion after access.

Preventing access during destructive or harmful motion is achieved by using fixed enclosing guards along the system's motion path, which does not require access, and movable guards with interlocking switches in the case where access is required. This is expected to be interlocked with the power source, in order to ensure switching OFF the hazard power routed through this switch. Some interlocking switches activate devices to lock the security guard door until the system is in a safe state. Typical industrial applications use combined solutions with movable guards and interlock switches. There are also other solutions, such as two-hand control, to prevent access while a system is in a dangerous operating condition (e.g., two start buttons pushed simultaneously to activate a painting robot).

Preventing destructive or harmful motion while allowing frequent access within the physical guards along the system motion path require the interlocking system. An interlock switch isolates the power source once any intrusion or external presence is detected. This is done using detective motion devices, such as *photoelectric light curtain emission* to enclose the perimeter guarding for the system motion path; *pressure-sensitive safety mats* to provide guarding of a floor area around the system motion path; *emergency stop devices* that are manually operated at the control panel level such that when they are actuated, they must latch in before a stop command is active; and *pressure-sensitive edge devices* mounted to the edge of the system motion path. It is important for the controller to have fail-safe measures and to be able to stop the system quickly after switching off the power source. Any activation of the interlocking system should lead to a hazardous event, but rather move the process at initial or normal state or an intermediary safe state.

4.7 Industrial Case Studies

In order to illustrate the process monitoring and control design methodology, three industrial case studies are developed in this section. In addition, features related to the implementation of the detection and diagnosis tools for induction motors are investigated. Some internal monitoring devices, such as watchdog, can be used to supervise the execution of the controller program operations, to report irregularities and difficulties, and even to stop the machine or process in case an emergency is not covered.

4.7.1 Bottle Washing Process

In the brewery plant, the washing bottle process aims to remove contaminants and organisms through a combination of biological, chemical, and thermal treatment. It produces empty bottles, safe to contain drinkable products from other brewery processes. This washing process starts through a loading wheel of dirty bottles in a handling chain. This chain is capable of carrying 500 trains with a capacity of 40 bottles each and moves through three successive washing tanks (two tanks of water maintained at 30°C and 55°C, respectively, and one tank of water mixed with soda maintained at 80°C). At the end of the washing process, an off-loading system removes the clean bottles from the handling chain for storage, as illustrated in Figure 4.8. All motors associated with bottle handling, loading, and off-loading systems must have a synchronized motion. All equipment involved in the washing bottle process is summarized in Table 4.7. Note that all pumps and motors associated with the handling system are ON/OFF controlled. There are two water temperature levels controlled by two electrovalves, EV4 and EV5, connected to motor pump MTR1 at the outlet of the hot water and steam tank. The hot water or steam is mixed with cold water to increase the water level in the respective tanks used for washing the bottles. Table 4.8 summarizes the process operating sequences. Thermostat-based temperature sensors (temperature sensors 1 and 2) are used to gather continuous temperature variations. Soda and water fill the tanks through electrovalves (EV1, EV2, EV3, and EV6).

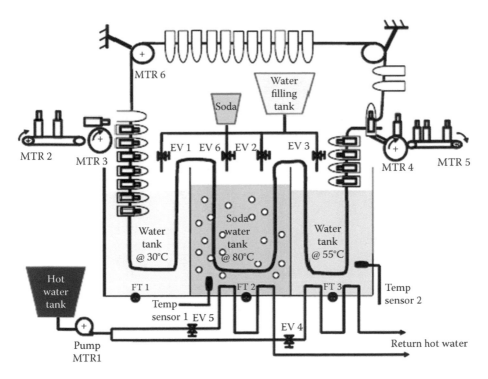

FIGURE 4.8
Simplified process schematic and description (P&ID).

TABLE 4.7

Process Equipment Listing

Equipment	Technical Characteristics	Operating Conditions	Associated Process Operations and Control Strategy
AC pump MTR1	$P = 18.5$ kW	Oil filled $n = 1500$ revolutions/min $p = 80$ bar	Star-based start-up of asynchronous motor for water tank injection at 80°C, 55°C ON/OFF controller
AC motor MTR6	50 Hz, 1.1 kW	1430 revolutions/min	Direct-based start-up of asynchronous motor for bottle handling chain in 30°C, 80°C, 55°C tanks ON/OFF controller
Conveyor MTR5	$P = 7.5$ kW $Q = 250$ m³/h	$n = 1500$ revolutions/min $p = 80$ bar	Asynchronous motor, 4 limit switches Conveyor speed $= 7.05 \times 10^{-2}$ m/s
On-loading/off-loading bottles MTR2, MTR3, MTR4, MTR5	50 Hz, 1.1 kW	1430 revolutions/min	Direct-based start-up of asynchronous motor for bottle ON/OFF loading handling system ON/OFF controller
Electrovalves EV4, EV5	Total volume of heated liquid: 93 L Total volume of heating liquid: 21 L	Heating liquid: Pmax = 6 bar Tmax = 100°C Heated liquid: Pmax = 7 bar Tmax = 170°C	For water heating stage of 80°C and 55°C
Thermostat	Range: 0–5 bar and 0°C–150°C	0–10 bar, 0°C–150°C	Transmitter 4- 20mA Maintain water temperature at 80°C and 55°C
Electrovalves EV1, EV2	$DN = 38.1$ mm 0°C–170°C, 0 to 10 bar	P = 2–4 bar 121°C–155°C	ASI-based pneumatics (ON/OFF regulation) Steam through heat exchanger for water tanks of 52°C and 62.5°C and command by pressurized air supply electrovalves
Manual valve for steam arrival	$DN = 125$ mm 0°C–40°C, 0–10 bar	P = 3–4 bar	Valve to supply steam for thermal heat exchanger

TABLE 4.8

Process Operating Sequence and Steps

Operating Mode	Operating Sequence	Variables	Variable Types	Description	Interlocks
Start mode	Filling tanks (1–3)	EV1, EV2, EV3	Bool, Bool, Bool	Activation valves for tank water inflow	If water top-level switch of tanks (Anti-overdrop) (LS1, LS2, LS3) and closure of valves EV1, EV2, and EV3
	Filling soda (tank 2) (5 min)	EV6	Bool	Activation valve for soda inflow tanks	(Anti-overdrop) LS (level detector) Soda solution Stop EV6, EV2, MTR1
	Activating hot water tank pump and electrovalves	Sensor_temp, MTR1, EV5, EV4	Integer, FBD	Opening of electrovalves for water cooling	Stop if Sensor_temp > (80°C, 55°C) for tanks 2 and 3
	Heating tank (80°C, 55°C) temperature control	Sensor_temp,	Integer		Close valves EV5 and EV6 if temperature of tanks 2 and 3 reached
	Activated uploading, off-loading, handling chain	PB_Start, MTR2, MTR3, MTR4, MTR5, MTR6	Bool	If entry/exit bottle misalignment, stop MTR6 (Anti-overdrop handling chain) Stop MTR6	Handling chain faults ZS 06X0, synchronization faults ZS 07EL
Stop mode	Stop filling water	LS_level	Bool	Deactivation valves for water tank inflow	
	Stop handling chain	PB_stop, MTR2, MTR3, MTR4, MTR5, MTR6	Bool		
	Stop hot water supply Emptying tanks	MTR1, EV5, EV4 FT1, FT2, FT3	FBD, Bool, Bool, Bool Bool		

Note: Bool, Boolean; FBD, function block diagram.

From the process audit, the following technical problems are encountered: (1) the soda tank could be overfilled with water due to a faulty limit switch for anti-overdrop (a drop on the floor causing a high risk of burning skin when in contact with shop floor operators); (2) the bottle is fragile due to a high temperature level and the poor synchronization between the bottle handling chain and the off-loading system; (3) finally, there is no monitoring system in place to allow real-time access or command and control of process operations. From the process operation analysis, it can be noted that (1) the temperature is maintained roughly around 80°C and 55°C using ON/OFF electrovalves that supply the hot water or steam, (2) the speeds of the conveyors (MTR1 and MTR5) are synchronized, and (3) all tanks are periodically emptied. Using the process devices, the equipment involved, and the corresponding process variables listed in Table 4.8, as well as the FAST method, the bottle washing process can be decomposed into the following primary subprocesses: tanks filling and emptying, water and soda tanks heating, tank temperature control, and bottles uploading and off-loading, as depicted in Figure 4.9. Then, the hierarchy of SFC can be filled in as shown in Figure 4.10a. Subsequent to the functional

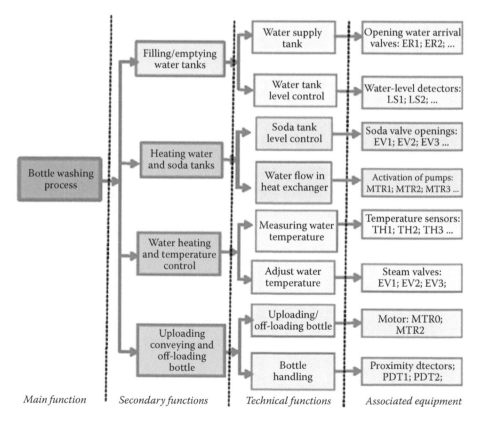

FIGURE 4.9
Functional analysis using the FAST method.

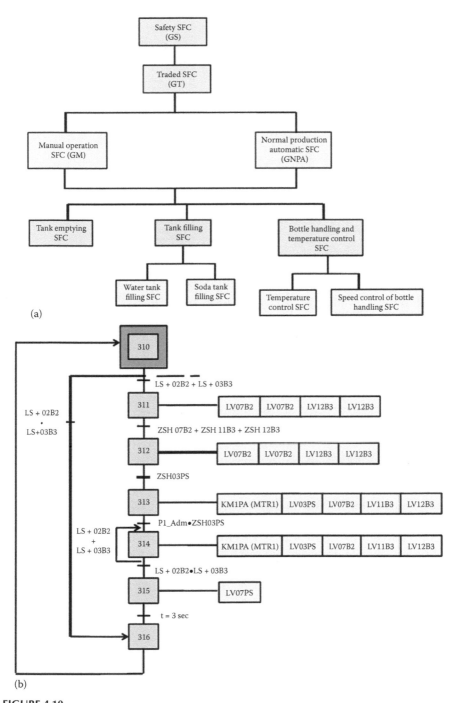

FIGURE 4.10
(a) SFC-based hierarchy of bottle washing process. (b) SFC for soda tank filling for bottle washing process. *(Continued)*

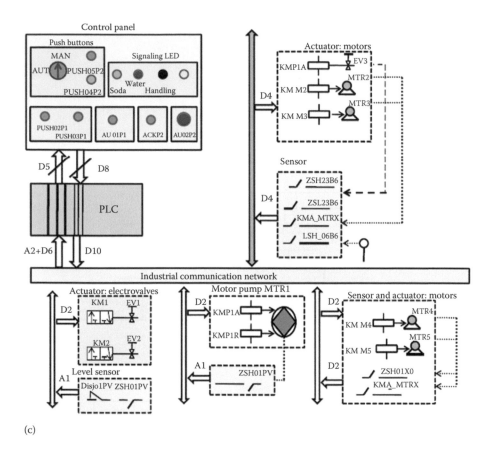

(c)

FIGURE 4.10 (CONTINUED)
(c) Control and command schematic for bottle washing process.

analysis and the SFC hierarchy, the graphical analysis of process operating sequences for start, stop, and failure modes, with their activation conditions, is summarized in Figure 4.11. Then, for logic control of the soda tank filling subprocess, an SFC is developed, as shown in Figure 4.10b, while the block diagram in Figure 4.12 is for closed-loop control of the coupling between the tank temperature and the conveyor speed. Figure 4.10c depicts the overall control and command wiring diagram of the bottle washing process.

4.7.2 Cement Drying Process

In a cement production plant, pozzolana with around 15% humidity is supplied from a quarry. A humidity of 3% is required for the subsequent grinding process. Hence, a mandatory drying process is performed. It consists of injecting heavy fuel and compressed air to be burned in a flame chamber. Then, the generated hot air between 700°C and 810°C is drawn into the kiln

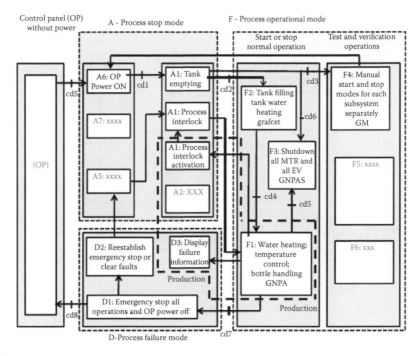

FIGURE 4.11
Graphical analysis of bottle washing process operating and stop modes. (Adapted from ADEPA, *Guide d'étude des modes de marches et d'arrêts*, GEMMA, Agence pour le development de la productique et l'automatique, Ministère de l'industrie, Paris, France, 1997.)

dryer tube to dry the fed humid pozzolana. The hot air leaves the tube with a temperature of around 100°C, while rotation of the kiln propels the dried pozzolana along the tube. Here, the dried pozzolana is extracted, while dust is removed by a separated filter, as illustrated in Figure 4.13a. Table 4.9B summarizes the pozzolana drying process devices and equipment, the corresponding process variable types, and the process operating sequences.

In order to generate a flame in the combustion chamber, two asynchronous motor-driven pumps (MTR1 and MTR4) inject fuel and compressed air toward a lighter that has ignited the flame. As an alternative fuel, butane gas can be used as a replacement combustible. Simultaneously, a motorized conveyor (MTR6) carries the humid pozzolana into a motorized rotary dryer (MTR5) tube connected to the flame chamber. In order to ensure complete combustion, the air-to-fuel ratio and the pressure in the flame chamber should be maintained constant by varying the opening of the dilution and combustion fans (MTR3 and MTR2). Electrovalve EV1 feeds the pozzolana from the silo and electrovalve EV2 extracts it from the rotary kiln. Table 4.9B summarizes the pozzolana drying operating sequences. From the process audit, a high-temperature variation within the flame chamber produces a frequent shutdown of the dryer tube. Such a temperature is reduced by

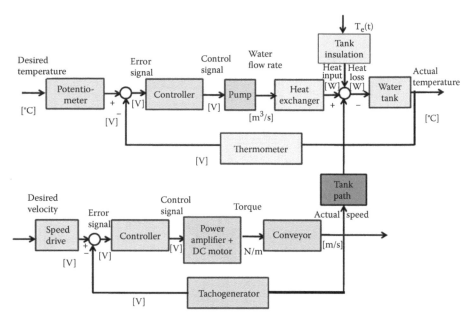

FIGURE 4.12
Subprocess feedback control block diagram for bottle washing process.

controlling the fuel inflow rate. Also, the fluctuation of the air-to-fuel ratio due to the combustion and dilution fans can cause incomplete combustion because of the zero pressure in the dryer tube. The performance of an ON/OFF control scheme of the pressure and temperature within the flame chamber is assessed and compared with the industrial standard, as shown in Table 4.9A. This can be used to set the design objectives of control loops from Figure 4.13c. There, it is required to migrate from an ON/OFF control scheme to a PID temperature and pressure control with a settling time and rise time of less than 450 and 150 s, respectively.

Using the FAST method and the process variables defined in Table 4.9B, the cement drying process can be decomposed into two subprocesses: (1) injecting pozzolana and (2) maintaining the flame within the chamber, as illustrated in detail in Figure 4.13b. The corresponding graphical analysis of the start and stop operating modes of the drying process is presented in Figure 4.13d. An example of a logic control program using the function block for the drying process is presented in Figure 4.13e. Figure 4.14 depicts the overall control and command wiring diagram of the pozzolana drying process. It can be noticed, for example, that there are eight digital output signals from the programmable logic controller toward the eight field contactors activating the motor and electrovalves (e.g., XVS AA09 and XSV BAC07 for MTR4) through the Modbus-based industrial network.

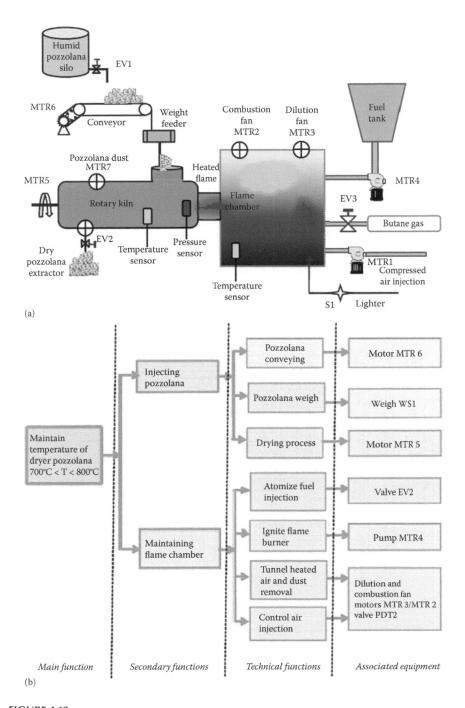

(a)

(b)

Main function *Secondary functions* *Technical functions* *Associated equipment*

FIGURE 4.13
(a) Simplified process schematic and description of the cement drying process. (b) Function analysis of the cement drying process using the FAST method. *(Continued)*

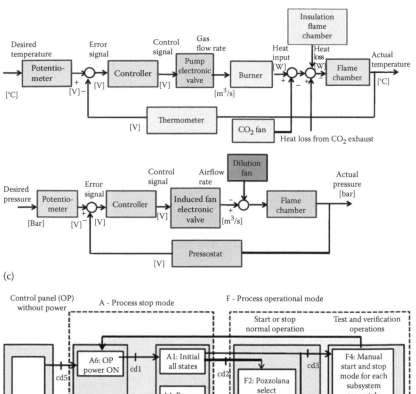

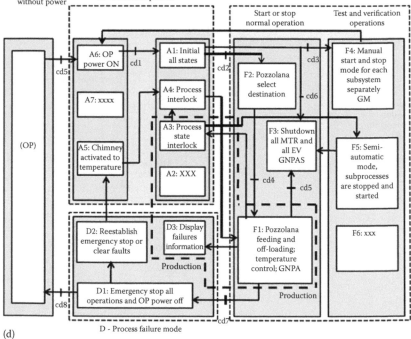

(d)

FIGURE 4.13 (CONTINUED)
(c) Feedback control block diagram of the cement drying process. (d) Cement drying process operating and stop modes graphical analysis. (Adapted from ADEPA, *Guide d'étude des modes de marches et d'arrêts*, GEMMA, Agence pour le development de la productique et l'automatique, Ministère de l'industrie, Paris, France, 1997.)　　　　　　　　　　　　　　*(Continued)*

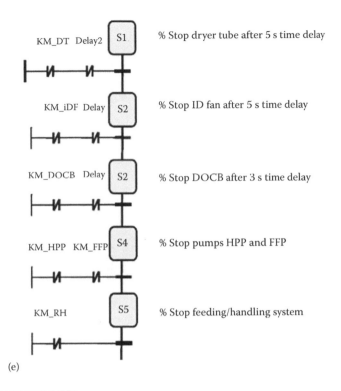

KM_DT Delay2	S1	% Stop dryer tube after 5 s time delay
KM_iDF Delay	S2	% Stop ID fan after 5 s time delay
KM_DOCB Delay	S2	% Stop DOCB after 3 s time delay
KM_HPP KM_FFP	S4	% Stop pumps HPP and FFP
KM_RH	S5	% Stop feeding/handling system

(e)

FIGURE 4.13 (CONTINUED)
(e) Example of function block for pozzolana drying process shutdown.

TABLE 4.9A

System Performance Audit and Specifications

Characteristics	Actual System Performance		Benchmarked Values	
	Temperature	Pressure	Temperature	Pressure
Set-point values	760°C	−2 mbar	760°C	−2 mbar
Precision	±20°C	±2 mbar	< ±2°C	±2 mbar
Steady state	0.578	0.19	1	1
Overshoot (%)	0	0	<2%	<2%
Peak time (s)	680	768	<400	<500
Settling time (s)	425	576	<300	<400
Rise time (s)	197	211	<100	<150

4.7.3 Induction Motor Condition Monitoring

Monitoring operations of the process equipment are highly expected to maintain a robust productive system. Among this equipment there are electric motors or generators whose unexpected failure could be very harmful and costly. Such equipment faults are either electrical-based operating faults

TABLE 4.9B

Process Operating Sequence and Steps of Cement Drying Process

Operating Mode	Operating Sequence	Variables	Variable Types	Description	Interlocks
Start mode	Select pozzolana destination	Direct grinding, Silo2	Bool, Bool	Select grinding machine or Pozzolana silo	Automatic mode
	Start pozzolana dust extractor	Start_Aux, Start_PDE	Bool, Bool		
	Dryer tube motor is started, horn is set on	Start_D, KMK, Elev, PB_Conv, KM_DT	Bool, Bool Bool	Activate motor and direction conveyor belt	
	Fuel feeding and high-pressure fuel pump	Start_RH, S1 EV3, MTR1	Bool, integer, integer	Heater activation Low fuel temperature: $90°C < T < 120°C$ Low compressed air pressure	Range: $5\ bar < P < 7\ bar$ Fuel thermostat Shut down the flame chamber and fuel pumps
	Burner of flame chamber lit, start flame chamber, dilution fan and combustion fan started	Start_FC, KM_DF, KM_CF	Bool	Fault such as excess flame chamber output temperature, absence of flame, or no compressed air	Flame fault from detector Flame chamber output temperature: $700°C < T < 810°C$ Shut down fuel pumps and flame chamber
	Start humid pozzolana extractor	Start_PE	Bool		

(Continued)

TABLE 4.9B (CONTINUED)

Process Operating Sequence and Steps of Cement Drying Process

Operating Mode	Operating Sequence	Variables	Variable Types	Description	Interlocks
Stop mode	Pozzolana extractor (weight feeder) stop	Stop_PE	Bool	Pozzolana extractor stopped	
	Flame chamber shut down, fuel electronic valves closed, combustion fan stopped, dilution fan stopped	Stop_FC, KM_DF, KM_PE, KM_CF	All Bool		
	Dryer shut down, dryer output motor and direction conveyor belt, ID fan	Stop_DT, KM_IDF, KM_DCB, KM_Elev	All Bool		
	Pozzolona dust extractor shut down	Stop_PDE, KM_PDE	Bool, Bool		
	Auxiliary shutdown sequence	Stop_Aux, KM_Aux	Bool, Bool		
	Reheater shut down	Stop_RH, KM_RH	Bool, Bool		
	Emergency stop	Automatic mode	Bool		Power supply to electrovalves, fuel and air supply system, and combustion fan

Note: Bool, Boolean.

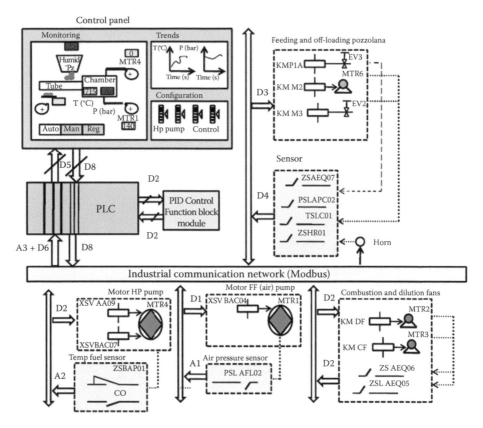

FIGURE 4.14
Cement drying command and control schematic.

(e.g., stator, and broken or loosened connecting wire at the armature motor faults) or mechanical-based faults (e.g., static and dynamic eccentricity, and bearing and gearbox faults). In the case of an induction motor, some common faults are vibration faults, which include broken or loosened connections of the motor, loose housing, poor lubrication, and roller elements, as illustrated in Figure 4.14. It has been found that faults are related to motor vibrations, and those vibrations can be characterized using stator current, axial flux, voltage, and even motor vibration signals. Hence, by collecting or estimating those signals on a motor, it is possible to process them in order to derive their frequency spectral and harmonic content using fast Fourier transform (FFT).

The signature analysis of the stator current signal, from the time domain in the frequency domain, reveals the presence of harmonics at specific frequencies in the current spectrum, related to the presence of some particular faults. Similarly, the vibration spectral analysis from the vibration signal of the motor under operation leads to the presence of certain harmonics in the signal (with amplitudes above some threshold). This indicates the presence of specific faults. Then, it is possible to compare the amplitudes and fault characteristic frequencies of healthy vibration signals and faulty vibration signals. Alternatively, signal trends, mean amplitude, and fundamental amplitude are compared with the thresholds specified by the manufacturer or with data recorded from a normal operating induction motor. The spectrum analysis is developed to circumvent motor testing under the same loading conditions (approximating by the sixth-order polynomial fit). Note that several induction motor faults can occur (bearing faults, eccentricity, etc.), but each fault has its frequency characteristics for each induction motor.

Mechanical-based faults are mainly due to motor vibrations. Hence, from the frequency analysis of the motor vibration signals, it is possible to capture individual fault features. In the case of a broken wire and a loose connection fault, it is suitable to perform an FFT of the vibration signal for a sampling rate, f_S, and then compare the faulty vibration frequency spectrum, F_v, with the supply frequency, f_S. There is the presence of a broken wire and a loose connection of the motor if there exists a frequency F_v such that

$$F_v \geq 2 \times f_S$$

In the case of a static eccentricity fault, its characteristic frequencies, f_S, would be given by

$$f_S = RB \times RS \pm nF_L$$

while in the case of a dynamic eccentricity fault, its characteristic frequencies, f_D, would be given by

$$f_D = b \times \omega \pm nF_L \pm \omega$$

In the case of broken rotor bar fault, its characteristic frequencies, f_B, would be given by

$$f_B = (1 \pm 2s)F_L$$

with b being the number of rotor bars, F_L the line frequency, ω the running speed of the motor, s the slip, and n the odd harmonics of line frequency. If f_D is greater than or equal to the threshold magnitude, this indicates the presence of fault eccentricity or broken bars.

In the case of a bearing fault, its characteristic frequencies can be determined on the basis that the failure is proportional to the inverse number of time between occurrences of the bearing impulses. The inner race frequency, $f_{B,in}$, is given by

$$f_{B,in} = 0.5\omega_r n \left[1 - \left(\frac{d}{D} \cos\beta \right) \right]$$

where n is the number of rollers, ω_r is the rotational speed of the rotor, d is the diameter of the bearing roller, D is the pitch diameter, and β is the slip (contact) angle of the roller in the bearing element. The roller frequency of the defect on the outer race, $f_{B,out}$, is

$$f_{B,out} = 0.5\omega_r n \left[1 + \left(\frac{d}{D} \cos\beta \right) \right]$$

The ball spin frequency, $f_{ball\ spin}$, is

$$f_{ball\ spin} = 0.5\omega_r \frac{D}{d} \left[1 - \left(\frac{d}{D} \cos\beta \right)^2 \right]$$

In addition, the loose housing fault frequency, $f_{housing}$, is

$$f_{housing} = 0.5\omega_r \left[1 - \frac{d}{D} \cos\beta \right]$$

It should be recalled that the synchronous speed is given by

$$\omega_s = \frac{60\ f_s}{number\ of\ pole\ pairs}$$

The slip is given by

$$s = (\omega_s - \omega)/\omega_s$$

And the rotor speed is given by

$$\omega_r = s f_s = \omega/60$$

In addition to mechanical faults, electrical-based fault features can be directly deduced from the stator current and voltages. Those variables need to be estimated or measured and are used to derive fault features from the variation of apparent power, the peak-to-peak value, the standard deviation of the currents, the supply impedance, and the mean power of the current. From Park's transformation, the current and voltage, $I_{s\alpha}$, $I_{s\beta}$, $I_{s\alpha}$, $V_{s\beta}$, can be expressed:

$$I_{s\alpha,\beta} = \sqrt{\left(I_{s\alpha}^2 + I_{s\beta}^2\right)}$$

$$V_{s\alpha,\beta} = \sqrt{\left(V_{s\alpha}^2 + V_{s\beta}^2\right)}$$

The active and reactive power would be respectively given by

$$P = V_{s\alpha}I_{s\alpha} + V_{s\beta}I_{s\beta}$$

$$Q = V_{s\beta}I_{s\alpha} - V_{s\alpha}I_{s\beta}$$

P and Q are normalized by the root mean square (RMS) value of the apparent power modulus $S^2 = \sqrt{P^2 + Q^2}$ and are called P_n and Q_n. Considering the mean values of P_n and Q_n being, respectively, m_p and m_q, the *apparent power deviation* called σ is given by

$$\sigma = \sum_{i=1}^{t_p} (V_i - \varepsilon_{p,q})(V_i - \varepsilon_{p,q})^T$$

with t_p being the discrete number of the signal points and V_i a point of the set. The *peak-to-peak values* of $I_{s\alpha}$ and $I_{s\beta}$, called δ_a and δ_b, are given by

$$\delta_a = \left|\max(I_{s,\alpha}) - \min(I_{s,\alpha})\right|$$

$$\delta_b = \left|\max(I_{s,\beta}) - \min(I_{s,\beta})\right|$$

The *standard deviations* of $I_{s\alpha}$, $I_{s\beta}$, and $I_{s\alpha,\beta}$ are respectively called σ_a, σ_b, and σ_s. The positive sequence components of current and voltage are given by

$$I_1 = \frac{1}{3}(I_{sa} + aI_{sb} + a^2 I_{sc})$$

$$V_1 = \frac{1}{3}(V_{sa} + aV_{sb} + a^2V_{sc}) \text{ with } a = e^{j\left(2\pi/3\right)}$$

The mean power of the current, P_1, is

$$P_1 = \frac{1}{N_p} \sum_{k=1}^{N_p} I_1(k)^2$$

The positive supply impedance at supply frequency is

$$Z_1 = \frac{V_1}{I_1}$$

Figure 4.15 represents the apparatus to capture and estimate the motor variables used to derive those features. From those features, the motor fault diagnosis decision can be modeled using the *sequential backward selection (SBS) method* on clusters with n samples during the fixed time period of each faulty operating mode (e.g., broken rotor bar mode, normal mode, and eccentricity). The induction condition monitoring block diagram is illustrated in Figure 4.16. The objective is to classify those samples' resemblance into fault type or feature. Consider that $X_{c,j}, j = 1...n_c$, are the samples for the cluster Ω_c with n_c samples and its center of gravity m_c, which is given by

$$m_c = \frac{1}{n_c} \sum_{j=1}^{n_c} X_{c,j}$$

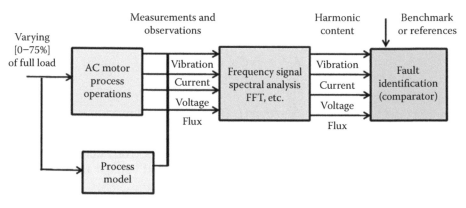

FIGURE 4.15
Typical apparatus for mechanical-based fault detection.

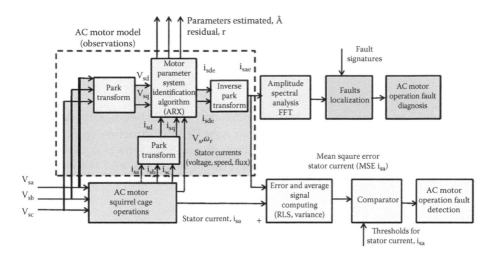

FIGURE 4.16
Block diagram and variables associated with AC motor fault detection.

Then, the overall center of gravity m for all the clusters, and thus for all the samples, is given by

$$m = \frac{1}{n} \sum_{c=1}^{n} \left(\sum_{j=1}^{n_c} X_{c,j} \right)$$

It should be recalled that the identified features have to maximize a criterion such that a weak point dispersion indicates minimal intraclass dispersion, while a maximum distance between the classes reflects a maximum interclass dispersion. Such a criterion can be quantified based on the within-class scatter SW and the between-class scatter SB matrices given by

$$S_w = \frac{1}{n} \sum_{c=1}^{M} \sum_{c=1}^{n_c} (X_{c,j} - m_c)(X_{c,j} - m_c)^T$$

$$S_B = \frac{1}{n} \sum_{c=1}^{M} (m_c - m)(m_c - m)^T$$

with M being the number of classes and n the total number of samples. In order to maximize the interclass dispersion while maintaining intraclass dispersion, the eigenvectors of $S_W^{-1} S_B$ can be estimated, assuming $(n - M)$

and M to be higher than d such that the criteria for feature selection are defined by

$$J_1 = trace\left(S_W^{-1} S_B\right)$$

In order to determine in which cluster a new sample belongs, the k-nearest neighbor rule uses is found through a membership function to define the degree to which the new sample belongs to clusters. This is done by measuring the Euclidean distance between the new sample and each training sample $X_i(X_1, X_2,...,X_n)$ (with n as the total sample number, $i = 1$ to n):

$$dE(X_u, X_i) = \left[(Xu - Xi) \cdot (Xu - Xi)T\right]^{1/2}$$

The distance between the new sample and its expected class is compared with a threshold T_c in order to apply a distance reject rule. The threshold value T_c is given by

$$T_c = 2 \cdot \max\left[dE(Xj, mc)\right], \text{ for } j = 1...nc$$

The severity of the fault can be derived by setting a 5% margin above the threshold value. From different types of induction motors, it is expected to have different mechanical characteristics. The induction motor threshold value can be derived based on (1) the type of motor, (2) the characteristics of motors, (3) the sequence of healthy (normal) motor data, (4) the sequence of faulty motor data, and (5) the characteristic frequencies of motor faults. Figure 4.17a and b depict typical monitoring screens and the fault detection and diagnosis results of a running induction motor.

At steady state, the error between the system response data and the model response is squared and passed through a discrete low-pass filter. This provides an average value of the MSE, which is then compared with a threshold value. The error detection filter subsystem outputs a 0 (meaning no error) if the MSE is below the threshold, and a 1 (meaning an error is detected) if the MSE is above the threshold. When the system changes state rapidly, it is more difficult to model the system, and more errors are used to prevent false detection. A remaining challenge is to detect errors and faults during transient response periods.

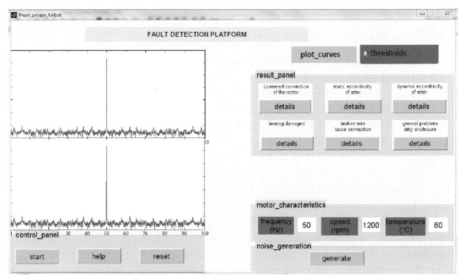

(a)

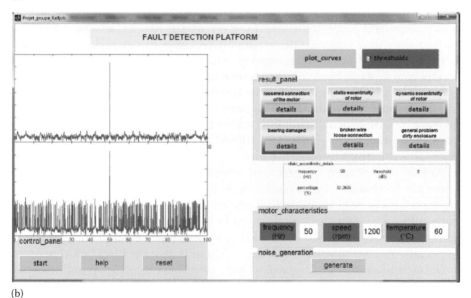

(b)

FIGURE 4.17
(a) Screenshots of condition monitoring of normal operating AC motor. (b) Screenshots of condition monitoring of faulty operating AC motor.

4.8 Distributed Control Systems

A DCS of process operations refers to the distributed configuration of a set of controller elements in order to achieve the synchronized operations of its network-based interconnected subprocess components. Such a configuration is used in a variety of industries, including electrical power grids, environmental control systems, and manufacturing. Usually, design methodologies of the distributed and cooperative control system include the following requirements: (1) definition of entities' or agents' decision objectives and cooperative constraints (e.g., sampling time), (2) definition of the process data (coordination variables) exchange mechanism or the coordination function, (3) mechanism of distributed problem solving through a suitable choice between centralized (hierarchical) and distributed (heterarchical) cooperation schemes, and (4) definition of a consensus-building mechanism around the decision criteria or the merit evaluation of expected process variables.

In the modeling and implementation of the distributed problem-solving mechanism for the process control system, centralized (hierarchical) problem-solving strategies are preferably used. However, in large-scale and geographically dispersed processes, converging consensus results have been found to be too slow and less reliable, especially in cases of network imperfection, such as changes in network topology due to node failures or time delays. Furthermore, it has been found to be difficult to predict their dynamics when they are relying on logic control or heuristic schemes. Therefore, there is a need to model and analyze interactions between autonomous and cooperative entities, by restructuring the centralized control system into distributed control architecture and the logic-based control technique into a feedback control of the discrete-event system.

Among distributed control and cooperative problem-solving methods there are *functional-oriented control methods*, where entities or agents grouped by their functional capability (i.e., one entity per functional task) communicate with each other through a common memory. They are suitable for systems with multiple functional tasks, but present some scalability problems, especially related to entity replication. There is also the blackboard or *operational (agent-based) control method*, where agents have process operation expertise in specific areas and share their expertise by posting partial solutions to the problem on a "blackboard." The blackboard provides a persistent channel of communication where past communications persist on the blackboard until they are modified by another agent. This architecture requires the use of global information storage, posing the problem of memory and scalability.

Those two methods use either a hierarchical-oriented data network structure or heterarchical-oriented data network structure. Within a hierarchical-oriented network, the command information flows downward and the feedback information flows upward, resulting in strong master–slave relationships. The tight coupling between a master and its slaves produces a short response time between them, and causes computing and information bottlenecks. The global information required by these master–slave relationships restricts the flexibility and scalability of this decision-making system and faces a high risk of information bottlenecks. Alternatively, within a heterarchical-oriented data network structure, the entities use (minimal) global information, which can be replicated many times and communicate as peers. Figure 4.18 depicts the various components of a generic cooperative agent with their communication and sensing or detecting capabilities. Among agent components there is a decision alternatives generation component, where the desired time of an event occurrence as an input objective is compared with a simulated time of occurrence (decision alternative) and processed into a regulation defined by the utility function of the subprocess. This leads to the subprocess time of occurrence without constraints (decision preference). When this is later assessed with the simulated process, based on the network information shared from other agents (information coupling function), the upgraded and suitable time of occurrence with constraints (decision alternative) is obtained. Then this time of occurrence is assessed through the entity component of the decision alternative merit ranking, where a decision evaluation function (merit) is used to apply a comparative decision merit evaluation function of upgraded time of event occurrence (decision alternatives). Commonly encountered industrial agents include (1) control and monitoring agent, (2) plant-specific processing agent, and (3) material handling and storing agent.

A key challenge for such a cooperative decision structure is to design coordination strategies that can be a logic-based time-driven control for

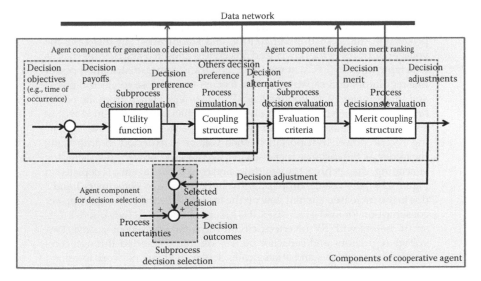

FIGURE 4.18
Generic cooperative agent schematic.

synchronous events or a logic-based action-driven control for asynchronous events. As illustrated in Figure 4.18, a time of event occurrence control scheme is retained, where the control objectives are expected to conflict with the objectives among cooperative agents. This is made independently of the coordination strategy chosen in order to ease the dynamical modeling of multiagent interacting systems. Using a time-driven control structure such as a first come, first served (FCFS) process queue model, those strategies could be implemented using either (1) continuous-time and discrete-time distributed averaging and consensus algorithms; (2) coordination algorithms for rendezvous, formation, flocking, and deployment; (3) distributed algorithms computed and optimized over networks; or (4) consensus design methodologies for distributed multivehicle cooperative control.

Example 4.1

Consider a distributed hybrid power system involving several geographically distributed nonrenewable energy sources, such as hydro-based and thermal-based power generation plants, as well as renewable energy such as solar plants, with AC and direct current (DC) networks and territorially dispersed electrical demands. The traditional power grid architecture is a centralized and radial topology with electric energy generated and delivered from one end to the other. Power dispatching strategies are an alternative to ease bidirectional power flow between power production sites and distributed energy consumption units. However, commonly encountered unidirectional power flow analysis cannot be used to control such disparate power topology when renewable energy sources are implemented near sites. As such, a distributed and cooperative control paradigm of power flow is suitable. Among possible power flow control objectives there are voltage control over the network, reduction of power transmission losses, production and consumption control, and stabilization. In this case, the voltage control over the power grid objective is chosen using devices such as load tap changers (LTCs) at the substation transformer, distribution power line regulators and capacitor banks, feeders (on or close to electric poles), and smart meters with photovoltaic (PV) inverters. Hence, from the power flow analysis, the voltage over network nodes has to be maintained within a specific range along the feeder by tuning capacitor banks. However, during electric power transportation, such voltage is expected to drop along the feeder as the distance increases. An agent-based power grid, including generating, dispatching, transmitting, and consuming agents, is depicted in Figure 4.19. Any voltage drop is discovered explicitly by consumers 1 and 3 due to the increased current flow on the feeder while the consumers' power consumption (or loads) increases. It is assumed that the network nodes (e.g., smart meters with PV inverters, circuit breakers, line sensors, convertors, voltage regulators, and capacitor banks) are interconnected through communication interfaces and strategically deployed at key network locations. In Figure 4.20a and b, the internal cooperative and control structure of production and voltage regulation, along with detailed control variables, is exchanged between them and other agents.

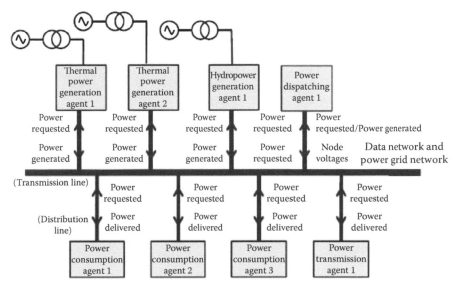

FIGURE 4.19
Hydro- and thermal-based power grid.

Example 4.2

Consider a fully automated factory environment consisting of (1) material handling equipment (MHE), such as robots and automated guided vehicles (AGVs); (2) transportation equipment (TE), such as conveyor belts and linear or rotary axes controlled by motors, trucks, and robots; and (3) processing equipment (PE), such as turns, lathes, and milling machines. Additional equipment related to measurements and process data gathering, such as that used for testing the quality of products, is discarded. Figure 4.21a depicts such a cooperative agent-based production control system, while Figure 4.21b–d illustrates the internal and cooperative structure for each agent for distributed manufacturing shop floor control.

From those block diagrams, distributed control algorithms can be derived where each agent recursively computes the best time to release (near optimal) or start processing a workpiece based on constraints such as buffer size and anticipated processing delays. With these distributed problem algorithms, a network communication system should enable (1) synchronous communications such that all agents are processing on exchanged data (process variables) at the same time (near real time) for all processors, (2) data (process information) to be exchanged between agents as messages (e.g., the processors' states) at the occurrence time, (3) agents in a network to possess only partial information about the network topology, and (4) network failure to occur intermittently or

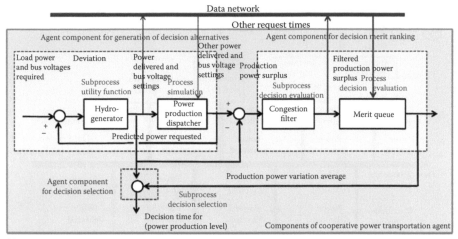

(a)

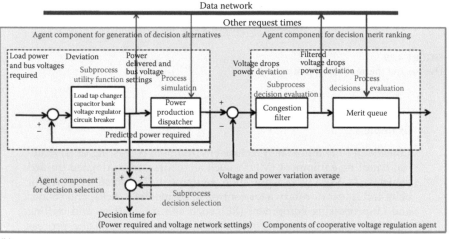

(b)

FIGURE 4.20
(a) Cooperative agent in charge of production of hydropower. (b) Cooperative agent in charge of voltage regulation.

permanently. The stability (consensus) of distributed problem-solving issues is dependent on the convergence theorem versus the amount of information exchanged. Usually, more the information that is exchanged, the higher is the consensus time delay. Hence, it is suitable to design an algorithm offering the fastest convergence. In addition, the system's relative stability can be defined when the distributed system is stable (converging) while some system agents are individually not stable, reaching unique decisions. Hence, in this case, if an agent is unstable,

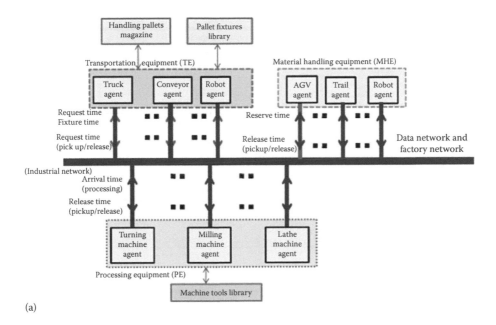

(a)

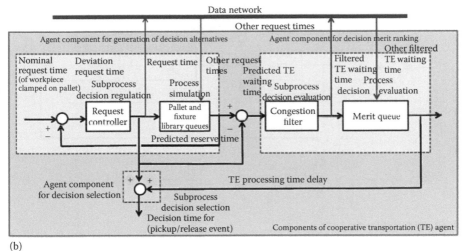

(b)

FIGURE 4.21
(a) Agent-based distributed production control. (b) Schematic of interaction between the components of the TE agent. *(Continued)*

the distributed system could be stabilized by the various decisions of an agent exchanged with other agents. Furthermore, network topologies (partial or arbitrary information sharing) are inherently unstable even if the agents are stable in reaching a unique decision. This is highly dependent on the quality of information received.

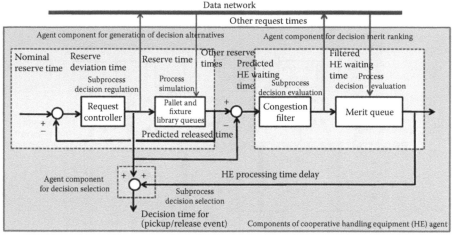

(c)

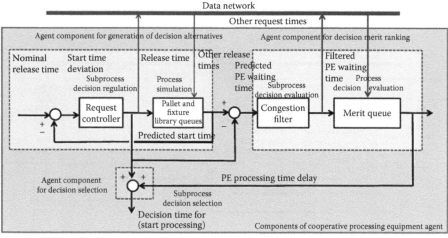

(d)

FIGURE 4.21 (CONTINUED)
(c) Schematic of interaction between the components of the handling equipment agent.
(d) Schematic of interaction between the components of the PE agent.

Exercises

4.1. Recall the bottle milk filling process, as illustrated in Figure 3.64. It is desired to develop a monitoring and control system for this process with the following operating conditions. In this process, a sensor detects the position via a limit switch and waits 0.6 s, and then the bottle is filled until a photosensor detects a filled condition. After the bottle is filled, an outfeed motor (M2) is energized and moves the filled bottle, while after waiting for 0.5 s, a fed motor (M1) is energized and conveys the next bottle. This logic control system includes start and stop push buttons for the process.

a. List in a table all actuating and sensing devices or detectors involved in this process, as well as their corresponding I/O variables.

b. From functional analysis using the FAST method, derive the operating cycles and sequences.

c. Update the process safety measures, along with start and shut-down operating modes, to derive interlocks.

d. Design the expected process sequential and feedback control loops.

4.2. Commonly used in the pharmaceutical industry, the granule-forming process (granulator) is shown in Figure 4.22. A feed stream of powder is sprayed between an outer stationary cylinder and an

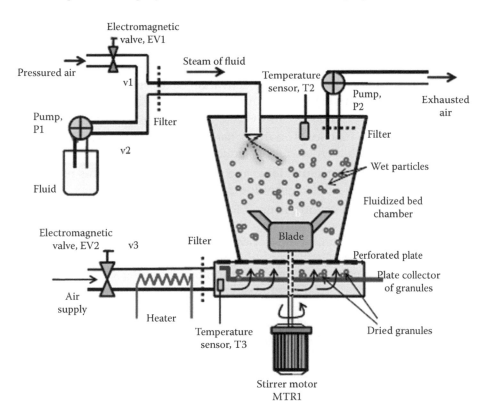

FIGURE 4.22
Granule forming process schematic.

inner rotating one. As the motor-driven cylinder rotates, the fluid-izing air drains out the particles of granules formed throughout the granulator. The rotational speed of the cylinder defines the final shape and size of granules. Valve EV1 is used to adjust the injection flow rate of the powder. Increasing valve EV1 and pump P1 (v_1 and v_2) on the edge by 1% causes an increase of the thickness of granules by 2 mm in diameter. A 1% increase in hot air supply EV2 (v_3) will increase the thickness in the center by 3 mm and reduce the edge thickness by 0.5 mm. A sensor measures the thickness of the formed granules. Once the particle powder is deposited into the chamber, the granulator process consists of injecting air with a starch-based fluid, heating the air, agitating (stirring) the wet particles, and form-ing granules from particle groupings. It is ideal to have a uniform, average size of granules (neither oversize nor undersize).

a. List all sensing and actuating devices involved in the granule-forming process and their corresponding I/O variables.

b. Perform a process decomposition using the FAST method.

c. After defining the controlled variables and process variables, draw the closed-loop block diagrams for the granule feed, shape, and size control.

d. Update the list of process binary variables with those from the control panel enabling the activation or deactivation of the pro-cess start and stop push buttons and indicating one of the two expected average granule sizes.

e. From the functional analysis, list all operating modes and cycles, along with their operating sequences.

f. Develop a sequential logic and feedback control for this process for various defined granule sizes.

4.3. The distillation column is used for the separation of liquid compo-sition (e.g., crude oil) through different boiling points. It is widely used in petrochemical industries. Figure 4.23 gives an overview of a typical crude oil distillation column and the equipment involved. To separate gasoline from asphalt, the reflux rate (at the top) is adjusted to control the distillate composition, while the rate of steam (at the bottom) to the reboiler is varied to control the bottom composition. Any change in the feed rate to the column acts as a disturbance to the process. Among the equipment involved there are a condenser, a vapor–liquid separator, a compressor, and a stripper.

When the composition of gasoline in the top distillate stream is below a temperature set point, the cold reflux flows into the column. This reflux is expected to increase the purity of the distillate stream. This permits an increase in the flow of gasoline out of the bottom stream. As the composition moves off the set point, the flow of steam

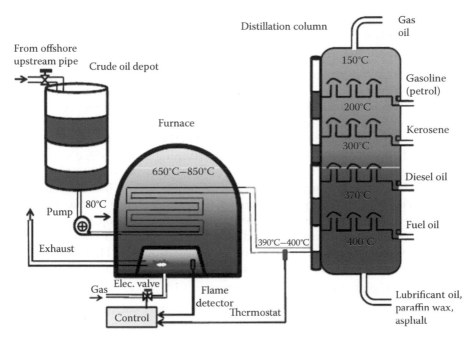

FIGURE 4.23
Distillation column schematic.

into the reboiler increases to heat up the bottom of the column, while causing an increase of hot vapors traveling up the column. This eventually causes the top of the column to begin to heat up. Similarly, as the top of the column heats up, the purity of gasoline in the distillate stream again becomes too low. In response, at the top, there is an increase of the flow of reflux into the top of the column.

a. List all devices (actuating and sensing) involved in the distillation column temperature and flow control and their corresponding I/O variables.

b. After defining the controlled variables and process variables, draw and label closed-loop block diagrams for pressure control, feed control, boiler temperature control, and reflux control.

c. Update the list of process binary variables with those from the control panel enabling the activation of process start, stop, and indicator lights when temperature threshold values are reached at each stage.

d. From the functional analysis, list the operating modes and cycles, along with their operating sequences.

e. List three types of safety measures and the corresponding interlocks that need to be implemented.

f. Develop a SFC normal production program integrating the logic control with the feedback control.

g. Sketch the command and control schematic.

4.4. It is desired to design the control and monitoring system of an AC motor with variable speed drives. This allows migration from a hardwired control to a programmable logic control and monitoring system. As such, the operator panel station performs the potentiometer speed control (speed regulator), forward and reverse direction selection, and selection of the manual or automatic operation running modes of the variable speed drive through run or jog switches, and even the process activation or deactivation through the start and stop push buttons. The incomplete wiring diagram in Figure 4.24 shows an operator station used to manually control a variable speed drive. Note that the start, stop, run and jog, potentiometer, and forward and reverse field devices are connected to the same names that are used in the control program. (Hint: Establish an SADT or process operating and stop modes graphical analysis graph).

a. Complete the I/O wiring diagram.

b. Design the sequential logic of the AC motor control and its speed feedback controller.

c. Write an assembly program for the logic control of this AC motor.

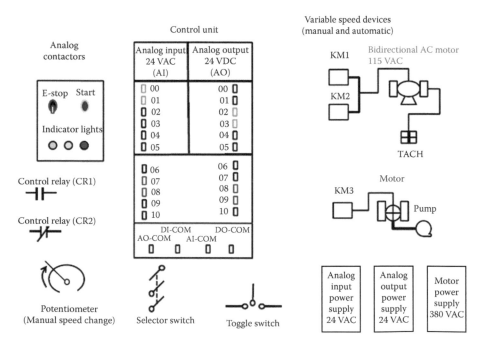

FIGURE 4.24
Wiring diagram of AC motor hardwired logic controller with variable speed drive.

4.5. Recalling the thermal plant and its power distribution network illustrated in Figure 3.59, it is desired to design a control and monitoring system. As such,

   a. Establish a list of voltage and power control devices and describe the operating sequences and associated constraints.

   b. Decompose the process operations using the FAST method.

   c. Based on operating conditions, establish the SADT (or GEMMA) graph of the start and failure modes, as well as interlocks.

   d. Define the visualization system for this power plant.

   e. Design the resulting sequential logic and feedback control system.

   f. Draw the control and command, power, and I/O wiring diagrams of a control and monitoring system for the thermal power plant.

4.6. It is desired to control the flow of water of a hydrodam, in order to maintain the generated electric power near the forecasted daily load.

   a. Based on the equipment listed in Table 4.10, list all major devices and equipment involved.

   b. Develop a monitoring and control system of the electrical hydrodam, as illustrated in Figures 4.25 and 4.26.

4.7. In a cement milling station, a portico scratcher with a capacity of 200 tons/h allows us to convey cement raw material, such as gypsum and dry and wet pozzolana, from the storage hall to the cement grinding or drying subprocesses. First, the speed transversal positioning of the portico scratcher above the cement raw material deposit is completed. Based on the two motorized arms (primary and secondary), the scratcher moves the deposit from right to left.

TABLE 4.10

Process Equipment List for Hydro Dam Water Flow Control

Equipment	Technical Characteristics	Operating Conditions	Associated Process Operations and Control Strategy
Generator GT1	Power 11 kW	Max. load arm: 35 tons Speed: 0.4 m/s	Asynchronous motor with squirrel cage for power generation
Gate motor drive MTR1	Power 6 kW	Max. load arm: 35 tons Speed: 0.4 m/s	Asynchronous motor with squirrel cage for the water gate motion
Water-level sensors LD1, LD2, LD3	Rittmeyer pressure-based sensor	Length of carrier: 80 m Speed: Low, 2.6 m/mm; high, 16 m/mm	

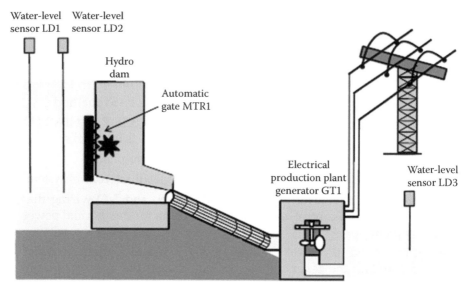

FIGURE 4.25
Simplified process schematic of electrical hydrodam water flow control.

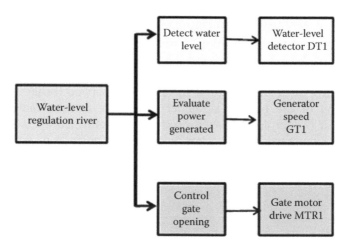

FIGURE 4.26
Functional analysis using FAST method of electrical hydrodam water flow control system.

From there, the conveyor base sends it to other cement subprocesses, as illustrated in Figure 4.27. Table 4.11A summarizes the process equipment involved, along with their operating characteristics, and Table 4.11B presents the process operating sequence from FAST decomposition.

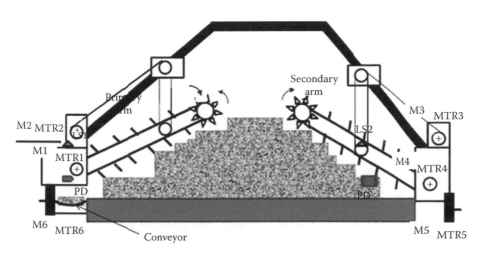

FIGURE 4.27
Schematic of cement pozzolana scratching process.

TABLE 4.11A

Process Equipment List for Cement Pozzolona Scratching Process

Equipment	Technical Characteristics	Operating Conditions	Associated Process Operations and Control Strategy
Primary robot arm scratcher MTR1	Power 11 kW	Max. load arm: 35 tons Speed: 0.4 m/s	Asynchronous motor with squirrel cage for primary robot arm motion
Secondary robot arm scratcher MTR4	Power 6 kW	Max. load arm: 35 tons Speed: 0.4 m/s	Asynchronous motor with squirrel cage for secondary robot arm motion
Vertical motion of secondary robot arm scratcher: MTR3	Force of 2000 kg	Length of vertical trip: 80 m Speed: Low, 2.6 m/mm; high, 16 m/mm	
Vertical motion of primary robot arm scratcher: MTR2	Force of 1600 kg Power: At low speed, 0.83 kW; at high speed, 5 kW	Length of vertical trip: 40 m Speed: Low, 2.6 m/mm; high, 16 m/mm	
Cross-motion portal MTR5 and MTR6	Power: 5 kW	Cross-motion: Scratching speed, 0.9 m/mm; high speed, 9.4 m/mm; brake torque max., 2 (10 mdaN)	

TABLE 4.11B

Operating Modes and Sequences of Cement Pozzolona Scratching Process

Operating Mode	Operating Sequence	Variables	Variables Types	Description	Interlocks
Start mode	Start transversal positioning motion	MTR6, MTR5	Bool, Bool, Bool	High-speed transversal displacement and positioning scratcher portal	Fault due to missynchronization of MTR5 and MTR6 during transversal motion (antitwisting)
	Positioning of primary scratcher robot arm	MTR1	Bool	Low-speed position of primary scratcher robot arm above pozzolana (72°)	
	Positioning of secondary scratcher robot arm	MTR4	Integer, FBD	Slow-speed position of secondary scratcher robot arm above pozzolana (68°)	Detect rotation of secondary scratcher robot arm
	Pozzolana scratching by winch of secondary scratcher robot arm (139 s)	MTR3	Repeat 1 successively	Slow-speed pozzolana scratching by winch of the secondary robot arm for 139 s	If primary scratcher robot arm is >23A, stop MTR3
	Pozzolana removal by winch of primary scratcher robot arm (139 s)	MTR2	Repeat 2 successively	Slow-speed pozzolana removal by winch of primary robot arm for 139 s	If primary scratcher robot arm is >73A, stop MTR2
	Repositioning of primary scratcher robot arm	MTR1	Repeat 3 successively until (72° − delta = 17°)	Low-speed position of primary scratcher robot arm above the pozzolana (72° − delta)	If collision between secondary and primary arms, stop MTR2 and MTR3
	Repositioning of secondary scratcher robot arm	MTR4	Repeat 4 successively until (68° − delta = 13°)	Low-speed position of secondary scratcher robot arm above the pozzolana (68° − delta)	

(Continued)

TABLE 4.11B (CONTINUED)

Operating Modes and Sequences of Cement Pozzolana Scratching Process

Operating Mode	Operating Sequence	Variables	Variables Types	Description	Interlocks
	Position referencing before transversal motion	MTR1, MTR2, MTR3, MTR4		Both primary and secondary arms to 72° and 68°, respectively	
	Start transversal positioning motion by 50 cm	MTR6, MTR5		Low-speed displacement of the portal by 50 cm	
	Repositioning scratching, etc.	All motors involved		Repeat step above until scratcher portal reaches pozzolana length of 60 m	
Shutdown mode	Stop secondary scratcher robot arm	MTR3	Bool		
	Stop secondary scratcher robot arm	MTR4			
	Stop primary scratcher robot arm	MTR2	Bool		
	Stop primary scratcher robot arm	MTR1	Bool		
	Stop transversal positioning motion	MTR6, MTR5			

Note: Bool, Boolean; FBD, function block diagram.

Based on the process information provided,

a. Create an SFC using the start and stop modes from the graphical analysis approach.

b. For each piece of actuating equipment, define the command types and parameters required for SFC.

c. Develop all required SFCs (especially safety, normal production).

d. Draw the resulting control and command schematic.

4.8. In the brewery industry, tunnel pasteurization consists of taking the thermal treatment of beer in a bottle through several water-based temperature media processes to reduce bacteria growth. This allows extension of the beer conservation. Here, as illustrated in Figure 4.28, the bottles filled with beer are conveyed at a constant speed of 7 mm/s through several water showers at respective temperatures of 30°C, 52°C, 62°C, and 30°C. Some motorized pumps are used to maintain water around the above mentioned temperatures with a flow rate of 250 m³/s in a closed-loop water flow between the water tank and heat exchanger at 170°C. Table 4.12 summarizes the pasteurization operating sequences.

The process function blocks for the temperature control can be decomposed into tank water temperature control, hot water flow control, and bottle conveying. Using the information provided,

1. Develop an interlocking logic (if necessary) for this process.

2. Create an SFC using the start and stop operating modes from the graphical analysis approach.

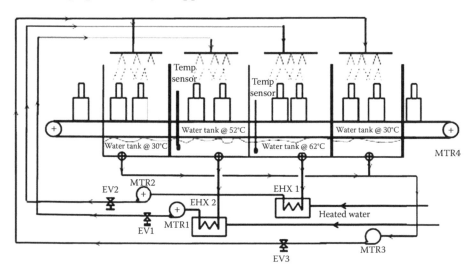

FIGURE 4.28
Simplified P&ID of beer tunnel pasteurization process.

TABLE 4.12

Pasteurization Process Operating Sequences

Operating Mode	Operating Sequence	Variables	Variable Types	Description	Interlocks
Start mode	Filling water	PB_START, all electrovalves (EVx)	Bool, Bool, Bool, Bool, etc.	Activation valves for inflow tanks	
	Turn on heat exchangers	EHX1, EHX2		2 bar $< P <$ 4 bar 121°C $< T <$ 128°C	
	Open steam electronic valves to supply water at temperature to the tanks	EV1, EV2, EV3			
	Starting pumps for heating water tanks to 52°C and 62°C in closed loop	MTR1, MTR2, MTR3, Sensor_temp, PID_temp_control	Integer, FBD		Pumps MTR2 and MTR3 start if tank water temperatures are reached
	Conveying bottle	LS1, LS2, LS3, LS4, MTR4	Bool, Bool, Bool, Bool	Move bottle filled with beer to pasteurization tunnel The conveyor speed is constant	MTR4 fails if temperature out of reference Four limit switches detect conveyor travel distance (stop and start)
	Temperature control	MTR2, EHX1/2, EHX2, EV1, EV2		Temperature control via steam flow through heat exchangers using electrovalve and thermostatic valve	Tank pressure ($P <$ 1.9 bar) temp_faults out of order (Temp_sensor) Stop conveyor
Stop mode	Stop filling water	LS_level	Bool		
	Stop conveyor	MTR4			
	Closed steam valve	EV1, EV, EV3			
	Stop pump	MTR1, MTR2			
	Removing yeast	Timer, EVx	Bool		
		Timer, EVx	Bool		

Note: Bool, Boolean; FBD, function block diagram.

3. Develop all required SFCs (especially safety, normal production, etc.).

4. Draw the control and command schematic.

4.9. Fermentation is a three-stage process. At the primary stage, yeast is added to malt seed in water-filled fermentation tanks. This mixture is maintained in a closed tank for 8–12 days. During this stage, the temperature must be maintained at 12°C and a pressure of 1.2 bar is applied, while the resulting CO_2 from the alcohol metabolism is evacuated, as illustrated in Figure 4.29. At the end of the CO_2 production, the temperature of the generated beer in the fermentation tank is reduced to 5°C and 1.2 bar to ease solidification of the remaining yeast in the tank. Then, the beer temperature in the fermentation is decreased to –1°C and kept at 1.2 bar to ease collection at the tank bottom of some solid-based deposits of yeast. After the –1°C fermentation stage, the solid-based yeast deposit is removed and the filtered beer is output through a dedicated electrovalve. Figure 4.30 illustrates the functional decomposition of this fermentation process, and Table 4.13 presents the fermentation process operating sequences. Based on the functional subprocess block

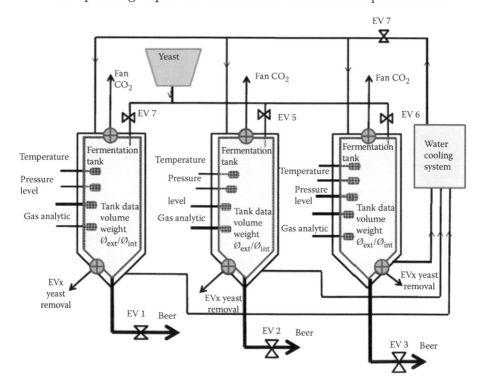

FIGURE 4.29

Simplified process schematic description of beer fermentation process.

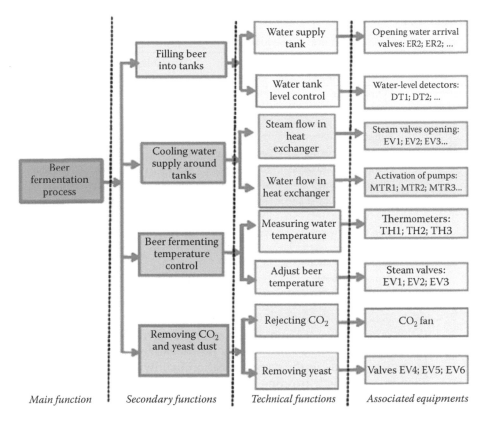

FIGURE 4.30
Functional analysis using FAST method for beer fermentation process.

diagrams for temperature and pressure control in the fermentation tank, and with information provided,

a. Develop an interlocking logic for this process.

b. Create an SFC using the start and stop operating modes from the graphical analysis approach.

c. Draw the power supply circuit schematics as well as sketch the control and command diagram.

4.10. Consider the implementation of a collision-free consensus decision making through a distributed, cooperative problem-solving mechanism for multiple mobile devices. Typical mobile devices could be autonomous underwater vehicles, unmanned aerial vehicles, or automated highway systems. In order to achieve such a consensus, each vehicle must communicate with its neighbors in a coordinated fashion in order to avoid collision, while adjusting its trajectory with respect to the targeted sequence points. Hence, the key characteristic

TABLE 4.13

Fermentation Process Operating Sequences

Operating Mode	Operating Sequences	Variables	Variables Types	Description
Start mode	Filling water	PB_START, EV4, EV5, EV6,	Bool, Bool, Bool, Bool, etc.	Activation valves for inflow tanks
	Filling yeast (20 min)	Timer	Bool	Yeast injection
	Cooling tank at 12°C	Sensor_temp	Integer	Stage 1 fermentation
	Fermentation (5 days)	Timer	Bool	Stage 2 fermentation
	Cooling tank at 5°C	Sensor_temp	Integer	Stage 2 fermentation
	Cooling tank at –1°C	Sensor_temp	Integer	Stage 3 fermentation
	Beer filtering	EVx	Bool	Removal of beer
Stop mode	Stop filling water	LS_level	Bool	Fermentation preparation
	Stop fermentation	PB_STOP		End fermentation
	Removing yeast	Timer, EVx	Bool	Collecting yeast
	Start cleaning (2 h)	Timer, EVx	Bool	Cleaning tank

of the cooperative scheme of the multiple agents is to have data exchanged with the nearby agents (i.e., localized communication) due to limitations in the communication bandwidth. In the case of autonomous vehicle running over an highway,

a. Identify through a schematic the cooperating structure and agents, the type of data exchanged, and the cooperative agent components.

b. Consider that each vehicle has a circular protection zone with a fixed radius around it. Determine how the time delay of the consensus convergence could be affected under the limited information exchange scheme and convergence on its real time positioning decision.

4.11. a. Describe any technical differences between a SCADA and a DCS.

b. For either the SCADA or DCS, give two reasons to separate the implementation and execution units of the management information system function and the online control system.

c. List and identify five major subsystems of a DCS.

4.12. A transmitter measures the difference in pressures, $p_i(t)$, as illustrated by the blood pressure transmitter with data monitoring in Figure 4.31. Depending on the pressure difference, the direction of the motion of the bellow could be either to the right or to the left, causing the magnet core to side over the coil accordingly. The motion of the magnet core over the coil causes an inductivity, as well as

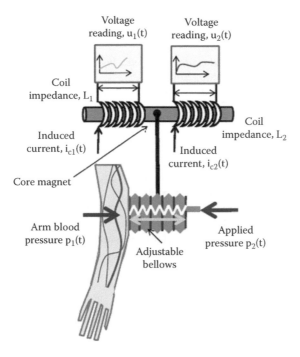

FIGURE 4.31
Blood pressure transmitter schematic.

voltage variation. Hence, it is possible to relate the pressure difference to the resulting voltage difference. In order to achieve a high reliable pressure measurement from this device, it would be suitable to upgrade its design to be fault tolerant according to the expected measurement faults.

a. In a table, identify and classify two types of faults causing measurement errors: the magnet core sliding out of the range and the bellow interface failure.

b. Perform a FMEA of identified potential failure and propose a fault-tolerant model of this blood pressure transmitter.

Bibliography

ADEPA, *Guide d'étude des modes de marches et d'arrêts*, GEMMA, Agence pour le development de la productique et l'automatique, Ministère de l'industrie, Paris France, 1997.

Arbel A., A. Seidmann, Selecting a microcomputer for process control and data acquisition, *IIE Transactions*, 16(1), 73–80, 1984.

Bai H., M. Arcak, J. Wen, *Cooperative Control Design*, Springer, Berlin, 2011.

Bailey D., E. Wright, *Practical SCADA for Industry*, Newnes (copyrighted Elsevier), Boston, 2003.

Barlett T.L.M., *Industrial Automated Systems: Instrumentation and Motion Control*, Cengage Learning, Boston, 2010.

Bertsekas D.P., J.N. Tsitsiklis, *Parallel and Distributed Computation: Numerical Methods*, Athena Scientific, 1997.

Bosh J., *Design and Use of Industrial Software Architectures*, Newnes (copyrighted Elsevier), Boston, 2003.

Bower J.M., *Control of Sensory Data Acquisition*, Elsevier, Amsterdam, 1997.

Boyer S.A., *SCADA: Supervisory Control and Data Acquisition*, 4th ed., ISA, 2009.

Bullo F., J. Cortes, S. Martinez, *Distributed Control of Robotic Networks*, Princeton University Press, Princeton, NJ, 2009.

Chiang L.H., E.L. Russell, R.D. Braatz, *Fault Detection and Diagnosis in Industrial Systems*, Springer, Berlin, 2001.

Cilliers A.C., E.J. Mulder, Adapting plant measurement data to improve hardware fault detection performance in pressurised water reactors, *Annals of Nuclear Energy*, 2012.

Ding S.X., *Model-Based Fault Diagnosis Techniques: Design Schemes, Algorithms, and Tools*, Springer-Verlag, Berlin, 2008.

Erickson K., J. Hedrick, *Plant Wide Process Control*, Wiley, Hoboken, NJ, 1999.

Francis B., A course on distributed robotics, Notes for CDC Bode Lecture, 2014.

Isermann R., *Fault-Diagnosis Systems: An Introduction from Fault Detection to Fault Tolerance*, Springer, Berlin, 2006.

Jadbabaie A., J. Lin, A.S. Morse, Coordination of groups of mobile autonomous agents using nearest neighbor rules, *IEEE Transactions on Automatic Control*, 48(6), 998–1001, 2003.

Jaksch I., Demodulation methods for exact induction motor rotor fault diagnostic, in *5th IEEE International Symposium on Diagnostics for Electric Machines, Power Electronics and Drives*, September 2005.

Kaltjob P., N.A. Duffie, Real time, cooperative, predictable decision-making in heterarchical manufacturing systems design: Requirements, generic methodology and applications, *Journal of Manufacturing Systems*, 34(2), 2005.

Komenda J., T. Masopust, J. Schuppen, Supervisory control of distributed discrete-event systems, in *Control of Discrete-Event Systems*, ed. C. Seatzu, M. Silva, J.H. van Schuppen, vol. 433, Lecture Notes in Control and Information Sciences 10, Springer, Berlin, 2013.

Luyben W., *Process Modeling, Simulation and Control for Chemical Engineers*, McGraw-Hill, New York, 1990.

Lynch N., *Distributed Algorithms*, Morgan Kaufmann, Burlington, MA, 1996.

Marlin T., *Process Control: Designing Processes and Control Systems for Dynamic Performance*, McGraw Hill, New York, 1995.

Mesbahi M., M. Egerstedt, *Graph Theoretic Methods in Multiagent Networks*, Princeton University Press, Princeton, NJ, 2010.

Ogunnaike B., W. Ray, *Process Dynamics, Modeling and Control*, Oxford University Press, Oxford, 1994.

Ondel O., Fault detection and diagnosis in a set "inverter–induction machine" through multidimensional membership function and pattern recognition, *IEEE Transactions on Energy Conversion*, 24, 431–441, 2009.

Santina M.S., A.R. Stubberud, Basics of sampling and quantization, in *Handbook of Networked and Embedded Control Systems*, Birkhäuser, Basel, 2005, pp. 45–70.

Seborg D.E., T.F. Edgar, E.A. Mellichamp, F.J. Doyle, *Process Dynamics and Control*, 3rd ed., Wiley, Hoboken, NJ, 2011.

Smith C., A. Corripio, *Principles and Practice of Automatic Process Control*, Wiley, Hoboken, NJ, 1997.

Trigeassou J.-C., *Electrical Machines Diagnosis*, Wiley-ISTE, Hoboken, NJ, 2011.

Wei R., Y. Cao, *Distributed Coordination of Multi-Agent Networks*, Springer, Berlin, 2011.

Wood R., Machine safeguarding protects employees and the business; the facts lead to one conclusion: Safety m, *Control Solutions*, February 2003.

5

Sensing and Data Acquisition Elements: Modeling and Selection

5.1 Introduction

As illustrated in previous chapters, process operations can be captured through binary logic or continuous-time variables. Depending on their time-varying characteristics, those variables are gathered by detection or measurement instrumentation and analyzed through a computer unit. The procedure of gathering, logging, and processing data is called *data acquisition* and consists of a sequence of operations, depicted in Figure 5.1.

Data gathering is achieved by the detection or measurement of process variables through transducers or sensors producing signals correlated to the variations of the physical input conditions. There are various types of signals, such as (1) analog voltage signals, (2) amplitude- or frequency-modulated signals, (3) pulse width–modulated signals, or (4) square-wave signals representing binary values of 1 or 0. Usually, analog process-level inputs (voltage) are transformed into digital output ranges (bits) in order to be processed by a computer unit.

The *data acquisition system* consists of sensory apparatus combined with signal filters, amplifiers, or modulators. Hence, typical components of such a system are

- *Transducers,* also called sensors, which transform one form of energy into another. There are two types of transducers: (1) active transducers, which generate their own energy (e.g., thermocouples), and (2) passive transducers, which require additional energy (e.g., strain gauge).

- *Filters,* to get rid of unwanted disturbances (noise), such as aliasing effects. Indeed, low-pass filters reduce or even eliminate the high-frequency content of noise signals.

- *Multiplexers,* which switch to connect multiple sensor input channels to one piece of data-measuring equipment (analog-to-digital

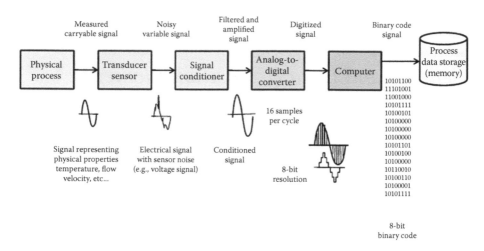

FIGURE 5.1
Sensing and data acquisition chain.

converter [ADC]). Multiplexers are characterized by their switching rate and cross talk.

- *Converters*, which transform either analog voltage signals into discrete signals, such as an ADC, or inversely in the case of a digital-to-analog converter (DAC).

Table 5.1 summarizes typical sensing methods for various types of variables.

However, the sensing challenge is to obtain a good estimate of a physical quantity. Indeed, converting physical phenomena into measurable voltages or current signals often requires signal conditioning, such as amplifiers that boost currents and voltages. Therefore, understanding the differences and limitations allows us to select suitable feedback devices for the design of a computer controller system.

TABLE 5.1

Typical Sensors and Associated Sensing Methods

Measurand	Some Sensing Methods
Displacement/position	Resistance, capacitance, magnetic, laser, optoelectric, Hall effect, ultrasonic, image vision, variable reluctance
Distance	Image vision, triangulation, ultrasonic, capacitance, inductance proximity
Temperature	Thermistor (NPC, PTC), infrared thermodiode, thermocouple
Pressure	Piezoresistance, capacitance, ultrasonic, strain gauge
Velocity	Hall effect, magnetic, variable reluctance
Luminance	Photoresistor, photoconductor, phototransistor

The *data logging and processing* system consists of four elements covering the following tasks and operations: (1) measuring the output signals, (2) recording the output signals (logger unit), (3) uploading or accessing recorded data, and (4) processing recorded data through data acquisition software. The performance assessment of the control system also requires the modeling of data gathering and processing time delays by computer interfacing.

In this chapter, sensor types and measuring principles are covered for various physical system variables, such as linear and rotational position, speed, acceleration, force, torque, chemical content, distance, flow, temperature, proximity, level, and pressure. In addition, signal-conditioning principles and categorization (zero-, first-, and second-order sensors, etc.), sensor and detector design specifications, and device characteristics, such as linearity, sensitivity, limitations, calibration, and saturation, are presented. Then, the sensing and operating principles of dynamic sensors, such as dynamics model, time domain, and frequency response characteristics, are developed. Computer interfacing and signal-conditioning issues, such as digitization, sampling theorem, antialiasing filtering, and noise characteristics, are discussed. Finally, methods for data mapping and rendering are explained.

5.2 Detection and Measurement Elements: Modeling and Selection

Detectors or sensors can be classified as active or passive devices capable of producing analog or digital (binary) output signals correlated to the measured variables. Digital sensors produce logic-level outputs that can be directly interfaced to a control computer, while analog sensors produce voltage outputs that require an ADC to interface with a computer. Furthermore, those sensing devices could be either contact (e.g., linear variable differential transformer [LVDT], tachometer, strain gauge, and rotary variable differential transformer [RVDT]) or contact-free (e.g., encoder, Hall effect, capacitance, and interferometer types) devices.

5.2.1 Sensor Classification

Sensors are characterized according to one of the following criteria:

1. *Power supply* requirements, which define if a sensor is active (i.e., it generates an electrical signal in response to an external stimulus, such as piezoelectric) or passive (i.e., with an external power, supply such as a thermocouple)

2. *Nature of the sensor output signal*, which can be digital (binary), such as a shaft encoder, or analog (continuous in magnitude and temporal content), such as object temperature or displacement

3. *Measurement operational mode*, which reflects the proportionality of the difference between the initial condition of the instrument and the measurand of interest, with the null mode being a case where the influence and measurand of interest are balanced (equal) but opposite in value

4. *Input/output (I/O) signal dynamic relationships* (zero, first, and second order, etc.) of the sensor

5. Type of *measurand* (mechanical, thermal, magnetic, radiant, chemical, etc.)

6. Type of *measured variable* (resistance, inductance, capacitance, etc.)

5.2.2 Sensor Performance Characteristics

The conversion of a physical phenomenon into electrical signals is made with limitations that are often characterized by

- *Input range*, which is the upper (maximum) and lower (minimum) sensed values of the physical process variable that can be measured.
- *Threshold*, which is the slightest change in input signal inducing a change in system output signal.
- *Transfer function*, which provides the physics-based relationship between the real phenomenon input stimulus and the resulting electrical output signal. As such, it describes the sensor characteristics through calibration curves.
- *Sensitivity*, which is the ratio between small variation in the electrical output signal and small variation in the physical input signal, and is defined by the slope of the calibration curve. The thermometer sensitivity is given by

$$S = \frac{\Delta V}{\Delta T} \text{ in V/}^{\circ}\text{C} \tag{5.1}$$

- *Accuracy* of a sensor, which is the difference between the real physical value of the variable and the measured value by the sensor. It is defined as the largest expected error.
- *Hysteresis*, which is the difference in output signals between repeated measurements of increasing and decreasing input signals.
- *Resolution*, which is the smallest variation of the process variable value detectable by the sensing device, taking into account the

accuracy, defined as the gap between the measured process value and the real physical value.

- *Precision*, which is the ability (in terms of accuracy and resolution) to derive similar values of a variable's repeated measurements under identical conditions.
- *Error characterization*, which is defined as an integration of the precision, accuracy, and resolution in the measurement.
- *Repeatability*, which expresses the variation over repeated measurements of a variable given the value under identical conditions.
- *Nonlinearity*, which is the deviation of the calibration curve K to a given straight line (transfer function).
- *Decay time*, which represents the time a sensor signal requires to return to its initial level after a step change.
- *Bandwidth* of a sensor, which is the range between these two frequencies.
- *Dynamic response*, which is the frequency range for regular operations of the sensor.
- *Slew rate*, which is the accuracy of the sensor to a changing input range.
- *Environmental* limiting factors, which affect measurement device (e.g., temperature, dust, magnetic interference, humidity, corrosives, and water).
- Sensor *calibration*, which determines the reference relationship between the phenomenon stimulus and the resulting electrical output signal.
- Identification of the *zero point or offset*, which is the resulting output from a zero input value, while the *saturation* occurs when any input change causes no significant change in output.
- *Measurement linearity*, also known as the *calibration curve*, which is the percentage of deviation from the best-fit linear calibration curve.
- *Impedance of the measurement equipment*, which is the ratio of voltage over the crossing current ratio such that a high input impedance is suitable for a low-power signal to the sensor, while a low output impedance is for the current provided by the signal-conditioning unit.
- *Operating temperature*, which defines the range in which the sensor performance is optimum.
- *Signal-to-noise ratio*, which is given by the proportionality between the magnitudes of the signal and the noise at the output.

Table 5.2 summarizes various sensor performance indicators.

TABLE 5.2

Commonly Encountered Sensor Performance Indexes

Sensor Performance Indicator	Unit	Description
Rise time	s	Time for signal to change from 10% to 90% of its peak-to-peak value
Settling time	s	Response time for signal to step change as input signal
Sensitivity	dV/dx	Ratio between small output variations resulting from small input variation
Offset voltage	mV	Output voltage obtained from zero-level input signal
Temperature coefficient	ppm/°C	Rate of reading change per degree Celsius
Repeatability	%	Assessment of output signal similarities between successive measurements under identical conditions
Reproducibility	%	Assessment of output signal similarities between successive measurements under various conditions
Temperature range	T (°C)	Allowable environmental operating interval for a given accuracy

5.2.3 Measurement Error Types

A typical analysis of the measurement error consists of identifying various types of errors and estimating their propagation. Measurement errors in the instrument chain can be classified as follows:

1. Systematic or reproducible errors, which are from different components of the measurement chain, such as
   a. Sensor accuracy
   b. Nonlinearity or linearity
   c. Hysteresis
   d. Zero point or offset (bias), drift
   e. Stochastic errors from random causes (noise)
   f. Repeatability
   g. Resolution
   h. Threshold
   g. Gain error
   i. Calibration, scaling
2. Application-related errors, which are those associated with the measurement process, including
   a. Spatial errors related to the number of measurement devices, as well as their positioning
   b. Interaction errors inherent to the measurement device size

    c. Probe errors due to the measurement device orientation

    d. Temperature effects related to the measurement device's resistance to environmental changes

3. Interface errors, which are those associated with the interfacing components, such as

    a. Connecting cable physical characteristics (resistance, impedance, etc.)

    b. Energy loading (electrical or mechanical)

    c. Common mode voltage

    d. Operating errors (electromagnetic interference)

4. Sampling and approximation errors, which are due to the discretization of analog signals, such as

    a. Sampling distribution error (related to an improper sampling rate, which could skip measurement data)

    b. Stabilization error (due to long measurement device response time)

    c. Approximation error (related to few measurement data)

5.2.4 Sensor Sizing and Selection Procedure

The selection procedure of a sensor consists of

1. Elaborating on process measurement requirements through a checklist derived from the analysis of the process behavior, especially on

    a. The *nature of the event* to be detected or captured.

    b. The *range of event values* representing the measurand set of values. The measurement component has to meet the dynamic range of the measurand.

    c. The *environmental conditions* in which the sensor or detector operates. For example, a detector has to resist (i) humidity or submersion, (ii) corrosion, (ii) wide temperature variations, (iv) soiling of any kind, and even (v) vandalism.

    d. The *selection of the measurement principle and the sensing method* based on the understanding of the physical phenomena involved and the knowledge of the built-in technology of available sensors. The measurement component can be continuous or periodic.

    e. The *human resources needed* in the functioning or maintenance of the sensor. Typical operating requirements are (i) ease of use and (ii) minimal level of maintenance.

    f. The *size of the sensing* device to meet the constraints of the space available and compliance within the automation chain.

g. The *device life span*.

h. The choice of the *nature of the signal* to be delivered by the sensors, which could be analog or digital.

2. Evaluating the *sensor performance and its energy consumption* by checking if the energy consumption and power supply provided in data sheets fulfill the network requirements from the client.

3. Selecting the sensors or detectors based on the *cost–performance ratio*. The cost depends on the technology used and on various technical criteria, such as (a) accuracy, (b) resolution, (c) repeatability, and (d) slew rate.

The key steps of the sensor selection procedure are (1) defining the system to measure, (2) choosing the function of the system and sizing it, (3) selecting the signal conditioning, (4) ensuring its reliability and robustness, (5) assessing the maintainability and operating conditions, and (6) estimating the cost–performance ratio.

5.2.5 Analog Measurement and Detection Instruments

A *transducer* transforms system energy generated by one physical source into a measurable energy form (i.e., electrical, mechanical, or thermal). Accordingly, transducers can be classified into two categories: electrical and mechanical. There are three types of electrical-based transducers: analog, digital, or frequency modulated. This section covers only the two major categories of analog output transducers:

1. *Active transducers* produce an electrical signal when activated by an external stimulus without using external power supply (e.g., piezo-electric sensors and photocells).

2. *Passive transducers* use external power supply to produce an electrical signal when activated by an external stimulus.

Table 5.3 lists commonly encountered sensors. Figure 5.2 summarizes the commonly encountered electrical transducers.

5.2.5.1 Relative Position or Distance Measurement

Relative position or distance sensors are either direct contact or contactless devices used to derive the presence or position of an object. Those sensors can be classified as follows:

- *Optical distance sensors (photoelectric) or detectors*, where the source emits an infrared light signal toward an infrared sensitive device that would (1) change its resistance based on the intensity of the

TABLE 5.3

Some Commonly Encountered Sensors

Sensor Type	Sensing Methods and Description
Linear and angular position and velocity sensors	Optical encoder (absolute and incremental angular position measurement)
	Electrical tachometer (generator or magnetic pickup velocity measurement)
	Hall effect sensor (position measurement with high resolution but low range)
	Interferometer (laser-based position measurement)
	Capacitive transducer (high-frequency dynamics position measurement)
	Magnetic pickup
	Gyroscope (angular position measurement)
	Vision (presence or absence, check position)
	Ultrasonic (flow or obstruction, counting, fluid or gas level, object shape)
Acceleration sensors	Seismic accelerometer
	Piezoelectric load cells
Flow sensors	Rotameter
	Ultrasonic type
	Electromagnetic flowmeter
	Pitot tube
Force, pressure, and torque sensors	Dynamometer
	Ultrasonic stress sensor
	Tactile sensor
	Strain gauge
Temperature noncontact sensors and detectors	Thermal imager
	Infrared thermometers
	Fiber optic
Temperature contact sensors and detectors	Phase change device
	Bimetallic thermometer
	Filled system thermometer
	Cryogenic temperature sensor
	Thermocouple (for operating range of 200°C to more than 1100°C)
	Thermistor (for operating range of more than 110°C)
	Resistance temperature detector
	Thermodiode and thermotransistor (nanodevices like chip)
	Infrared type (wavelength resolution)
	Infrared thermography (temperature distribution measurement)
Smart sensors	Optical fiber (use as strain sensor, temperature sensor, high-resolution and range force sensor, level sensor)
	Magnetoresistive (use as force sensor, torque sensor)
	Piezoelectric (accelerometer, force sensor, strain sensor)

(Continued)

TABLE 5.3 (CONTINUED)

Some Commonly Encountered Sensors

Measured	Sensing Methods and Description
Proximity sensor	Photoelectric
	Capacitance
	Hall effect
Light sensor	Photodiode
	Photoresistor
	Phototransistor
	Photoconductor
Binary noncontact detector	Inductive proximity to detect a metal object
	Capacitive proximity to detect a dielectric object
	Optical presence to detect an object or a subject (use for counting; checking the passing or obstruction of objects; monitoring the presence, absence, or position of an object)
Binary contact detector	Mechanical contact (switch) to detect an object

Physical phenomenon

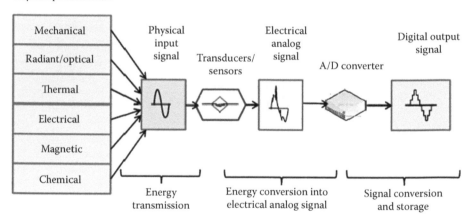

FIGURE 5.2
Typical electrical transducers and their computer interface.

received signal (i.e., photoresistor or photocell), (2) allow the current to flow in a circuit (i.e., photodiode), or (3) change the flowing current amplitude based on the intensity of the received signal (i.e., phototransistor). Those photoelectric sensors can have either (1) an opposite-side design (the transmitting element emits a light toward the receiving element physically at the opposite side, and the signal is interrupted when an object passes between them), (2) a same-side (retroreflective) design (the transmitter and receiver elements are on the same side, but a mirror is used to reflect the emitted signal back), or (3) a diffuse design, where similar to the retroreflective design, the signal is returned back to a wide surface rather than to a single point of interest. Phototransistors can be used in distance (linear position) measurement applications. Photodiodes are suitable for binary applications where the photointerruption or circuit breaker activation/deactivation occurs when an object passes between the emitter and the detector. Photoresistors are interferometer linear-type sensors that can be categorized according to the light intensity. It is required to differentiate the ambient light from the emitter infrared light by using a commutator switching the emitted signal into a high-frequency signal that is decoded at the receiver side. Those sensors offer a high resolution (μm) and wide range (up to several meters). Their typical applications include manufacturing part counters and door access and security systems.

- *Resistive-based position sensors,* such as rotary potentiometers or rheostats, where conductive materials connect two electrical terminal contacts. A wiper rotates between the two ends of the resistance material such that each contact position of the wiper corresponds to a specific total resistance of the circuitry, and consequently to the observed voltage. It is possible to have single- and multiturn rotary potentiometers.

- *Magnetic position sensors and detectors,* using the Hall effect voltage principle, where a conductor carrying a current crosses a magnetic field and induces a small voltage proportional to the distance between the sensor and the object, given by

$$V_h = \frac{R_h I B}{t} \tag{5.2}$$

with V_h, R_h, I, t, B being, respectively, the Hall voltage, the material-dependent Hall coefficient, the current, the element thickness, and the magnetic flux density. Hall effect magnetic sensors are sensitive to vibration and are mainly used within security systems, position, and speed robot guide path, as well as motor commutation. Position measurement using the Hall effect offers a high resolution and a smaller sensing range (from 0.2 to 5 mm). There are also Hall effect detectors that turn ON and OFF if the intensity of the magnetic field exceeds the threshold value. Those devices are mainly used to detect the presence or absence of a moving part, or a particular position or reference position when combined with an incremental encoder.

- *Inductive position sensor,* where the variation of the inductance is proportional to the gap distance between the metallic object and the magnetic field and generates a probe oscillator voltage variation. Here, an induced eddy current in the metal object surface causes a secondary magnetic field interfering with the probe magnetic field. Those devices are used for the detection of metal objects over short distances.

- *Optical, ultrasonic, or microwave distance-measuring sensors*, using the principle of wave propagation (absorption, reflection, and passing) over air or a liquid medium. As such, those devices can be (1) radio based or radar (using time-of-flight [TOF], frequency modulation, and phase shift measurement techniques), (2) laser light based (using time-of-flight or phase shift measurement techniques), or (3) sound based or sonar (using time-of-flight measurement techniques). However, those devices are sensitive to temperature and target obstruction. Table 5.4 summarizes the distance formula using the time-of-flight or frequency technique based on the propagation time (or elapsed time) of the signal (ultrasonic, radiofrequency, or optical energy) from the reference to the target point.

Among other reflective sensing techniques, there is *interferometry*, which is based on the returning signal intensity. Here, an emitted laser light is split into two different paths (one directed to the detector giving a flight path of fixed length and another directed toward a retroreflector attached to the moving target). The estimation of the relative travel wavelength of both signal paths gives the distance to the moving target. Those sensors are sensitive to vibration, object size, shape, reflectivity, speed and directions of motion, and so forth. Typical applications include maritime guidance and referencing systems, robotics, object detection (collision avoidance), security systems (sensing presence or intrusion), and system diagnostics. There are also *microwave position sensors*, where the emitted microwave frequency signal reflects on the target and returns back to the sensor, thus indicating an object's presence or absence within the radiofrequency field. Similar to microwave position sensors, there are *ultrasonic position sensors*, where the emitted signal is ultrasonic and capable of distance measurement from 1 cm up to a few meters within an accuracy of 1%. Figure 5.3 illustrates the ultrasound-based distance measurement principle for automatic car parking.

TABLE 5.4

Distance Formula Using Time-of-Flight Technique

Measurement Setup	Distance Formula
Emitter and receiver with same location (same sensor)	$distance = \dfrac{TOF}{2}.speed\ of\ signal$
Emitter and receiver with different locations (receiver attached to targeted object)	$distance = TOF\ .speed\ of\ signal$

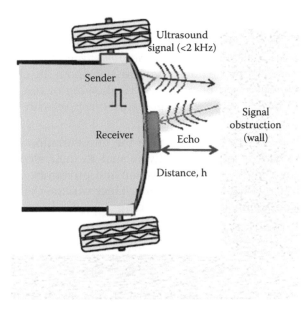

FIGURE 5.3
Ultrasound-based distance measurement operation for car parking.

In the phase shift measurement technique, a frequency modulation unit at the receiver side is used to derive the frequency or the time shift with the emitted signal such that the distance to the object is given by

$$distance = \frac{(f(t) - f(t-T)) \times speed\ of\ light}{4 f_{modulated} f_{deviaton}} \tag{5.3}$$

where $f_{modulated}$ is the signal-modulated frequency, $f_{deviation}$ is the total modulated frequency deviation, and $f(t), f(t-T)$ are, respectively, the emitted and received signal frequencies. A typical ultrasonic signal is above a frequency of 16 kHz. The ultrasound travel time to the target and back to its emitter source corresponds to the distance to the target. This is given by

$$h = \frac{Ct}{2} \tag{5.4}$$

with C, t being, respectively, ultrasound speed on the media and signal travel time from sender to receiver.

The ultrasound signal is sensitive to the ambient temperature and humidity. Those contactless sensors are suitable for applications such as the measurement of contaminated fluid levels in tanks and distances up to 10 m, regardless of target shape and material.

- *Position measurement sensors using electromechanical switches* (e.g., limit switch), which provide binary information about the closed or open status, when pressed or not pressed, allowing or not allowing the current to flow. Each switch has at least one pole with two points of contact.

- *Capacitance distance sensors*, where the gap variation between two metallic plates is proportionally related to their electrical capacitance, thus allowing a linear position and proximity measurement. As illustrated in Figure 5.4, the capacitance variation of those passive and noncontacting sensors is given by

$$C = \frac{q}{V} = k\epsilon_0 \frac{A}{d} \tag{5.5}$$

with C being the capacitance, q the charge (in C), V the potential difference between the two plates (in V), ϵ_0 the dielectric constant, d the permeability constant, A the effective plate area (in m^2), and d the plate separation (in m). This capacitive sensor delivers a high-resolution position measurement. However, it is highly sensitive to temperature and humidity variation. Those contactless devices are typically applied to detect a metallic object's presence or absence and for liquid or powder with a dielectric material constant higher than that of air.

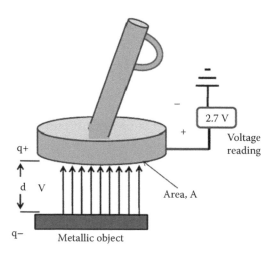

FIGURE 5.4
Capacitor-based distance measurement principle.

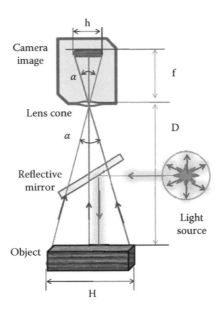

FIGURE 5.5
Measurement principle using a vision system.

- *Distance measurement sensors using a computer vision system* to gather an object photograph and process it by deriving its physical characteristics, such as dimension, position, appearance (e.g., shape and color), and specific signs (brand name). Figure 5.5 illustrates how the computer vision system is used in the distance measurement of an object. Depending on the light positioning, the distance is given by

$$D = \frac{fH}{h} \tag{5.6}$$

- *Position measurement sensors using the force of gravity* through mercury conductivity measurement to close or open a switch. However, such sensors are sensitive to temperature.
- *Position measurement sensors using the alternating current (AC) inductive (magnetic) principle,* which can be described as an LVDT consisting of two coil layers (a primary internal one and a secondary external one) wrapped around a ferromagnetic tube in the measurement of a surface finishing process, depicted in Figure 5.6. The second coil sliding around the inner magnetic core is used for linear displacement measurement from a few millimeters up to 1 m with micron-level accuracy. The AC voltage input signal applied to the primary winding coil induces the magnetic field to generate a current in the two external coils. Subsequently, the core drags the

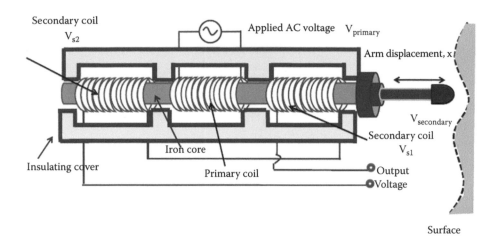

FIGURE 5.6
LVDT measurement principle in surface finishing probe.

magnetic field toward it, and a current is generated in the left-hand coil. Conceptually, the secondary outer winding coils have an equal number of turns. The axial displacement of the iron-based magnetic core within the tube cylinder generated a variation of the magnetic flux linking the primary coil to each external one. The magnitude and phase of the AC applied to the excitation secondary coil are compared with the detected signal from the pickup primary coil. Any deviation is related to the position of the ferromagnetic tube and can be captured through a synchronous detection electronic circuitry. From Faraday's law, the electromagnetic induction is given by

$$V = -N\frac{d\phi_B}{dt} = -N\frac{d(BA)}{dt} \tag{5.7}$$

where V represents the induced voltage, N the number of turns for the coil, ϕ_B the magnetic flux, B the magnetic field, and A the cross-sectional area of the coil. A current variation produces a magnetic field and consequently changes the orientation of the source.

The position of the core is proportional to the level of the output signal voltage. This is only valid for small motion corresponding to any motion around the center. Otherwise, the relationship is nonlinear. The secondary coil's connection in series opposition allows us to measure and set the voltage difference between the two coils to zero at the center core position. The windings are designed to allow a linear relationship between the core position and the output voltage, such that the primary winding voltage is given by

$$V_{primary} = V \sin \omega t \qquad (5.8)$$

The voltage difference results in

$$V_{secondary} = k\,Vx \sin \omega t \qquad (5.9)$$

where k is a constant and x is the core linear position. By its value and sign, the level of output voltage indicates the magnitude of the induced winding voltage differences, as well its in- or out-phase status with the primary coil's applied signal. It is given by

$$V_{Output} = k\,Vx \qquad (5.10)$$

LVDT sensors capable of measuring up to a few micrometers offer more accurate measurements than linear potentiometers and are applied to quality control. While they are less sensitive to friction, their measurement displays a small residual voltage at the output when the coil is in the central position due to small variations in the windings.

- *Position measurement sensors using magnetostrictive time.* Such sensor components consist of a moving magnet (target) and a magneto-strictive wire. Here, a current pulse passes through the magnetostrictive wire over the acting magnet's field and generates an ultrasonic pulse in the wire. The position of the magnet is related to the time delay between the current pulse and the detected ultrasonic pulse.

- *Position measurement sensors using a potentiometer,* which consist of a resistive element sliding along a wiper, as depicted in Figure 5.7. When a voltage is applied across this resistive element, the resulting output AC voltage magnitude is found linearly proportional to the position of the brush over the resistive element. Although such sensors have the same operating principle as the LVDT, potentiometers do not require signal conditioners. They measure the absolute rotational position and can rotate without limits (multiple 360°), making the wiper jump from one end of the resistor to the other.

For linear displacement measurement, x, the output voltage can be derived from

$$V_0 = kVx \qquad (5.11)$$

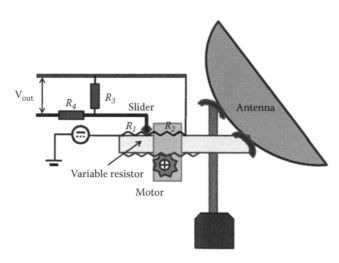

FIGURE 5.7
LMP electric circuitry for antenna positioning system.

For rotary displacement measurement, θ, the output voltage would be

$$V_0 = kV\theta \tag{5.12}$$

A linear motion potentiometer (LMP) consists of (1) a wiper attached to a mobile object whose position is monitored, (2) a potentiometer whose resistance is related to the displacement of the wiper, and (3) a direct current (DC) voltage source, as illustrated in Figure 5.7. Using the voltage division properties, the relationship between the input voltage signal, V_{signal}, and the output voltage signal, V_{out}, can be captured by the equation

$$V_{out} = \frac{R_0 + \dfrac{R_2}{R_L}}{R_1 + \dfrac{R_2}{R_L}} V_{cc} \tag{5.13}$$

This linear relationship expresses that to every position, there is a given value of electrical output voltage. Table 5.5 summarizes some characteristics of a typical LMP.

LMP-based displacement measurement is limited by the friction during the wiper displacement, as well as the speed level of the moving object due to the contact mechanism. Furthermore, the resistance value of the potentiometer is sensitive to temperature variations. Typical applications of LMPs include (1) use as a solenoid positioning sensor for automated controllers, (2) satellite dish positioning, (3) automatic valve positioning, and (4) robotics.

TABLE 5.5

Example of LMP Sensor Characteristics

Power supply	5–40 VDC
Dynamic range	0-15.621 mm
Repeatability	0.5–1 mm error
Accuracy	0.02%–15%
Sensitivity	4 mm/V
Resolution	Infinite (<0.01 mm)
Hysteresis	0.02%
Operating temperature	−65° to 105°

Note: VDC, volts DC.

5.2.5.2 Angular Position Measurement: Resolver and Optical Encoder

Angular position measurement is usually achieved through resolver and optical encoders. Optical encoders convert the motion into an electrical pulse train by using the photointerrupter principle. Such generated electrical pulses can then be counted or "decoded" by the circuitry. Hence, the components of a rotary optical encoder consist of a light emitting diode (LED), a light guidance and shaping device (optical mirror), a photosensitive receiver, and a coded rotating wheel (disk) directly in contact with the motor controller through its shaft. The series of transparent (binary level 1) or opaque (binary level 0) angular or linear slots on the surface of the disk are designed according to Gray code binary number patterns. The light emitted by the LEDs through the transparent zones of the disk is detected by photodetectors (photodiodes). The disk rotation induces a succession of square-wave signals corresponding to the encoder output signal. This output signal is normally a binary or Gray code number. This code is then converted into the displacement measurement within the decoder through a pulse counting device. The number of pulses per revolution is directly related to the number of slots on the disk or a multiple thereof.

There are two basic configurations of optical encoders: (1) incremental encoders where generated pulses are incremented or decremented according to the angular position of a motor in order to monitor its displacement and (2) absolute position encoders that allow estimation of the shaft angle position with precision. These two families of encoders include variants such as absolute multiturn encoders and tacho-encoders of processing data to provide speed information. Overall, the average velocity can be derived using the time between pulses.

With *incremental optical position encoders*, an initial angular position is arbitrarily set to zero and the direction of the rotating object attached to the coded disk is derived by using two 90° out-of-phase sinusoidal shape-like signals (channels A and B) generated from photodetectors. The two channels, A and B, which can be divided into several intervals with equal angles, are alternatively opaque and transparent. Here the number of division corresponds to the resolution or number of periods. The light beam passing through a transparent slot activates two out-of-phase photodetectors, which subsequently generate a pulse-like signal that will be squared as follows: if the track A signal and track B signal are above the reference signal, then the square circuit signal is set to 1; otherwise, it is set to 0. The signal decoding is done through the pulses' counter device. Here, the reference point can be lost when the power supply is temporally interrupted. The motion direction is given by the phase relationship between the A and B pulse trains, that is, the indication of which signal leads the other. The number of slots and the disk circumference define the encoder resolution. The total count (pulse) defines the total distance.

The direction of motion of the motor shaft can be derived from the relative 90° ($\Delta\theta/4$) phase shift between signals A and B. The opaque area arrangements on the disk of the incremental optical linear and angle encoders are illustrated in Figure 5.8 for an inkjet printer motor positioning system and a rotational motor position, respectively. This has the advantage of simplifying the encoder because only two photodetectors are required. However, an external counter hardware has to be used in order to develop an absolute position measurement from the output of the incremental encoder. The initial or reference position is given through a Z signal in comparison with signals A and B. It is designed by a transparent slot in the gray coded disk, and after each revolution, it allows us to reset the motor position as depicted in Figure 5.9a and b. The angular spacing of the opaque areas is $\Delta\theta$.

Angular position measurement can also be performed with a counter hardware. At each transition in A and B, a pulse is generated and sent to the counter. The binary output of the counter can be as many bits as needed to represent the entire range of motion required. Each pulse received during a positive rotation causes the binary counter output to increase by one (count up). Provided that the counter is properly initialized so that its output is zero at the desired initial angular position, it gives an absolute reading of the angular position. Similarly, a separate set of pulses can be generated for a negative rotation and sent to the counter. In this case, each pulse causes the counter output to be decremented by one (count down) along the bidirectional encoder angular position. The output of the counter changes by 4 for an angular rotation of $\Delta\theta$, so the effective resolution is $\Delta\theta/4$. Incremental optical angle encoders are available with 100,000 or more opaque areas or "lines" on the disk. However, the dynamic response of the photodetectors limits the rotational velocity of the sensor. If the maximum frequency at which the photodetectors can operate is known, the maximum rotational velocity of the encoder can be found from

$$\omega_{max} = \frac{f_{max}\Delta\theta}{6} \ in \ rpm \tag{5.14}$$

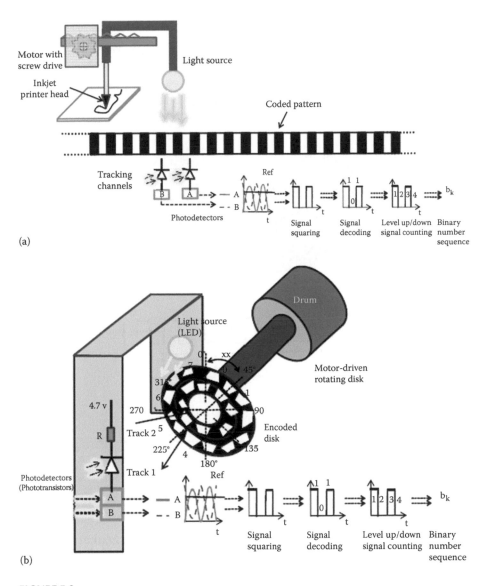

FIGURE 5.8
(a and b) Principle of incremental encoder with two tracks.

where f_{max} is the photodetector cutoff frequency (in Hz) and $\Delta\theta$ is the incremental angle. Figure 5.10 illustrates some binary signals generated by an incremental encoder.

The accuracy of such an angular measurement device depends on the latch-based pulse counting circuit, which very sensitive to unpredictable noise interference. Hence, in order to detect possible measurement errors, it is suitable to continuously compare generated signals A, B, and Z.

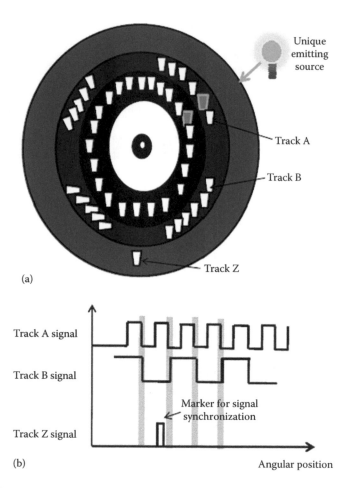

FIGURE 5.9
(a and b) Detection principle for direction of rotation and synchronization.

Such a technique also allows regeneration of the original signal in the case of counting error occurrence. Figure 5.11a and b summarizes the mitigation of encoder signal interference.

Absolute rotary encoders are similar to incremental position encoders, except that the disk is designed around three, rather than two, tracks and arranged such that the direction of rotation is embedded with the angular position signal, which corresponds to a unique binary word (a binary Gray code) that increments or decrements accordingly. For example, an increment from 7 to 8 in decimal corresponds to a change from 0100 to 1100 in Gray code. Hence, those encoders can produce the actual position image of a moving part by producing a set of logic-level signals representing a binary word proportional to its angular position. Figure 5.12 illustrates an absolute optical angle encoder with 3 bits of binary position output. The angular position encoder does not require DAC.

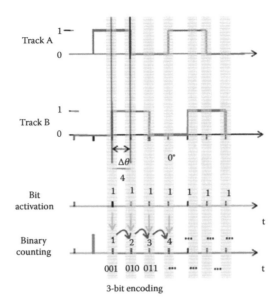

FIGURE 5.10
Binary signals generated by incremental encoder.

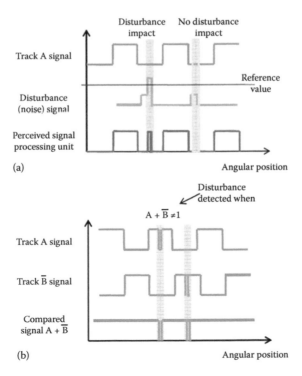

FIGURE 5.11
(a) Case of a counting error on encoder channel A. (b) Case of a noncounting interference signal.

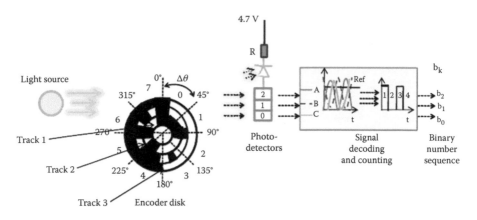

FIGURE 5.12
Principle of absolute rotary encoding.

Each bit is generated by one photodetector according to the arrangement of opaque areas on the disk. As shown in Figure 5.12, the half opaque and half transparent track gives information on the direction of motion, while the others provide information on the angular position of a quarter of the disk motor shaft. The higher is the number of windows per ring, the higher is the encoder resolution, which varies from two to thousands of windows per ring. Collectively, these bits generate the binary code for any angular position, as shown in Table 5.6.

At boundaries between some regions on the disk, more than one bit changes simultaneously in the binary output, in order to avoid large errors in position measurement. For example, when the disk rotates from 3 to 4, the binary output changes from 011_1 to 100_2 through intermediate combinations, such as 111_2, if they are not synchronized.

Figure 5.13 depicts a signal encoding example. It could be the number of parallel outputs that correspond to the number of bits or tracks on a disk. Also, the code generated by an absolute encoder can be either natural binary (pure

TABLE 5.6

Example of Binary Codes Generated by Absolute Rotary Encoder

n	θ [deg]	b_1	b_2	b_3
0	0–45	0	0	0
1	45–90	0	0	1
2	90–135	0	1	0
3	135–180	0	1	1
4	180–225	1	0	0
5	225–270	1	0	1
6	270–315	1	1	0
7	315–360	1	1	1

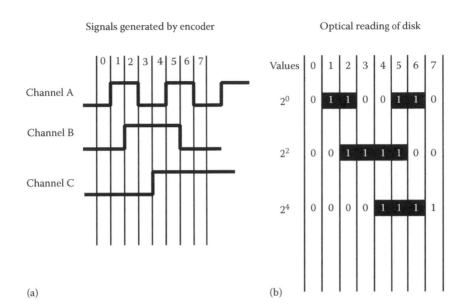

FIGURE 5.13
(a) Bit-based optical reading of disk. (b) Binary signal-generated encoder.

binary) or reflected binary (Gray binary). Overall, with the absolute encoder, the derived code is repeated after one revolution, and several geared encoders can be used to increase the resolution or range of angular position sensing. However, absolute angle encoders are useless when a large number of output bits are required to meet range and resolution requirements. Hence, usually incremental optical angle encoders are used for position sensing applications, while absolute encoders are applied to movement and position monitoring.

Linear optical encoders are similar to optical angular encoders, with the coded wheel (disk) replaced by a coded bar and a slider that carries the optical and electronic components. The distance measurement is operated along the bar, allowing a resolution from a few micrometers to a maximum reading scale of about 3 m at an operating speed of about 600 mm/s. There are also *angular resolvers*, which are shaft-oriented angular sensors based on the AC magnetic principle. Here, two coils (primary and secondary) provide sinusoidal-like outputs that are compared.

5.2.5.3 Velocity Measurement

The direct measurement of velocity can be performed using analog transducers, such as

- *Magnet-and-coil velocity* sensing devices, which are designed using the Faraday law of induction. Here, the magnet is surrounded by a coil with n turns of wire such that the voltage generated in the coil is a function of the linear velocity of the magnet. Those sensors are suitable for a displacement of less than 0.5 m and above 10 mm.

- *Tachometer AC (or DC) generators*, which are attached to a rotating shaft such that its rotation induces a voltage directly proportional to its angular velocity. If ω is the armature speed and k_t is the tachometer gain, the DC commutating output voltage, V_0, is

$$V_0 = k_t \omega \qquad (5.15)$$

Note that this generator voltage device and its hooked armature produce a noise over the DC output of the tachometer with a fundamental frequency as a function of the angular velocity. These tachometer generators are commonly used for speed measurements above 5000 rpm. This technique introduces some drag into the system, and it is not appropriate for motion tracking applications.

- *Counter-type velocity* sensors, which consist of counting generated pulse signals during a set time and estimating the velocity as the ratio of the resulting count per time unit.

- *Linear velocity transducers* (LVTs), which are inductive devices consisting of a rod called the core (a permanent moving magnet) and two electrical coils connected in series opposition, as illustrated in Figure 5.14. Based on Lorentz's law, the core moving through the coils generates a current flow through the wire, which is proportional to the magnet velocity transducer. The voltage generated by the transducer is given by

$$V_0 = Blv \qquad (5.16)$$

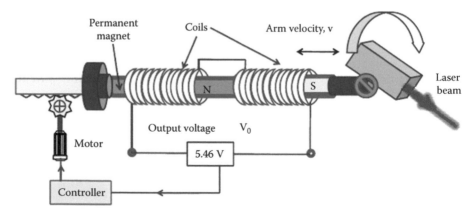

FIGURE 5.14
Linear velocity transducer device for laser-based cutting process.

with B being the component of the flux density normal to the velocity, l the length of the conductor, and v the magnet linear velocity. The magnet can be directly mounted on a moving mechanical element without electrical connections. The output voltage, V_0, must be converted from analog to digital.

Using any position sensor, the velocity measurement can be estimated through *digital differential position algorithms*, such as the Euler backward approximation:

$$\omega_i = \frac{\theta_i - \theta_{i-1}}{T} \tag{5.17}$$

where ω_i is the average velocity (rather than the instantaneous angular velocity), θ_i is the current position, θ_{i-1} is the previously sampled position, and T is the sampling interval. It is suitable to use a trapezoidal approximation for a refined velocity estimation. Note that the velocity resolution is highly dependent on the sampling period T and on the position resolution. These sensors' operating interval is around 50 mm.

5.2.5.4 Acceleration Measurement

Acceleration measurement is derived either through

- Digital differential velocity algorithms, in the case of angular motion.
- Accelerometers based on the Newton principle, in the case of linear motion. This is achieved through
  - The displacement measurement of a seismic mass moving freely within a cage surrounded by a spring–damper support structure with its relative motion being recorded and converted into voltage levels. Such devices are called electromechanical accelerometers. Here, each motion of this structure generates a magnetic force that is opposed to and proportional to its displacement from the zero position. This force depends on the length of the conductor within the field, the density of the magnetic field, and the value of the angle between the conductor and the magnetic field.
  - The force–balance principle applied to the core of an electromagnet, such as the high-speed train accelerometer illustrated in Figure 5.15
  - Using one electrode of a piezoelectric crystal as the vertical vibration sensor (accelerometer), illustrated in Figure 5.16. Such devices are called piezoelectric accelerometers due to the charge generated when force is applied on the crystal induced from an acceleration.

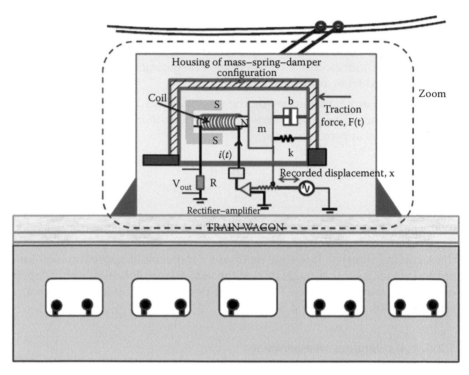

FIGURE 5.15
Accelerometer measurement device mounted in high-speed train.

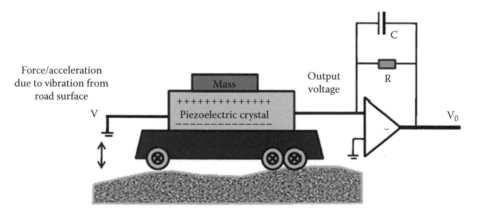

FIGURE 5.16
Piezoelectric-based accelerometer circuit for vertical vibration measurement.

- Using the plate of a differential capacitor as the seismic element. For seismic accelerometers based on resistive, capacitive, inductive, or piezoelectric material, the acceleration motion of a specific mass is captured through its resulting force, $F(t)$, given by

$$F(t) = m\frac{d^2x(t)}{dt^2} + b\frac{dx(t)}{dt} + kx(t) \qquad (5.18)$$

where m is the mass, b the damping ratio, and k the spring constant of the mass–damper–spring apparatus. Hence, the cage apparatus dynamics equation relating $x(t)$ the mass motion, $y(t)$ to the train motion, and α to the angle between the gravity and the motion axis is

$$m\frac{d^2(y(t)-x(t))}{dt^2} + b\frac{d(y(t)-x(t))}{dt} + k(y(t)-x(t)) = mg\cos\alpha - m\frac{d^2x(t)}{dt^2} \qquad (5.19)$$

Note that the natural frequency of the fixed mass bounds the frequency application interval. In addition, such a sensory system is highly sensitive to environmental conditions (temperature, dust, magnetic field effects, humidity, etc.). Typical applications of accelerometers are to reduce the system vibration and noise by adjusting speeds or to actively eliminate chatter.

For *piezoelectric accelerometers*, the piezoelectric material is compressed or shear stressed over the cross-sectional area, A, of the material crystal during any motion (vibration) of the apparatus. Its motion is proportional to the electric charge given by

$$q(t) = k_i m\frac{d^2x(t)}{dt^2} \qquad (5.20)$$

with k_i being the piezoelectric coefficient of the piezoelectric materials (quarts, ceramic crystals, etc.). It is possible to estimate the deformation in the crystal and the resulting voltage, V, across it, such as

$$V = k\frac{F}{A} \qquad (5.21)$$

where k is the piezoelectric voltage constant of the crystal. Applying Newton's law, the acceleration of the mass, a, is a function of the voltage produced across the crystal, such as

$$a = V\frac{A}{kM} \qquad (5.22)$$

The voltage produced by the piezoelectric accelerometer is detected by a charge amplifier that is proportional to the acceleration. Those accelerometers are sensitive to temperature variations causing nonlinearity in the measurement, and angular acceleration causing rotation-induced errors. Typical applications include vibration detection and navigation system, gyroscope, or vehicle acceleration monitoring.

Other accelerometers are strain gauge piezoelectric accelerometers where the piezoelectric crystal material is connected to the mass apparatus. Here, during any motion of the apparatus, a change in the length of the strain gauge causes a variation of the conductors' resistivity, ρ, such as

$$\frac{dR/R}{dL/L} = 1 + 2\delta + \frac{d\rho/\rho}{dL/L} \tag{5.23}$$

where $\dfrac{d\rho/\rho}{dL/L}$ indicates the resistance change due to piezoresistivity. There are also electrostatic accelerometers that consist of an electrode pivoting around fixed electrodes. Hence, based on Coulomb's law, any variation of the gap between the fixed electrodes and the pivoting electrode produces an electrical signal.

5.2.5.5 Force Measurement

A force is any action on an object (or a subject). It is considered dynamic when it results in a deformation and/or a displacement, and is considered static when it does not result in an accelerating motion. Among the methods of measuring a force there are those using

- *Levers* to compare the applied weight force with the displacement of a known mass attached over a spring as the counterweight. This displacement is measured using a gyroscope, a position sensor, or a velocity sensor, sometimes within a simple spring balance scale.

- *Accelerometers*, when the unknown force is balanced by the gravitational force of a known mass such that its acceleration yields

$$a_c = \frac{F}{m} \tag{5.24}$$

The unknown force can also be balanced by an electromagnetically developed force such that $F = F_{em}$, with F_{em} being the electromagnetical force at equilibrium. Hence, converting the force to a fluid pressure and measuring that pressure yields

$$P = \frac{F}{A} \qquad (5.25)$$

where P is the pressure to be measured, and F is the acting force on an object surface, A.

- The correspondence, defined by Hooke's law, between the deflection of an elastic object (strain gauge, piezoelectric crystal, or force sensing resistor) and the applied static force. The strain gauge length increases inversely proportional to the area of the cross section causing an increase of its resistance, as depicted in Figure 5.17. Hence, the resistance is given by the relationship between the resulting strain gauge length and its cross-sectional area, which is given by

$$R = \frac{V}{I} = \rho \frac{L}{A} = \rho \frac{L}{wt} \qquad (5.26)$$

with R, l, V, A, t, w, ρ, being, respectively, the resistance of wire, the length, the voltage, the cross-sectional area of the conductor, the thickness, the width, and the resistivity of the material. Commonly encountered strain gauges are (1) wire strain gauges, (2) foil strain gauges, and (3) semiconductor strain gauges. Usually, those sensors transform any change in resistance into a small voltage scale through

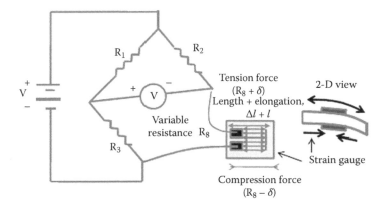

FIGURE 5.17
Quarter bridge strain gauge circuit.

a Wheatstone bridge circuit. For example, this could be achieved by connecting a wire strain gauge to a leave of electric sensitive material (e.g., Teflon). The sensitivity of a material to the strain or gauge factor (K_{GF}) is given by

$$K_{GF} = \frac{\Delta R/R}{\Delta l/l} \tag{5.27}$$

with ΔR being the variation in the initial resistance in Ω, R the initial resistance in Ω, Δl the elongation in m, and l the initial length in m. The strain gauge is highly sensitive to temperature.

Piezoelectric materials relate elastic crystal deformation to electrical potential. Hence, among the piezoelectric materials used to measure the applied force there are quartz, Rochelle salt, lithium sulfate, and barium titanate ceramic.

- *Capacitive force transducers*, where the membrane small elastic deflection from the applied force induces a variation of capacitance, which is converted into a voltage through an electronic circuit. This could cause either a variation of distance, surface area, or dielectric. Here, in the case of the sliding motion of two plates separated by an air gap, the capacitance of the sensor is given by

$$C(x) = \varepsilon_0 x \frac{A}{d} \tag{5.28}$$

with C being the capacitance, ε_0 the permittivity of the dielectric, x the relative overlapping or plate gap positioning of the dielectric, A the overlapping area for two plates, and d the gap between the two plates. Hence, based on the voltage, v(t), the force could be given by

$$F(x) = \frac{1}{2} \frac{C(x)}{dx} v(t)^2 \tag{5.29}$$

Typical applications of force sensors include multiple-axis robotic handling, manufacturing assembly operations, and impact shock forces in automotive crashes.

- *Magnetoresistive force sensors*, where temperature-sensitive sensors change their resistivity within a magnetic field (such as Bismuth).
- *Tactile force sensors*, where the pressure distribution across the closed-space array of the robot arm fingertip is measured. Typical industrial applications of such sensors are manufacturing, robot handling, or the surface identification of a product shape.

5.2.5.6 Torque Sensor

A torque is the action causing a momentum around a rotational axis. Thus, in a torque sensor using a shaft made of ferromagnetic material, the torque can be derived from the variation in magnetic properties caused by the strain stress. Such sensors generate voltage output signals that are magnetoelastically or magnetoresistive dependent from either solid–solid or liquid–solid interactions. Among those sensors there are torquemeters (noncontact strain gauges) or torkducers (magnetoelastic transducers).

5.2.5.7 Flow Measurement

Gas, liquid, or mixture of gas–liquid can be sensed in terms of mass or volumetric flow rate and flow direction (turbulent or laminar, compressible or not, free of particles or nonhomogenous, and time varying). Commonly encountered devices that measure fluid flow rate are called flowmeters. They are designed from various measurement approaches, such as (1) the fixed-point fluid flow rate measurement at a pipe cross section, (2) the average fluid flow rate measurement at a pipe cross section, and (3) the measurement of the mass flow rate or volume of fluid flowing through a pipe. The flow velocity profile is related to the fluid instrumentation design (pipes, valves, etc.) and to the fluid properties (laminar, turbulent, etc.), which are characterized by the Reynolds number. Hence, the selection of a flowmeter requires knowledge of the fluid type (gas, solid, or liquid), its properties (viscosity, density, corrosive, etc.), the fluid piping and other installation conditions (pipe diameter, pipe length, Reynolds number, valve type and characteristics, etc.), the dynamic characteristics expected (dynamic response, accuracy, etc.), and the operating conditions (temperature, pressure, etc.). Typical encountered flowmeters can be classified into **contact-based sensors** and **contact-free sensors**.
Contact-based sensors include

- *Differential pressure flowmeters*, which consist of deriving a differential measurement of the flow from at least two points along the fluid flow direction. The sensing device configurations include venture; an orifice plate, as illustrated in Figure 5.18; a flow nozzle; an elbow; and a pitot-static tube. Such flowmeters are suitable for flow rate measurement of low-viscosity fluids.
- *Positive displacement flowmeters*, which consist of deriving the flow rate proportional to a shaft rotation. Such a sensing device configuration includes a rotameter (variable-area in-line flowmeter) or a turbine meter, a cylinder or a plug, and a variable aperture.
- *Turbine flowmeters*, which consist of measuring the fluid flow through their rotor speed measurement. Hence, the magnetic-based sensor converts the average rotor speed into voltage levels proportional to the fluid average velocity or the volumetric rate in a pipe.

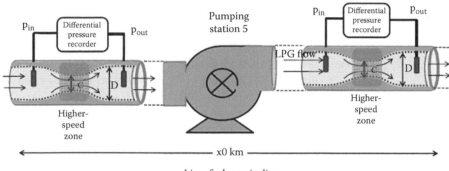

FIGURE 5.18
Displacement pressure flowmeter with LPG flow rate within a pipeline.

The sensing device configuration includes an impeller, an axial turbine, and a propeller.

- *Mass flowmeters*, which are fluid flow rate measurement devices based on the heat transfer principle, where an electrical heating element raises the temperature of the fluid flowing through a pipe proportional to the applied voltage. Hence, the quantity of voltage required to maintain a constant fluid temperature can be measured using a thermal anemometer. Such flowmeters are suitable for slow-motion fluids going through large pipes.

Contact-free sensors include

- *Electromagnetic flowmeters*, where a fluid behaving as a conductor moves along a pipe containing electrodes. This produces a magnetic field across its section wall. The fluid is ionized such that its motion induces a voltage proportional to the fluid mass flow rate, which can be measured by the electrodes, based on Faraday's law. Electromagnetic flowmeters (pulsed DC magnetic, AC magnetic, etc.) are suitable for low-speed flows of liquids.
- *Ultrasonic flowmeters*, which are fluid flow rate measurement devices based on the Doppler effect, where the propagation and reflection of the signal are proportional to the medium characteristics (temperature, viscosity, fluid flow rate, etc.). All other characteristics being constant, the transmission delay is directly proportional to the flow rate. These flowmeters are suitable for clean (particle-free) subsonic fluids, as small particles could disturb the signal.
- *Displacement pressure flowmeters*, which consist of measuring a fluid's (e.g., liquefied petroleum gas [LPG]) flow rate through the pressure

difference across an orifice, as illustrated in Figure 5.18, such that the volumetric flow rate is given by

$$Q = \frac{C}{\sqrt{1-\beta^4}} \, \varepsilon \frac{\pi}{4} d^2 \sqrt{\frac{2(p_{in} - p_{out})}{\rho}}$$

(5.30)

with p_{in}, p_{out} being the downstream and upstream pressures of the orifice plate, ρ the density of the fluid upstream of the orifice plate, d the downstream hole diameter of the orifice plate, and β the ratio d/D, with D being the upstream internal pipe diameter.

5.2.5.8 Pressure Measurement

Considering pressure as a force acting on a surface, it is possible to measure it as a force per unit area. The pressure measurement is usually achieved within a chamber and can lead to (1) an absolute pressure when the chamber containing the fluid is sealed at 0 Pa, (2) a gauge pressure when this chamber is vented to the atmosphere, (3) a differential pressure when another pressure is acting on this chamber, or (4) a sealed pressure when this chamber is sealed at a given pressure (different from 0 Pa). Usually, fluid pressure transducers consist of elastic diaphragms dividing two chambers (the working chamber and the reference chamber). The measurement is given by the difference in pressure between the two chambers. Pressure transducers transform a device motion or deformation proportional to an applied pressure.

Commonly used transducers are (1) the Bourdon tube, where an elliptical metal tube becomes deformed when there is a differential between the inside and the outside pressure; (2) the Bellows tube, which is a metal tube that is flexible under applied pressure; and (3) the diaphragm, which is a circular plate that is elastic under pressure. As summarized in Figure 5.19,

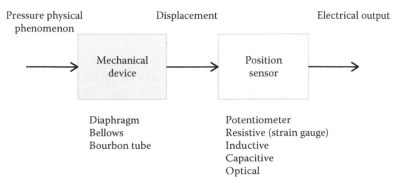

FIGURE 5.19
Pressure measurement principle.

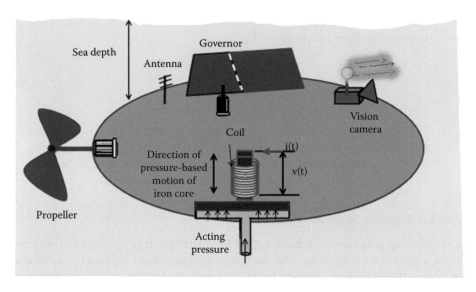

FIGURE 5.20
Baffle-based pressure measurement within an unmanned underwater vehicle.

some of these sensors can used a resistance-based strain gauge or the outer walls as a differential capacitor to derive the corresponding electrical signal output. As illustrated in Figure 5.20, a baffle-based pressure measurement device is embedded into an unmanned underwater vehicle.

Other types of pressure sensors are

- Measurement devices using a variable capacitance pressure, which consists of changing a physical property of one of the variables (e.g., distance, d). This varies the capacitance, C, as depicted in Figure 5.21 in the case of a water dam. The capacitance is given by

$$C = \frac{KS(N-1)}{d} \tag{5.31}$$

  with K, S, N and d being, respectively, the dielectric constant of the material between the plates, the area of a plate side, the number of plates, and the distance between two adjacent plates.

- Variable resistance pressure measurement devices capable of detecting large resistance variations. These devices usually operate in potentiometer circuits or bridge circuits (strain gauge transducers).

- Measurement devices using variable inductance pressure, which operates on the principle that the voltage drop across a coil generates a magnetic field proportional to the rate of current variation.

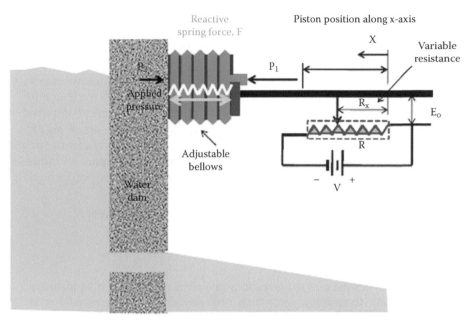

FIGURE 5.21
Capacitance pressure sensor based on a Bellows transducer for a water dam.

- Measurement devices using piezoelectric pressure are based on the material property capable of generating an electric potential once it is subjected to an applied strain force. Some materials exhibiting such characteristics are quartz, Rochelle salt, and ammonium dihydrogen phosphate.

5.2.5.9 Liquid-Level Measurement

Liquid-level sensors can be either contact or contact-free and ensure continuous or threshold value measurements. Typical liquid-level sensors include

- Classic *float-type liquid-level sensors*, which measure either continuous or discrete points of a liquid level.
- *Hydrostatic pressure liquid-level sensors*, where the weight of the liquid in the tank is proportional to the differential pressure between the surface and the bottom.
- *Electrical capacitance liquid-level sensor*, where a capacitor is a metallic bar within a cylindrical measurement vessel that determines the level of liquid in a tank with respect to the liquid level in vessel walls. Such a sensor is usually combined with signal-conditioning devices for applications such as average fuel-level indicators.

Commonly encountered contact-free liquid-level sensors include

- *Capacitive proximity switches,* which are used for point measurement of a liquid level.
- *Ultrasound sensors,* which consist of liquid-level point measurements, where an attenuated ultrasound energy signal sent from a transmitter over the liquid is collected by the receiving unit and compared with the reference signal when the gap is filled with air. A signal-conditioning circuit is usually required. Typical applications of such sensors are seawater depth measurement and fish locators on industrial fishing boats using echo-ranging transducers, or powder and grain level measurement in brewery silos.
- *Radiofrequency sensors.* Radiofrequency using TOF principle.
- *Electro-optical-based sensors.*

Example 5.1

Here, it is desired to evaluate the flow rate of a corrosive fluid in a pipeline. For safety compliance, contactless measurement devices are preferred. Typical measurement methods are summarized in Figure 5.22.

Hence, selecting the appropriate sensing method consists of choosing among the following candidates:

1. Differential pressure in two pipeline points
2. Rotor speed of attached turbine-based measurement
3. Electromagnetic-based measurement
4. Ultrasound Doppler effect–based measurement

For this pipeline, the ultrasonic method appears to be the most suitable and cost-effective, as it uses a contact-free device, as required.

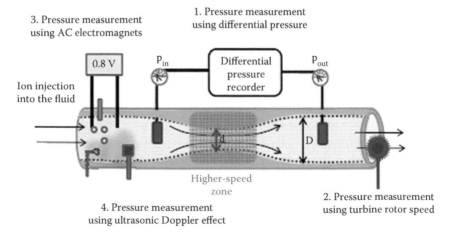

FIGURE 5.22
Example of four types of pipe-based fluid flow rate measurement.

5.2.5.10 Temperature Measurement

Temperature can be measured through direct contact or contact-free devices. Among contact-based temperature measurement instruments there are

- *Thermistors*, which are devices with an internal resistance that varies with temperature. Unfortunately, due to their nonlinear relationship, a current is required to reduce their sensitivity.
- *Thermostats*, which are thermistors integrated with a potential divider to improve the temperature measurement. Here, the temperature variation is inversely proportional to the resistance of the thermistor change.
- *Resistance temperature detectors* (RTDs), which are similar to thermistors connected to a Wheatstone bridge. Here the temperature material can be either copper, nickel, or platinum.
- *Thermocouples* consisting of two thermo-electric-sensitive wires connected in a circuit. The two wires are made of different materials that induce a voltage any time they are in contact. Those sensors offer measurement within 1% accuracy.

Contact-free thermal measurement devices include thermal imagers and radiation thermometers.

5.2.5.11 Radiofrequency-Based Level Measurement

RF-level measurement is based on the capacitance measurement principle. Here, the device consists of an electrical capacitance located between two conductors normally positioned at a specific distance from each other. The medium between the two conductors is insulated, meaning that the level measurement does not involve a conducting material. The capacitance is given by

$$C = E\frac{KA}{d} \tag{5.32}$$

with C, E, K, A, and d being, respectively, the capacitance in picofarads (pF), the free space permittivity, the insulating material dielectric constant, the conductor's effective area, and the distance between the conductors. The value of K is directly proportional to the system charge storage capacity. Such device is also suitable for level measurement at more than one discrete point within a tank.

5.2.5.12 Smart Sensors and Nanosensors

Smart sensors and nanosensors are usually low-scale devices based on electrostatic, piezoresistive, piezoelectric, and electromagnetic sensing principles.

These sensors are made of different materials, such as silicon or polymer. More specifically, there are

- *Optic fiber–based smart sensors* (magnetorestrictive, piezoelectric, optic fiber, etc.), which use the proportionality between the intensity or spectrum of their material (glass and silica) physical characteristics and the variations in the sensed system strain, force, temperature liquid level, vibration, and so forth.

- *Electrostatic or capacitive-based smart sensors*, which consist of transducing a force, vibration, pressure, or temperature variation into the displacement of a membrane or beam. This displacement then induces variations of the membrane capacitance. These electrostatic sensors are used as accelerometers, sound sensors, gyroscopes, and pressure and even tactile sensors.

- *Piezoresistive-based smart sensors*, which transduce the applied mechanical strain or deformation into a magnitude electrical field, through variations of the resistive material dimension. Typical materials used include alloy and silicon. Some applications include piezoresistive or capacitive-based micropressure sensors where the diaphragm deformation is transduced into pressure differences, or using polymer-based material to monitor the high temperature of an internal combustion engine. Other examples of applications are accelerometers used in vibration measurements or in vehicle anticollision systems.

- *Thermal-based smart sensing devices*, which produce heat from a current flowing through a resistive element over an ambient fluid. Hence, when the fluid with lower temperature passes through the device, by heat convection, it reduces the temperature of the heated element. The temperature variation of the element provides the flowmeter (anemometer) with information on the fluid direction and its flow rate. The mechanical momentum transfer principle is an alternative flow sensing technique.

- *Light-based smart sensors*, which use phototransistors, photodiodes, or photoresistors such that their exposition to light intensity is proportional to their resistance variation, which is then converted into voltage change. These sensors can also be used as photodetectors that are activated by a thermal light source, such as diode, quartz halogen, fluorescent, or daylight. Such a light source is capable of producing various electromagnetic signal wavelengths: 30 nm for ultraviolet (invisible light), between 400 and 700 nm for visible light, and up to 0.3 mm for infrared light. Here, the absorbed radiation capacity of the detector is not proportional to the emitted signal wavelength (i.e., pyroeletric). Typical applications of such light sensors include use in fire detectors, infrared-based intruder alarms, and object shape capture.

- *Image sensors*, which consist of an image recording component combined with image data processing software. First, the image generation and image encoding are achieved through a light irradiance source of the image. Then, the image is projected over a plane on which the charge of the image spatial distribution is presented. The charge of the image is accumulated, stored, and converted into a voltage signal. Here, a charge sample is called a picture element or pixel. Color sensors are achieved by placing color filters over the individual pixels. Typical applications include use in machine vision detecting systems in manufacturing operations or public event surveillance, as well as robotic guidance.

- *Radiant-based detectors*, in which the absorbed photons' capacity is proportional to the induced mobility of electrons (i.e., electrical charge), causing change in their resistance values (i.e., photoresistors) or change in current or voltage values (i.e., photodiodes or phototransistors). Using the semiconductor functioning principle, photodetection is the conversion of the thermal excitation of electrons into the current. This signal is proportional to light irradiance and is sensitive to temperature. Typical applications of phototransistors include coin detectors in vending machines, level sensors, and proximity sensors.

Other customized measurement devices are microelectromechanical-micromechatronic systems and nanoelectromechanical-nanomechatronic systems.

5.2.6 Binary Instruments

Binary instruments as bistable devices are used to detect two-state physical phenomena or energized and de-energized actuators. This can be done either by direct contact or by proximity between the detector and the object. Examples of methods used to detect physical phenomena are (1) the inductive proximity to detect if a metal object is nearby, (2) the capacitive proximity to detect if a dielectric object is nearby, (3) the optical presence to detect if an object passes through a light beam, and (4) the mechanical contact applied to the detection of a contact force between an object and a switch. Typical binary instruments are

- Sinking and sourcing switches, which switch the current flow ON or OFF. When the current flows out of the device, it is called a sourcing switch.
- Plain switches, which switch the voltage ON or OFF.
- Solid-state relays, which switch AC outputs.
- Transistor–transistor logic (TTL), which signals logic levels from 0 to 5 V applied.

Typical applications of such binary devices include use in material handling operations, the detection of a vehicle's presence in car parking, the detection of a gas or liquid level, an object's shape and position using infrared-based shape recognition tool, and the tracking of a tagged object using the radiofrequency identification (RFID) tool.

5.2.6.1 Binary Device: Electromechanical Limit Switches

The electromechanical limit switch is a detection device made of an actuator and electrical contact components. The detection is performed through direct contact with a moving entity. The resulting electrical signal is sent to the processing system. These limit switches are commonly encountered in automated systems.

5.2.6.2 Binary Device: Photoelectric Sensors

The photoelectric sensors detect various opaque to nearly transparent entities. Here, a LED generates infrared light signals. The generated signals allow the current to flow through a photodiode or a phototransistor. The comparison of the current signal with a reference signal results in a binary value indicating whether the presence of the object or subject is detected. There are several factors affecting the performance of these detection systems, such as (1) the distance between the sensor and the targeted object or subject; (2) the characteristics of the targeted object or subject, especially in terms of material reflectivity, transparency, color, and size; and (3) environment conditions, such as the ambient light or background. Such devices are used within security access management for automatic doors or barriers.

5.2.6.3 Binary Device: RFID-Based Tracking and Detection

RFID is a set of automatic identification devices that use protocol-based radio wave signals exchanged between a targeted subject or an object embedded within an electronic tag and a remote reader (transponder) component. Initially, an electromagnetic signal is sent from the transponder to the tag antenna, where the flowing current charges the integrated capacitor. When the capacitor threshold voltage value is reached, it releases back to the transponder, over the tag coils, a time-varying amplitude radio-based signal corresponding to the encoded electronic tag information. The transponder demodulates and stores this encoded signal. Figure 5.23 summarizes the RFID operating principle.

There are three types of tags: (1) passive tags, which produce surface acoustic waves upon reception of a request signal from the transponder; (2) semipassive tags with embedded capacitive diodes that enable the production of a small current; and (3) active tags with an integrated circuitry capable of data processing. In either type of RFID communication protocol,

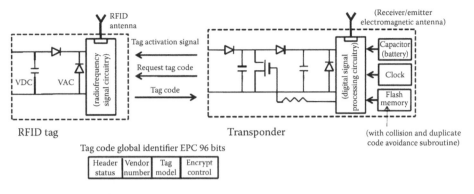

FIGURE 5.23
RFID detection principle.

TABLE 5.7

RFID Signal Properties and Characteristics

Frequency Range	Suitable Operating Conditions	Application
Less than 135 kHz	Signal propagation attenuation acceptable in water and metal	Animal identification
13.5 MHz	Standard installation	Book tracking Contactless payment
850–950 MHz	Long-range remote data transfer	Large-scale supply chain Data warehouse
Above 2.45 GHz	High data rate	Tracking object device in motion (vehicle)

tag talk first (TTF) or interrogator talk first (ITF), it is suitable to involve a subroutine, to avoid tag data collision when the reading of simultaneous tag codes is required. The tag code standard is defined by the global identifier Electronic Product Code (EPC) (96 bits). RFID detection technologies are suitable for counting and tracking various products in dirty environments and counting various products remotely. The low cost of RFID tags offers a opportunity to replace traditional barcodes by product tracking in several sectors, especially in the e-commerce supply chain, along with some new applications, including security and payment. Some radio wave frequencies define the distance coverage and determine RFID possible applications, as summarized in Table 5.7.

5.2.6.4 Binary Device: Pressure Switches

Pressure switches consist of a diaphragm, a piston, or a Bellows tube, which activates actuating electrical contacts mechanically when pressure-driven action is applied to them. This enables the generation of an electrical signal proportional to the applied pressure level.

5.3 Signal Conditioning

Active transducers are energy converters without requiring external excitation to operate. This is the case, for example, for a thermocouple in temperature measurement, in contrast to passive transducers, which change the electrical property of a device (capacitance, resistance, etc.) according to magnitude variations of the measured physical phenomenon. For example, with strain gauges, the resistivity varies proportionally to the applied stress, and with LVDTs, the inductance changes proportionally to the resulting displacement. Usually, transducer output signals require a signal-conditioning unit, in order to be properly read by other control devices. Thus, the signal conditioner consists of processing a transducer output by demodulating, amplifying, filtering, linearizing, range quantizing, and isolating the signal from noise, before transmitting it to the comparator in a suitable form. The functions performed by signal conditioners can be described by

- Scaling (amplification or attenuation), in order to use a reasonable portion of the analog-to-digital (A/D) range and to avoid the A/D saturation effect
- Filtering (noise elimination or isolation), in order to ensure a better restitution of the initial signal received from analog filters or digital filters
- Phase shifting or reference shifting, in order to ensure proper wave shaping
- Signal buffering, in order to capture useful output signal content within an analog input board
- Bridge balancing, in order to calibrate output signals or detect signal traffic overloads
- Demodulation through mathematical manipulation of the signals, such as summation, squaring, and integration
- Linearization of the nonlinear relationship to the magnitude of the measured physical phenomenon

In summary, signal conditioning is achieved by variation of the input signal's amplitude and frequency. Such signal conditioner output could be directly used by the ADC unit. Table 5.8 summarizes the functions required from typical signal conditioners for some measurement systems.

Typical transducers require external voltage or current excitation from the signal-conditioning unit, such as strain gauges, thermistors, and RTDs. The connection between the analog measurement device and the A/D device can be single ended (to avoid offset) or differential (to measure signals with a large voltage range). In order to minimize the effects on the automation

TABLE 5.8

Types of Transducers and Associated Signal-Conditioning Functions Required

Temperature	Amplification
	Cold junction
	Filtering
Strain gauge	Voltage excitation
	Filtering
	Linearization
	Bridge configuration
	Excitation
High voltage	Isolation
Absorption transducer	Offset removal
	Amplification

system and the process performance, the selection of a signal conditioner should take into account the sample rate through (1) the sensor response speed and range, (2) the conversion time, and (3) the frequency of sample signal itself.

Example 5.2

An 1 mV/°C resolution of a thermometer output signal requires a minimum of 2.5°C before a 12-bit A/D device indicates a change in digital value. Hence, by amplifying the output signal of a thermometer by 1000×, each bit on the A/D device is equivalent to 0.001 mV, which corresponds to 0.0025°C, thus avoiding A/D device saturation. If the temperature variations are measured with 1000× amplification, 22°C is measured as 22 V, which is much greater than the A/D input range. Therefore, it is possible to adjust the output of the amplifier with an offset of −22 V such that the output temperature would be 0 V.

5.4 Signal Conversion Technology

Typical signal conversion includes DC to AC, AC to DC, frequency to voltage, voltage to frequency, D/A, and A/D. In this section, only the two last conversion types are covered.

5.4.1 Digital-to-Analog Conversion

A DAC consists of a resistive ladder network connecting switches within an analog circuitry, and an operational amplifier receiving an equivalent set of input digital values to produce a corresponding analog voltage signal. Depending on the binary-driven switches' position (logic 1/closed or logic 0/open), the current flows across specific resistances within this ladder network. The operational amplifier is used to sum all voltage potentials across the set of those specific resistances and to generate an output voltage. For example, using operational amplifiers and D flip-flop, a set of switches' position equivalent to a binary word is converted into an output voltage signal. Figure 5.24 depicts an 8-bit unsigned binary word DAC. Consider an 8-bit binary word given by $b_0 b_1 b_2 b_3 b_4 b_5 b_6 b_7$ such that $b_i \{1, 0\}$ is sent out of the computing unit toward the D flip-flop through its I/O address buses. Then, the D flip-flop I/O data bus collects this binary number. Depending on the binary word, the switches' status is closed or open, allowing the current to pass through their connected resistor. This generates voltages $v_0 v_1 v_2 v_3 v_4 v_5 v_6 v_7$ of the D flip-flops. These voltage outputs of D flip-flop are fed into the summing operational amplifier. The resistor values on the operational amplifier are chosen such that their output voltage is given by

$$V_{out} = v_h \left(\frac{R_B}{R/2^{k-1}} b_{k-1} + \frac{R_B}{R/2^{k-2}} b_{k-2} + \ldots + \frac{R_B}{R/2} b_1 + \frac{R_B}{R} b_0 \right) \qquad (5.33)$$

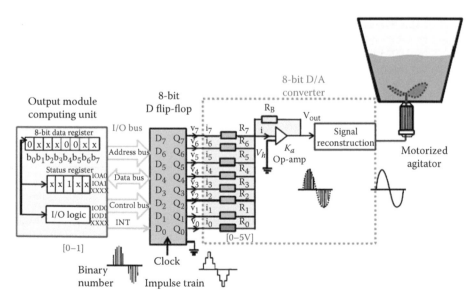

FIGURE 5.24
Example of an 8-bit resistive ladder DAC.

with $v_h = v_0 = v_1 = v_2 = v_3 = v_4 = v_5 = v_6 = v_7$ being the logic 1 output voltage of the D flip-flops. Equation 5.33 can be rewritten as

$$V_{Out} = v_h \frac{R_B}{R} N \tag{5.34}$$

with the binary number output to DAC given from $N = b_{k-1} 2^{k-1} + b_{k-2} 2^{k-2} + \ldots + b_1 2^1 + b_0 2^0$.

In addition, the gain of the DAC is $K_u = \frac{R_B}{R}$, with $R = R_0 = R_1 = R_2 = R_3 = R_4 = R_5 = R_6 = R_7$ being any resistor connected at the output of the D flip-flop. The voltage levels produced by the DAC are given by the counting numbers derived from the number of binary bits generated. In this case, b_0 is the most significant bit (MSB) and b_7 is the least significant bit (LSB).

Example 5.3

As shown in Figure 5.25, a resistive ladder network is used as a 4-bit DAC. The binary input is sent continuously across four separate data lines, v_4 through v_7, which results into the targeted discrete value of the DAC output voltage.

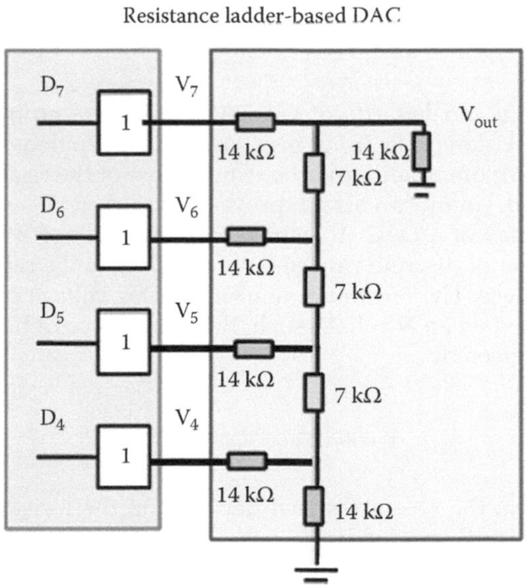

Resistance ladder-based DAC

FIGURE 5.25
Resistive ladder of a DAC.

The DAC sampling rate is the number of points per second that a DAC can generate as output analog values (up to a million samples per second). Among the characteristics of a DAC there are

- *Range*, which is the voltage interval of values that the DAC can output, given by $[V_{min}, V_{max}]$. The maximum output voltage, V_{max}, is produced when all bits are 1 ($N = 2^k - 1$) and is given by

$$V_{max} = V_h \frac{R_f}{R}(2^k - 1)$$

(5.35)

The minimum output voltage, V_{min}, is achieved when all bits are zeros ($N = 0$) and it results in $V_{min} = 0$. When a signed analog output signal is obtained from the DAC, the number of point areas derived for k bits is estimated by inverting the sign bit and adding a bias voltage of $-V_h (R_f/R) 2^{k-1}$ to the output v_o. Here, the interval is given by

$$V_{max} = V_h \frac{R_f}{R}\left(2^{k-1} - 1\right)$$

(5.36)

and

$$V_{min} = -V_h \frac{R_f}{R}\left(2^{k-1}\right)$$

(5.37)

- *Accuracy* and *linearity* of the DAC, which are proportional to the precision of the input resistors on the operational amplifier. Furthermore, stabilizing the temperature of the resistor ladder network circuit can also improve the accuracy.
- *Resolution* of a DAC, V_q, which is the number of k digital bits (number of discrete values) corresponding to the range of analog values. The minimum nonzero analog voltage produced is obtained when $N = 1$. As such, the resolution as a binary number is given by

$$resolutions/combinations = 2k$$

(5.38)

Then, the resolution that depends on the resistor network ladder configuration is given by

$$V_q = V_h \frac{R_f}{R}$$

(5.39)

When the resolution and the range are known, the lowest number of bits required in the DAC is given by

$$k = \frac{\log\left(\frac{V_{max} - V_{min}}{V_q} + 1\right)}{\log 2} \qquad (5.40)$$

with k being an integer or resulting from Equation 5.40 rounded up to the next highest integer. An ADC resolution of 8 bits would allow an encoded analog input in $2^8 = 256$ different levels. This corresponds to 0–255 or 128–127 ranges for unsigned integers and signed integers, respectively, accordingly to the application. It is also possible to express the resolution in volts. In the case of encoder pulse signal counting, the resolution is given by

$$\bar{\omega}_{res} = \frac{\theta_{res}}{T} \qquad (5.41)$$

while in the case of clock pulse signal counting, it is given by

$$\omega_{ccp} = \frac{\theta_{res}}{N_x T_{CLK}} \qquad (5.42)$$

Then, the resolution yields

$$\bar{\omega}_{res} = \omega_{ccp_{pos+}} - \omega_{ccp_{pos-}} = \frac{\theta_{res}}{N_x T_x} - \frac{\theta_{res}}{N_{x \mp 1} T_x} \qquad (5.43)$$

Consider a sampling rate, T; the average velocity sensed (or weighted moving average) is given by

$$\bar{\omega} = \frac{\theta(k) - \theta(k-1)}{T} \qquad (5.44)$$

By increasing the sampling period T, $\bar{\omega}$ appears to be closer to ω with a degraded resolution. This may cause *quantization*. However, using clock pulse counting, the resolution yields

$$\bar{\omega}_{res} = \omega_{pulse} = last \frac{\theta_{pulse}}{T} \qquad (5.45)$$

- *Settling time (slew time)*, which is the conversion time delay of the input DAC binary word within the expected accuracy voltage level.
- *Offset error*, which is the deviation of the real value from the expected DAC output analog signal and when the digital value of 0 is applied. The *gain error* is the linear deviation from the ideal transfer line of a DAC, which corresponds to a change of slope from the ideal converter.

5.4.2 Analog-to-Digital Conversion

An ADC is a device that samples an analog input voltage and encodes it into a binary number (1 or 0). ADC is implemented through using either an ADC software conversion process or an ADC hardware conversion process. In the case of hardware conversion, the voltage-level input signal is the reference voltage. Hence, a two-step converter consists of a sampling and holding (S/H) unit, along with a quantizing and encoding (Q/E) unit, as illustrated in Figure 5.26. The quantization consists of the partitioning of the reference voltage signal range into several discrete quanta, and then the input signal is matched to the corresponding quantum. During encoding, each quantum is assigned a specific digital code, which is then allocated to the input signal.

The ADC allows the computer to proportionally sample the analog feedback voltage from the transducers into a digital voltage. This is achieved through a sample and hold circuitry, used by the flip-flop memory. The ADC operates by using a comparator between the input voltage, V_i, and a variable digital voltage, V_0. The time from V_o to V_i is the conversion time. The main characteristics of an ADC are the speed and resolution of the conversions.

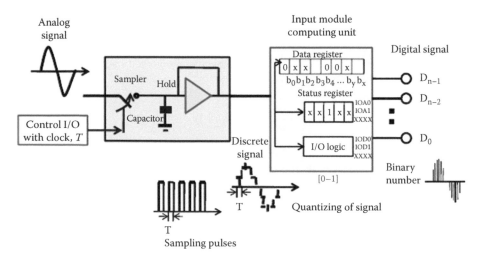

FIGURE 5.26
Analog-to-digital conversion technique.

The ADC *resolution* is the smallest detectable change in analog voltage signal resulting in a variation in the digital output, given by

$$\Delta V = \frac{V_r}{2^N} = \frac{V_{RH} - V_{RL}}{2^n} \qquad (5.46)$$

with V_r being the reference voltage range, N the number of bits in binary output, 2^N the number of possible states, n the number of digits, V_{RH} the higher value of input voltage, and V_{RL} the lower value of input range. The resolution represents the discretization of the error related to the conversion of the signal. For example, consider a 6-bit ADC corresponding to $2^6 = 64$ different values such that in a 1 V scale voltmeter with 0.000001 V steps, it could measure from −0.999999 to 0.999999 V. The resolution of an ADC is degraded by an increase in temperatures and resistances within its components. A voltage divider or amplifier is used to fit the input signal without changing the resolution. The *input range* of an ADC is the interval of voltage values that are properly converted. The ratio of the highest to the lowest voltage values converted by an ADC is called the *dynamic range* and is expressed in decibels (dB) by

$$Dynamic\ range = 20\log\left(\frac{S}{N}\right) \qquad (5.47)$$

with S being the highest voltage level and N the noise level. The *rate of accuracy* of an ADC is the smallest resolution of analog voltage (LSB). The A/D conversion consists of a successive S/H process of a chain of digital values (bits), which is achieved by a switch and a capacitor. The minimum sample rate is obtained when the capacitor is discharged (no value held), while the maximum sample rate is derived from the speed at which the capacitor is charged. The time required to commute the sample switch is called aperture time or delay (t_a). Any deviation between the analog voltage value and the current voltage value input while the switch is opening causes an error in accuracy. The maximum input frequency, $f_{MAX\ LSB}$, is given by

$$f_{MAX\ LSB} = \frac{2}{2\pi t_a 2^n} \qquad (5.48)$$

The accuracy of A/D conversion can be improved either by increasing the resolution or by increasing the sampling rate, which causes an increase in the maximum frequency that can be measured. A pulse signal from a clock is used to synchronize the timing of the ADC conversion process.

Commonly encountered ADCs are ramp, successive approximation, dual slope, and parallel or "flash."

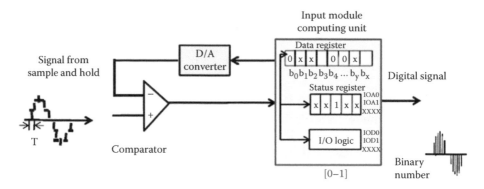

FIGURE 5.27
Successive approximate A/D conversion circuitry.

Digital ramp ADC generates a toothed voltage signal that raises (ramp-like) before dropping to zero. The timing is set ON at the beginning of the ramp segment of the voltage signal until it reaches its initial value. At that stage, the comparator signal output value changes and resets the timer, as well as assigns the corresponding analog voltage value as a digital value.

Successive approximation ADC uses a counter circuit commonly referred to as a successive approximation register (SAR). It is an *n*-bit DAC iterating a series of levels and comparing them to the input voltage, as depicted in Figure 5.27. Within this register, all bits are counted from the most significant to the LSB, instead of being counted in binary sequence. As the successive approximation ADC converges on the closest input voltage value, its processing time is shorter than the digital ramp ADC. Such an *n*-bit ADC requires 2^n cycles to perform a conversion.

Example 5.4

It is desired to convert a 0–12 V range command input signal into an 8-bit ADC, or $2^8 = 256$ binary counts (or values) from 0 to 255, with a count increment of 1/256. The 0000 0000 binary number is 0 V corresponding to the 0 value; similarly, 6 V corresponds to 128 count values, and the binary number 1111 1111 for 11.95 V corresponds to 255 count values. A 10-bit ADC offers greater resolution, as there are $2^{10} = 1024$ binary counts, which is equivalent to a count increment of 1/1024. Hence, the case of a 0–5 V range command input signal with a 10-bit ADC corresponds to a resolution of 4.88 mV per LSB, while for an 8-bit ADC it would be 19.5 mV.

Example 5.5

Consider an 8-bit resolution A/D device used in the measurement of a 0–10 V signal. The input signal can be measured in steps of $10/2^8 = 39$ mV (i.e., the ADC can only detect changes greater than 0.039 V). Thus, a 10 V analog input corresponds to the digital number 255 and an 0 V analog input corresponds to 0. Any input signal increase or decrease causes a change by 1 from the previous number (i.e., 9.961 V is digitally represented by 254). A 12-bit A/D device is more sensitive to variations in the input voltage since its minimum resolution is given by 10 V/4096 = 0.00244 V. If the input signal exceeds 10 V, the A/D produces a flat line corresponding to the saturation from the analog data acquisition device. Another ADC device is the *dual-slope ADC* (DS-ADC), which consists of an amplifier for integration, a signal comparator, a counter, a clock, some controlled switches, and a capacitor, as illustrated in Figure 5.28.

The ADC derives the voltage by measuring time and using digital logic to compute the input voltage. Initially, switch S_1 is set to ground, S_2 is closed, the counter is set at 0, and the capacitor is discharged. Then the conversion starts, S_2 is open, and S_1 is set such that the input to the integrator is V_{in}. At the end of the time delay T_{d1}, the input is applied across the capacitor and starts to charge for a period of T_{d1}. After T_{d1}, the capacitor commutes to a negative reference voltage, V_{ref}, and the discharge rate is proportional to the reference voltage. The timer is started when S_1 is set such that the counter begins to count clock pulses for a period of T_{d1}, after which the counter resets to zero. A comparator is used to determine when the output voltage of the integrator crosses zero.

An advantage of a DS-ADC over the single-slope one is that errors in component values cannot affect the conversion result, as they are compensated during the integrate and de-integrate phases. This method is insensitive to noise.

In the case of *flash ADCs* with n bits of resolution, there are $(2^n - 1)$ comparators, (2^n) resistors, and a control logic unit, as depicted in Figure 5.29. The 2^n resistors divide the reference voltage into 2^n equal intervals (i.e., one for each comparator), which correspond to the input range, while the

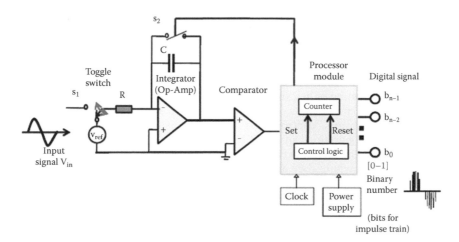

FIGURE 5.28
DS-ADC circuitry.

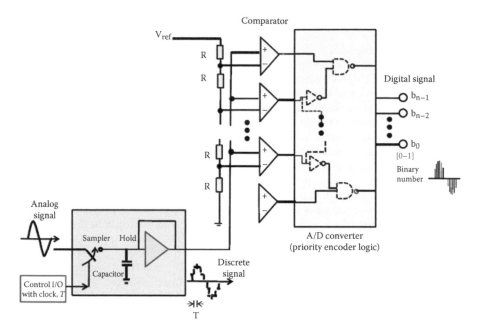

FIGURE 5.29
Flash A/D conversion circuitry.

$(2^N - 1)$ comparators indicate the voltage intervals corresponding to the input voltage, V_{in}, within these 2^N intervals. Then, V_{in} is compared with each voltage level simultaneously such that the comparator can set the output to 1 for all voltages lower than the input voltage and to 0 otherwise. The resulting chain of digital values is fed into an encoder with combinational logic to translate comparator outputs into corresponding n-bit values. The flash ADC conversion time is bounded by the comparators' settling time and the combinational logic propagation time.

Pipeline ADC or multistage ADC is similar to the flash converter, with fewer comparisons. It uses a coarse conversion such that the difference with the input signal is estimated by a DAC. At the last step, all results are combined. This type of ADC is faster and provides higher resolution while requesting only a small die size.

The main components of a *delta-encoded ADC* are the resistors, capacitors, comparators, control logic, and DAC, as shown in Figure 5.30. During its operation, an input signal is oversampled and goes to an integrator. The integrated output signal and the DAC are channeled to a comparator, where they are compared with the ground. The counter is iteratively adjusted by the comparator through negative feedback, running until the DAC's output signal approximates the input signal. This delta modulator adjustment of the output signal ensures that the average error at the quantizer output is zero. After iteration, it results in a digital signal output whose sequence of binary numbers is proportional to V_{in}.

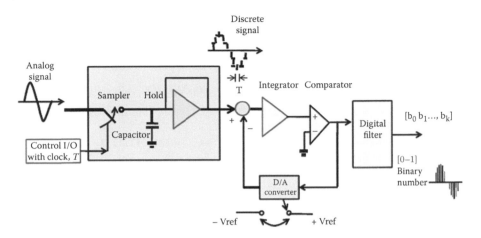

FIGURE 5.30
Delta-encoded ADC block diagram.

In a delta-encoded ADC, the converter is fed by an up–down counter. This device has a high resolution and very wide ranges and does not require precision components. The conversion time varies according to the input signal level. Furthermore, it requires a high sampling rate.

Some ADCs successfully combine the delta and successive approximation approaches in the case of a signal with high frequencies and low magnitude.

Sigma-delta (SD) ADC is a high-sampling-rate 1-bit ADC that combines a low-pass filter (to cut or shape higher-frequency electronic noise) with a DAC in a feedback loop. The resulting converted 1-bit sequence is digitally filtered through a low-pass filter to improve resolution and reduce the data flow rate.

Example 5.6

Consider an n-bit ADC receiving from a speed sensor an analog voltage signal of $y_a(t)$ in the range [Ymin, Ymax], and convert it into a digital signal of y_d number of bits, such as

$$y_d \approx b_{n-1}b_{n-2}b_{n-3}\ldots b_1 b_0$$

where each bit b_i value can be 0 or 1 and the last one is the LSB. In case 2, it corresponds to

$$y_d \approx b_{n-1}2^{n-1} + b_{n-2}2^{n-2} + \ldots + b_1 2^1 + b_0 2^0$$

Example 5.7

Consider a 12-bit ADC used to convert a signal in the [0 V, 10 V] interval. The digital signal is given by

$$y_d = Y_{min} = 0.2^{11} + 0.2^{10} + \ldots + 0.2^1 + 0.2^0 = 000000000000_2 = 0_2 = 0\ (decimal)$$

$$y_d = Y_{max} = 1.2^{11} + 1.2^{10} + \ldots + 1.2^1 + 1.2^0 = 111111111111_2$$
$$= 10000000000000_2 - 1 = 2^{12} - 1 = 4095\ (decimal)$$

The resolution is given by

$$y_d = \frac{Y_{max} - Y_{min}}{Number\ of\ intervals} = \frac{Y_{max} - Y_{min}}{2^n - 1} = \frac{10V - 0V}{4095} = 2.44\ mV$$

5.5 Data Logging and Processing

5.5.1 Computer Bus Structure and Applications

Besides digital-to-analog (D/A) conversion and signal conditioning, knowledge of the computing unit architecture and interface to process equipment and sensing devices is highly valuable for the process control system design. Indeed, the computing unit is either sending instructions encapsulated into voltage signals toward the process actuating devices, or receiving binary encoded signals from the sensing devices. Such instructions or encoded signals can handle a set of bits corresponding to I/O field device connections or addresses. The exchange of instructions and process data occurs between the computing unit components. The three main components of the computing unit architecture are the processor, the memory, and the bus structure, as depicted in Figure 5.31. While the processor component is responsible for the execution of program instructions using data stored in the memory component, similarly, the bus structure ensures the communication links for data exchanged between the processor, the memory, and the I/O devices. This is achieved through the storage or execution of interrelated I/O logic-level data or programs. More specifically, there are synchronous buses using a clock-based protocol to activate and deactivate exchanges, and asynchronous buses using a handshake-based protocol to activate and deactivate exchanges. Furthermore, during process data and instruction transfer between the computing unit components and the field devices through the bus structure, there are communication lines or function-specific buses: address, data, and control. The address line specifies which I/O (with a

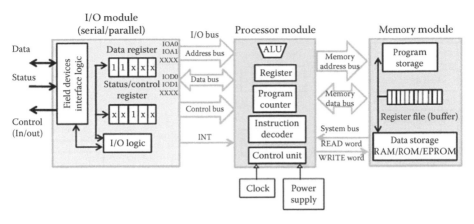

FIGURE 5.31
Computer data and instruction addressing structure.

unique address) device communicates with the computing unit at a given time. It consists of a number of separate signals, each specifying 1 bit in the address of the I/O device to be accessed by the processor. Data I/O devices are transferred to or from the computing unit using a data line based on IN and OUT signals representing write or read requests on the control line. As presented in Table 5.9, IN and OUT signals are used to indicate when the addresses and data specified by the I/O address bus and the I/O data bus are valid. The I/O data bus is bidirectional, depending on whether it is an input or output operation performed by the processor. The number of lines in the I/O data bus is usually the same as the computing unit word size (e.g., 8, 16, 24, 32, 64, or more bits). Note that the bus creates a bottleneck, limiting the I/O data throughput.

The I/O device interfacing mechanism can be captured through the memory register or the buffer mapping activities, including synchronizing signals, system clocks, interrupt lines, and status lines. For example, in order to alert the processor when an I/O operation is complete, interrupt circuitry and programs are used. As such, READ and WRITE instructions from and toward I/O devices are executed using techniques for I/O service operations, including programmed I/O (polling), interrupt drive I/O, or direct memory

TABLE 5.9

Logic-Level Signals of I/O Data Bus

IN Data Bus	OUT Data Bus	Description
0	0	No read or write operation
0	1	Execute write operation to specific address
1	0	Execute read operation from specific address
1	1	Undefined

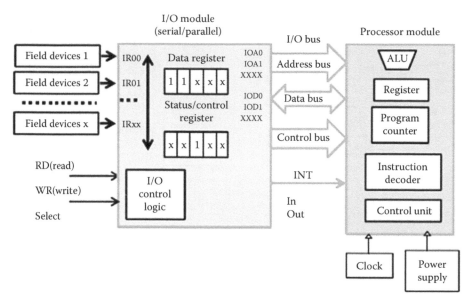

FIGURE 5.32
I/O field devices interfacing with computing unit.

access. With the programmed I/O technique, a program routine requests an I/O service operation by periodically checking the status of the I/O devices. In the case of the interrupt-driven technique, an interrupt communicates with the processor through a dedicated control bus based on an interrupt service routine program for enabling or disabling the active signal from or toward field devices. This allows the processor to interrupt normal program execution momentarily in order to process I/O service operation instead of waiting for the completion of an I/O operation, as is done with the polling of the I/O port. The direct memory access technique uses a specialized processor to ensure the execution of I/O service operations. Figure 5.32 illustrates a generic I/O interface for an x-bit computing unit.

Example 5.8

Consider push-button (PB) and limit switch (SW) input devices used to activate a LED by checking their status (open or closed) through the computing unit, as illustrated in Figure 5.33. From the execution of the READ instruction, the LED can be activated or deactivated when the ID_0 bit 1 or 0 is loaded over the data bus (bus driver) onto the lowest bit of a register in the processor. Then, after the status is checked, the computing unit interface generates a SELECT signal to transfer the ID_0 bit data toward the I/O data bus and the address of the LED interface on the I/O address bus. The processor can set the OUT signal to logic 1 or 0, indicating a change of the LED status through variation of both signals (OUT

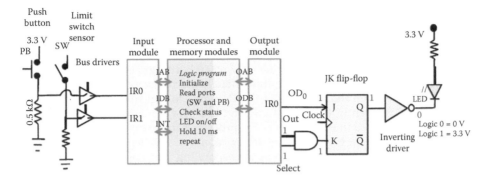

FIGURE 5.33
Computing unit interface for the LED activation.

and SELECT signals). This causes a JK flip-flop to sample and hold the level (0 or 1) of the OD_0 at the output Q connected to the RS-232 driver. If the logic level at the output of the driver connected to the LED is 1 (3.3 V), the current through the LED is zero, as there is no voltage differential across it, and the LED is OFF. Otherwise, the LED is ON, with the power source and the resistance defining the LED brightness.

Consider another example of a pneumatic-driven cylinder whose travel is bounded by two limit switches, LS3 and LS4, as depicted in Figure 5.34. The solenoid activation causes the pneumatic piston rod.

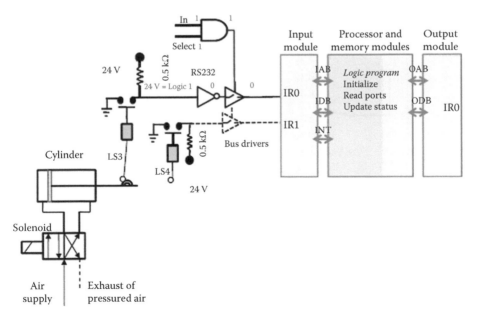

FIGURE 5.34
Pneumatic-driven process with limit switch I/O interface circuitry.

Any contact between either switch closes the corresponding switch contact. An open switch would correspond to the lowest bit 0 sent to the processor, and a closed switch to bit 1. Here, the bus driver (RS-232) can connect all the corresponding bit data from all limit switches to the I/O interface over the data line only when both signals (SELECT and IN) are at logic 1. Then, it is immediately disconnected from the I/O data line upon the resetting of the IN signal to logic 0 by the processor. Subsequently, the loaded LS3 and LS4 data are checked and their status updated in the register of the processor. In the case where a limit switch is open, the inverting driver receives 24 V (logic 1), which is inverted into logic 0. Then, the corresponding input bit data are sent over the data bus in the register of the processor. In that case, any switch is closed; the bus driver is set to logic 1 to be sent to the lowest bit of the processor.

Example 5.9

By using an asynchronous flip-flop, it is possible to store and hold the output value for a time interval when the input conditions (limit switch changes) have varied. This is called debouncing circuitry in reference to the mechanical contact bounce effect when the switch rapidly changes position (i.e., repeatedly opens and closes). It is useful to avoid a reactivation output value despite an input change, as illustrated in Figure 5.35. Here, a toxic liquid has to interlock its input control valve to avoid

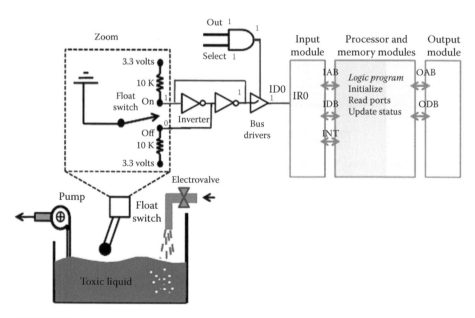

FIGURE 5.35
Double-switch position interface with debouncing.

harming personnel or destroying the equipment in a facility. A safe design uses a float switch with two positions: ON and OFF. When the switch is in the OFF position, the logic level 0 is set to 1 in the register of the processor. Otherwise, logic level 1 corresponds to the switch in the ON position. This is done using two successive inverting drivers to eliminate the contact bounce effects. When the switch is in the ON or OFF position, the output of the inverter in the series flip-flop is logic 1 or 0, respectively. For a time interval, any subsequent variation of switching contacts would not affect this logic output. Such debouncing circuitry cand be implemented through D flip-flop. This design is useful for systems having inputs with more than two positions.

Exercises and Problems

5.1. a. Derive the number of bits a DAC is capable of delivering for the control of a process in the ±8 V range with a resolution of 0.05 V.

   b. From a process signal range of ±8 V and a resolution of 0.05% of this range, derive the number of bits a ADC requires.

   c. Derive the allowable number of bits a DAC can have for the velocity command of a bidirectional motor between 2.5 m/s and 0.1 m/s (Hint: Nonzero velocity is 0.1 m/s).

   d. Derive the number of bits a DAC can deliver for the temperature command profile of a heater system such that the highest temperature is 280°C and the minimum nonzero temperature is 45°C.

5.2. a. Consider a motor shaft operating in the velocity range of 0.90 – 2.30 m/s. Determine the number of bits provided by an ADC with a 150 Hz sampling frequency to ensure a 0.02 m/s resolution.

   b. Consider an 8-bit ADC with V_{ref} = 5 V. For input signals of 1.5 V and 5 V, derive the resulting digital signals (codes) in the cases of bidirectional and unidirectional operations, respectively.

   c. From the list of physical phenomena below, list two (contact and contactless) sensors to be used to measure them:

      i. Flow rate of liquids in pipelines

      ii. Speed of elevator motor

      ii. Pressure of underwater vehicle

      iv. Fermentation tank temperature

5.3. Design and sketch the circuitry of a 10-bit binary-signed DAC capable of outputting a ±10 V signal.

5.4. Consider an 8-bit ADC with $Vref = 5$ V and a differential input channel. The input voltages V+ and V− are in the range of [0.99, 1.03] V and [1.02, 1.025] V, respectively. Answer the following questions:

    a. What would be the corresponding input of the differential input?

    b. What is the percentage of the input range?

    c. What is the quantization error, in percent of differential input range?

    d. Among the ADC gains 2, 5, 7, 10, 15, 20, 50, 110, 180, and 220, which gain would be suitable?

    e. How does the selected gain affect the quantization error (as a percentage of the differential input range)?

5.5. Consider a motor rotating at a nominal speed of 1500 rpm, with each revolution of the motor shaft involving 49 pulsation signals (commutation) toward the velocity sensor. What will be the maximum allowed sampling period of the ADC to avoid the aliasing effect?

5.6. a. Among the system bus, the I/O address bus, the I/O data bus, and the control bus, derive (i) the bus that sets the computing unit processor size and (ii) the bus that sets the memory size.

    b. Define the following terms from the computing unit: (i) clock cycle, (ii) clock frequency, (iii) interrupt lines, (iv) status lines, (v) polling of the I/O port, and (vi) asynchronous bus and synchronous bus.

    c. Which bus carries the computing unit dynamic instructions?

    d. List all buses carrying information between the processor, I/O, and memory.

    e. Which bus specifies the direction of the information flow? Give the name of the lines used.

    f. Among the process operating characteristics listed below, indicate which require the use of interrupts:

        i. The process events occur frequently.

        ii. The process is fully automated.

        iii. The state change is important.

        iv. The time intervals between two successive events are short.

    g. Among the process operating characteristics listed below, indicate which require the use of I/O polling:

        i. The process events occur randomly.

        ii. The state change is important.

        iii. The process output signal is noisy.

5.7. Consider a 100 Hz ADC with the resolution on the angular position sensor being 0.005 *rpm*. Derive the velocity resolution. If the A/D sampling frequency is doubled, what will the new velocity resolution be?

5.8. Consider a 1024-element linear photodiode array used as a feedback device for laser beam positioning. Each output of each element corresponds to a voltage proportional to the light intensity, and the element outputs appear serially at the output of the array. It takes 20 *μs* to perform the A/D conversion of voltage values generated by each array element and another 1.2 ms to locate the center of the beam image on the detector. What is the minimum response time that is expected for the closed-loop control of the beam position?

5.9. It is desired to design a small autonomous vacuum cleaner including a 16-key multiplexed keypad and an LCD as the user interface for motion control, depicted in Figures 5.36 and 5.37. It has an antenna to communicate with the base station (remote control). It also includes four dirt sensors, each measuring the number of dirt particles per square centimeters. The range of the dirt sensor, D, is 0–10,000, and it has a resolution of 100 dirt particles (DP) per square centimeters. The transducer has a maximum slope of 1000 DP/s and produces a voltage between 0 and 30 mV. In addition, there are two DC motor–driven wheels. Due to pin limitations, only one 5-bit DAC is used to the control the speed of both motors. The microcontroller provides four additional control bits to determine the motor directions (00, both motors stopped; 01, left motor in reverse and right motor forward; 10, left motor goes forward while the right motor goes in

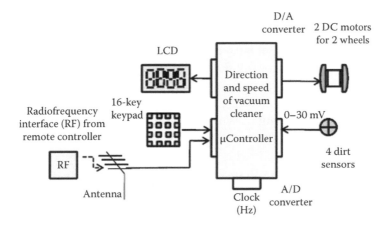

FIGURE 5.36
Vacuum cleaner system.

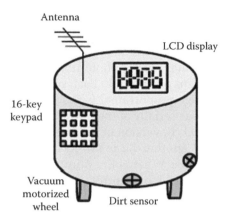

Antenna

LCD display

16-key keypad

Vacuum motorized wheel

Dirt sensor

FIGURE 5.37
Computer interfaces for vacuum cleaner motion control.

reverse; 11, both move forward). Finally, the vacuum is turned ON and OFF by using a relay interface. Considering the information given about this vacuum system, answer the following questions:

a. What is the required ADC precision? How many bits does the ADC need to be?

b. Assuming an ADC conversion time of 64 μs, is a sample and hold necessary? Why or why not?

c. What is the maximum allowable noise at the input of the amplifier? What could be the amplifier's gain?

d. The motor controller has a voltage range between −12 and +12 V. A positive voltage corresponds to spinning the wheel attached to it in a forward direction, while a negative voltage corresponds to spinning it in a reverse direction. What is the resolution of the DACs used to control the motors?

e. Draw a block diagram for the vacuum cleaner, including as much detail as possible, including port of I/O devices connected to external circuitry, sensors, amplifiers, filters, DACs, and logic. Add any component as long as its functions are described.

f. Derive a required value (min/max) of the interrupt or clock.

g. The output of the motor (direction control and speed) is updated every 1 ms. Write an assembly language–based program and the interrupt handler required to update the values presented those values required by to the motor controllers. Be sure to use the I/O port address from the diagram. Assuming that the values to be output are determined elsewhere, (i) derive the global variables required and (ii) give the assembly program for the controller.

h. Designing a logic controller that searches for dirt is considered, using the input of the dirt sensors to determine the control signals for the motor. The controller adjusts the motor speed based on the amount of dirt. When there is a lot of dirt, it moves slowly to make sure it collects all the dirt. Hence, it tries to maximize the difference in dirt measurements by its front and back sensors. From your own creativity, determine what are the control input and output variables.

5.10. a. A sinusoidal signal $u(t) = 2 \sin (2\pi157000t)$ is sampled through an ADC consisting of a 100 kHz data acquisition card. Is there any aliasing effect on the signal?

b. By justifying your answers, choose the suitable device to measure the temperature in the following processes and systems:

i. Beer fermentation vessel

ii. Preheating furnace in an oil distillation process

iii. Combat airplane turbo engine

iv. Crude oil flowing through a pipeline

5.11. A 10 kΩ potentiometer is used to measure the linear displacement while a voltmeter with an internal resistance of 100 kΩ to measure the sensor output. Determine the loading error when the potentiometer slider is at a 50% position for an excitation of 20 V.

5.12. In the following statements, select the correct answer:

a. A measurement system includes an instrument, a signal processing unit, a signal filter, and a signal recorder. The uncertainties of the four units are ±2.5%, ±1.5%, ±0.5%, and ±2.0%, respectively. The overall uncertainty of measurements is approximately (i) ±2.1%, (ii) ±6.2%, (iii) ±3.5%, or (iv) none of the preceding.

b. A seismic-type accelerometer provides accurate dynamic measurements (i) at frequencies well below its natural frequency, (ii) at frequencies above its natural frequency, (iii) at any frequency, or (iv) none of the preceding.

c. A force transducer is used to measure a true force of 200 N. The measurements, repeated five times, result in the following readings, as shown in Table 5.10.

i. The bias of the instrument is (difference between mean and true value) (1) 199.0 N, (2) 1.0 N, (3) 0.5 N, or (4) none of the preceding.

ii. The accuracy of the above instrument based on the previous readings is (1) 1.5%, (2) 2.0%, (3) 0.5%, or (4) none of the preceding.

TABLE 5.10

Measurement Data

Sample	1	2	3	4	5
Force N	197	201	200	198	199

    d. The linearity of a measurement instrument defines (i) the maximum deviation of the output from the true value, (ii) the minimum allowable variation in input required to reflect a variation in the output, (iii) the maximum deviation of the output from the best-fit straight line, or (iv) none of the preceding.

    e. The precision of a measurement instrument expresses (i) the closeness of the data to the mean value, (ii) the closeness of the data to the true value, (iii) the closeness of the data to the median value, or (iv) none of the preceding.

5.13. a. Derive the corresponding DAC voltage signal output from an input binary number of 10011001.

    b. List three advantages of the resistive ladder network in the signal conversion. What is its major drawback as a conversion tool?

    c. Draw the circuitry of a resistive ladder 8-bit DAC capable of delivering in the 256 V analog range.

5.14. a. Derive the resolution of an absolute 5-binary track optical encoder.

    b. Derive the voltage-level output of a thermocouple at 195°C, if it produces a voltage of 30 mV at 155°C and 195 mV at 175°C.

    c. A 5-volt signal is supplied to a potentiometer directly measuring the angular position of the rotating joint ankle of a humanoid robot. Considering that the rotational potentiometer has a wiper travel range of 280°, derive the corresponding potentiometer output voltage for 43°, if there is a voltage offset of 4.3 mV.

    d. A position encoder is hooked up to a motor combined to a 0:1 ratio-reducing gearbox. Derive the minimum allowable incremental encoder resolution when it is required to control the position of the attached load with a precision of 0.2°.

    e. How the following devices can be converted for velocity measurement purposes:

      i. A permanent magnet (PM) DC motor into a tachometer

      ii. An incremental encoder into a relative position counter from which the average velocity is estimated

    f. For position measurement purposes, compare the advantages and inconveniences of incremental or absolute encoders and potentiometers?

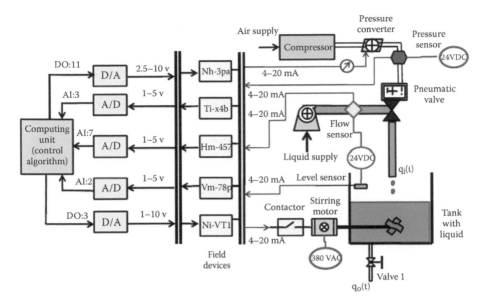

FIGURE 5.38
Apparatus used for liquid flow into tank process.

5.15. Consider a mixing tank filling process control system as depicted in Figure 5.38.

A 4-bit AD signal conversion unit converts each process analog signal output in 38 µs, while the computing unit executes the algorithm of control loops in 96 µs. The time required to convert the digital signal into its corresponding analog is 5% of the algorithm execution time.

a. Based on Figure 5.38, identify all process control loops.

b. Considering the time delays from all process control loops, estimate the maximum content (i.e., the range of process signal frequencies that could be entirely captured by this ADC and associated computer control system).

c. Derive the process frequency response and suggest the sampling period for each fluid-level control algorithm strategy (hint: characteristics for A/D conversion time and computing time).

5.16. Consider two linear position measurement devices (D_1, D_2) connected in series. The sensitivity of D_1 is 2.5 mV/mm. A system linear motion of 1.5 mm induces a measurement system output voltage signal of 7.5 mV. Derive the sensitivity of D_2 when there is an offset voltage at D_1 of −0.2 mV. Determine the value of D_2 when the offset voltage of D_1 is zero.

5.17. a. Consider a temperature measurement device with a resolution of 0.01°C and a sensitivity of 5 V/C°. Derive the smallest voltage variation the device can display.

   b. A 10-bit ADC with a range of nominal values of ±10 V is used to record data from this measurement device. Find the range of voltages that would generate an ADC digital value of 450. (Hint: Only positive integer codes are generated by the ADC).

   c. A 250 Hz sinusoidal input signal is sampled by a digital data acquisition system. Derive the suitable sampling period between 4 and 20 ms.

   d. Derive an increment for a 3-bit ADC with a 0–10 V range.

   e. Determine the maximum linearity error.

5.18. a. Indicate three different methods for a continuous temperature metering system installed inside the tanks.

   b. Sketch and explain the principle of resistance thermometers.

   c. For each of the measurement devices listed in Table 5.11, select the corresponding process applications.

5.19. Consider a 4-bit resistive ladder network DAC, as depicted in Figure 5.39.

   a. In the figure, identify the MSB and LSB.

   b. From Figure 5.39, derive the voltage out, V_{out}, equation using all bits D_i (including MSB, LSB), V_s, and R.

   c. From Figure 5.39, derive the voltage out, V_{out}, equation using all bits D_i (including MSB being Y, LSB being X), V_s, and R.

5.20. Consider a 4-bit ADC that operates such that after reading 5 V, it activates a micropump when the level switch is closed (binary level 1). Otherwise, the open switch would be read as the input level 0, corresponding to 0 V to be sent over the micropump circuitry.

   a. Complete the circuit diagram in Figure 5.40.

   b. Replace the level switch by a digital input device equipped with an internal pull-up resistor.

5.21. Consider a three-dimensional (3-D) sonar sensor for an underwater vehicle, as depicted in Figure 5.41. It is desired to connect it to a computing unit using a 60 Hz-sampling-rate ADC.

   The sampled data will be used to find $x = f(z)$ and $y = g(z)$. The maximum x, y, and z velocities are 0.04 m/s. The ADC has one analog input, and its conversion time is 140 µs. Complete the information flow through the A/D conversion system shown in Figure 5.41 and justify each additional component required.

TABLE 5.11

Measurement Devices and Typical Process Applications

Measurement Device	Process Application
Resistance thermometer	Temperature measurement in high flow rates
Oscillation piston flowmeter	Flow metering of water
Orifice plate	Flow metering in corrosive fluid
Thermocouple	Mass metering
Ball valve	Temperature measurement inside tank
Inductive-based flowmeter	Flow metering in pipe with large diameter

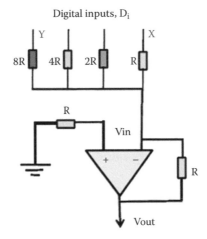

FIGURE 5.39
Four-bit DAC circuitry.

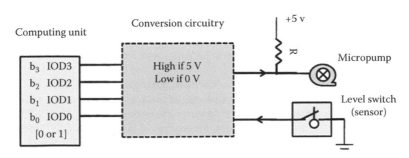

FIGURE 5.40
Computer interfacing and A/D conversion for level-controlled pump.

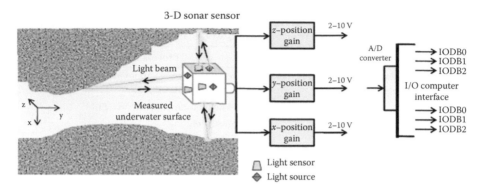

FIGURE 5.41
3-D sonar sensor and ADC.

Bibliography

Alciatore D., M. Histand, *Introduction to Mechatronics and Measurement Systems*, 2nd ed., McGraw-Hill, New York, 2003.

Areny R.P., J.G. Webster, *Sensors and Signal Conditioning*, 2nd ed., Wiley, Hoboken, NJ, 2001.

Baker R.C., *Flow Measurement Handbook*, Cambridge University Press, Cambridge, 2000.

Ball D., Sensor selection guide, *Sensors*, 50–53, 1986.

Bishop R., *Mechatronic Systems, Sensors, and Actuators: Fundamentals and Modeling*, 2nd ed., CRC Press, Boca Raton, FL, 2008.

Baxter L.K., *Capacitive Sensors: Design and Applications*, IEEE Press, Piscataway, NJ, 1997.

Bollinger J.D., N.A. Duffie, *Computer Control Machines and Processes*, Addison-Wesley, Reading, MA, 1989.

Carr W.W., Eddy current proximity sensors, *Sensors*, 23–25, 1987.

de Silva C.W., *Mechatronics: A Foundation Course*, CRC Press, Boca Raton, FL, 2010.

Elgar P., *Sensors for Measurement and Control*, Prentice-Hall, Upper Saddle River, 1998.

Fraden J., *Handbook of Modern Sensors: Physics, Designs, and Applications*, Springer, Berlin, 2004.

Garfinkel S., H. Holtzman, *Understanding RFID Technology*, eds. S. Garfinkel and B. Rosenberg, *RFID Applications, Security, and Privacy*, pp. 15–36, Addison-Wesley, Reading, MA, 2005.

Isermann R., *Mechatronic Systems: Fundamentals*, Springer, London, 2003.

Jarvis R.A., A perspective on range finding techniques for computer vision, *IEEE Transactions on Pattern Analysis and Machine Intelligence*, PAMI-1(20), 122–139, 1983.

Lenz J.E., A review of magnetic sensors, *Proceedings of the IEEE*, 78(6), 973–989, 1990.

Lynnworth L.C., *Ultrasonic Measurements for Process Control: Theory, Techniques, Applications*, Academic Press, Boston, 1989.

Lyshevski S.E., *Nano- and Microelectromechanical Systems: Fundamentals of Nano- and Microengineering*, CRC Press, Boca Raton, FL, 2000.

Manolis S., Resolvers vs. rotary encoders for motor commutation and position feedback, *Sensors*, 29–32, 1993.

Miller R.W., *Flow Measurement Engineering Handbook*, 3rd ed., McGraw Hill, New York, 1996.

Pinney C.P., W.E. Baker, Velocity measurement, in *The Measurement, Instrumentation and Sensors Handbook*, ed. J.G. Webster, pp. 16.1–16.15, CRC Press, Boca Raton, FL, 1999.

Ramsay D., *Principle of Engineering Instrumentation*, Arnold, London, UK, 1996.

Regtien P.L., *Handbook of Measuring System Design*, Wiley, Hoboken, NJ, 2005.

Regtien P.L., *Selection of Sensors*, Wiley, Hoboken, NJ, 2005.

Regtien P.P.L., *Sensors for Mechatronics*, Elsevier, Amsterdam, 2012.

Santina M.S., A.R. Stubberud, Basics of sampling and quantization, in *Handbook of Networked and Embedded Control Systems*, Birkhäuser, Basel, 2005, pp. 45–70.

Schuermeyer F., T. Pickenpaugh, Photoconductive sensors, in *The Measurement, Instrumentation and Sensors Handbook*, ed. J.G. Webster, pp. 56.1–56.6, CRC Press, Boca Raton, FL, 1999.

Spitzer D.W. (ed.), *Flow Measurement: Practical Guides for Measurement and Control*, 2nd ed., ISA, Research Triangle Park, NC, 2001.

Taylor H.R., *Data Acquisition for Sensor Systems*, Chapman & Hall, London, 1997.

Xi N., M. Zhang, G. Li (eds.), *Modeling and Control for Micro/Nano Devices and System*, CRC Press, Boca Raton, FL, 2014.

Webb J., K. Greshock, *Industrial Control Electronics*, 2nd ed., Macmillan, New York. 1993.

Wojcik S., Noncontact presence sensors for industrial environments, *Sensors*, 48–54, 1994.

6

Data Transmission System

6.1 Introduction

Industrial data transmission systems are mainly the serial bus–based multi-drop data networks interconnecting the automation system components (sensors, actuators, human–machine interface [HMI], etc.). Commonly encountered physical media used for such data transmission systems include copper-based cables, optical fiber, and even air for wireless communications. Furthermore, these data transmission networks use protocols defining the message format and specifying how the message is transmitted. Among commonly encountered industrial network protocols there are Modbus, Profibus, Actuator Sensor Interface (AS-I), DeviceNET, Industrial Internet, and WorldFIP. Hence, in industrial environments, the data transmission systems are expected to provide some exceptional data rates, a higher data transmission range, power availability on the wire, and the highest degree of data integrity for the microcontroller, programmable logic controller (PLC), and supervisory control and data acquisition (SCADA) systems. Therefore, specialized protocols of automation system integrators have to adapt communication networks and protocols between various automation devices (logic controllers, actuators, sensors, etc.) according to some of the industrial benchmarks and compliance requirements.

This technological-oriented chapter aims to guide automation field engineers into the quantitative design, analysis, and performance assessment of a wide range of data transmission networks. This especially covers how to (1) size and select the technology related to the data transmission network (topology specifications, network devices, and protocols) for specific industrial applications and (2) perform audits on the quality of service (QoS) of industrial data transmission networks. First, a general overview of commonly encountered industrial transmission networks, as well as their technical characteristics and protocols, is presented. For each protocol, the estimation of transmission delay, the estimation of data losses, the modeling of the congestion rate, the throughput, and the network efficiency are developed. Based on the network benchmarks and the compliance check, some guidelines are reviewed for selecting the transmission media and protocol, as well as choosing the industrial data network architecture.

6.2 Network Topology

A simple form of an industrial transmission network consists of a master command and control device (e.g., PLC) sending requests to read or write data to the slave-related field automation devices (e.g., electrovalves). Generally, the industrial network aims to interconnect several automation devices into a single link to form a multidrop network. As illustrated in Figure 6.1a–d, the commonly encountered interconnection patterns (topology) of these automation devices (nodes) on an industrial network are

- Star topology, where a centralized node enables individual connection with several types of terminal equipment. This topology is suitable for large networks with a high density and rate of data flow, but it is central unit failure dependent.
- Ring topology, where each unit can only transmit messages to its direct neighbors. It is for networks with high reliability requirements.

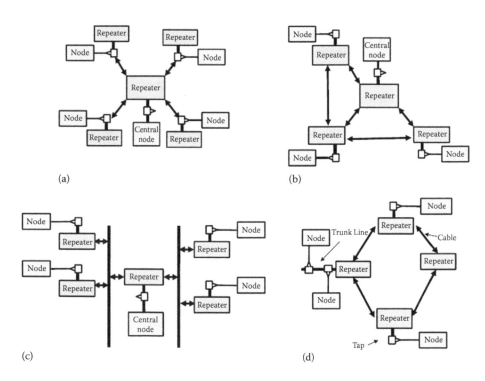

FIGURE 6.1
(a) Star network topology. (b) Net (full-mesh) topology. (c) Bus network topology. (d) Ring network topology.

- Full-mesh network topology, which consists of interconnecting all other nodes in the network through a plethora of possible alternative transmission routes. It is fitted for short-reaction-time industrial applications.
- Hierarchical network topology, which is a star network type with multiple network layers and where the highest layers are centers and the lowest layers are devices (e.g., switches).
- Bus topology, where only one participant (central node) at a time is able to send or receive simultaneous information to or from all active nodes. It is suitable for high-flow-rate networks with geographically distributed systems (e.g., the automatic handling system of airport luggage).

Similarly, among networks buses there are

- Serial buses, where segments of messages (bits) are transmitted one after another for longer distances, from 20 m up to 15 km
- Parallel buses, where message addressing of data and control signal transmission is done in parallel, through a set of wires, for distances up to 20 m

6.3 Components of Industrial Automation Networks

In order to implement those topologies, the network devices such as repeaters, tap or cable length should be selected to enable efficient connection between all automation field devices. Furthermore, the selection of network topologies and protocols should be based on the performance requirements, as well as the isolation constraints. In addition, network data have to be transmitted across dissimilar network types, it is requiring the efficient integration of those network devices listed in Table 6.1.

Unless otherwise specified, in common industrial data network, the required number of repeaters, R, could be derived from

$$R = integer\ of \left(\frac{number\ of\ master - slave\ connecting\ devices}{max.\ number\ of\ devices\ allowed\ per\ network\ segment} \right) \quad (6.1)$$

while the terminator, *Term*, is a resistor of 75 Ω, mounted in connectors plugged into any network segment head to prevent any disturbance to

TABLE 6.1

Definitions of Some Industrial Automation Network Devices

Term	Network Device Description
Node	Field automation components (devices) (PLC, HMI, actuators, sensors, etc.) connected to a transmission network
Tap	Network device connecting a node to the network's trunk cable
Repeater	Network device enabling us to extend the network beyond its amplifying electrical signals
Bridge	Network device enabling network link data traffic
Trunk cable	A physical link (or bus) connecting any two network devices
Network	All devices involved in data transmission between at least two automation devices (nodes)
Router switch	Network device defining the transmission route of some data packets over several networks
Hub	Network device acting as a central network link where its connecting wires are used to transmit some communication packets to geographically dispersed network nodes
Gateway	Network device providing the interoperability of different buses and protocols by allowing some applications to communicate
Segment	Network link with terminators at both ends without repeaters

electric power diffusion in the signals transmitted. The number of *Term* could be derived from

$$Term = 2 \times number\ of\ network\ segment \tag{6.2}$$

The number of digital signal converters in a, industrial network, $n_{converter}$, is given by

$$n_{converter} = n_{sensor} \times other\ analog\ equipment \tag{6.3}$$

with n_{sensor} being the number of analog sensors in the network. Similarly, the number of bus connectors, n_{bus}, in such a network is given by

$$n_{bus} = number\ of\ equipment\ to\ be\ connected + 2 \times number\ of\ repeaters \tag{6.4}$$

6.3.1 Typical Transmission Interface and Transmission Media

There are two types of data transmission: (1) serial data transmission, where the packets of bits (in bit format) are transmitted one after the other and require three wires (send, receive, and earth), and (2) parallel transmission, where the packets of bits are transmitted simultaneously in byte format and are only used for short distances. A serial data transfer scheme could be either synchronous or asynchronous serial transmission. In synchronous serial transmission, the data signal is transmitted continuously and simultaneously to a clock-based synchronizing signal or a handshake signal. Here, the handshake signals, with data, are sent simultaneously over the network, synchronizing the future data packet to be sent. In the case of asynchronous serial transmission, the data are transmitted aperiodically, although the interval between 2 bits is fixed. Here, the synchronizing bits (START, STOP) encapsulate the data. Indeed, in addition to process information, the data packet contain information on the transmitter and receiver, as well as the start and end bits, to structure for the receiver when the transmission starts and when it is completed. This data packet transfer mechanism is based on identical frequency or the synchronized clock between the transmitter and receiver. Among types of transmission networks there are

- Baseband networks, where data are transmitted based on gate ON/OFF switching to indicate bit status
- Carrier band networks, where data are transmitted based on signal frequency-based switching (frequency shift keying [FSK]) corresponding to bit status
- Broadband network, where the data are transmitted over multiple frequencies through multiple channels (carrier)

The design of all transmission networks requires compliance with the specific interface. A typical interface or transmission medium is supported by various types of cables, including coaxial cables and optical fiber cables. The coaxial cable consists of a copper conductor surrounded by plastic insulating the layer and grounding shielding. The coaxial cable transmits the electrical signals and is suitable for high-speed transmission. The optical fiber cable, which consists of a glass or plastic fiber, transmits data through light signals. It is used for harsh environmental conditions and is reliable over long distances.

6.3.1.1 Transmission Support Media

A transmission support medium is used to interconnect some network nodes. It is made of either (1) twisted-pair copper wiring (10baseT) terminated with an RJ-45 connector, (2) optical fiber (10baseF), (3) thin shielded coaxial wiring (10base2) with BNC (Bayonet Neill–Concelman) style connectors, or (4) a microwave (wireless). The choice depends on the protocols in the network.

6.3.1.2 RS-232 Interface Standard (EIA232)

The recommended standard (RS) for serial communication (232) is a three-wire interface (with separate ground, receive, and transmit wires) capable of receiving and transmitting data simultaneously. RS-232 has a voltage wire signal-carrying bit (1 and 0) to transmit or receive mode setting information. This is suitable for the initial setup of systems or for some diagnostics applications. Its physical layer has the following characteristics: (1) peer-to-peer connection, (2) signal levels that can be +3 to +25 V corresponding to logic level 0 or –3 to –25 V for logic level 1, and (3) the low traffic speed.

6.3.1.3 RS-485 Interface Standard

RS-485 is a three-wire interface such that two of them with a voltage signal are used for setting the transmit and receive mode. The RS-485 interface is used with industrial protocols, such as Profibus, Interbus, and Modbus, as well as for applications requiring immunity to noise, bidirectional and multiple-driver capability, and a high transmission speed (9.6 Kbs to 12 Mbs). RS-485 allows interconnecting 10–32 devices.

6.3.1.4 RS-422 Interface Standard

RS-422 uses a differential electrical signal with two lines for transmitting and receiving signals. It displays a high noise immunity.

Among other transmission media there are also USB interfaces (USB 1.0, 2.0, and 3.0) and infrared device attachment (IrDA).

6.3.2 Serial and Parallel Data Communication Format

Data packets are transferred in a bit format. In the case of serial data communication, and depending on the protocol, data packets or messages can be transmitted in the format of series of bits or bytes within a frame structure. Hence, a typical frame structure includes the data packet, the address transmitter or receiver, an error detection or data collision prevention mechanism, and the start and stop of the bits. Sometimes, it also contains the length of the message, as well as the time of stamping. With this generic digital data signaling structure, the transmitted bit data stream, $m(t)$, is mapped to a digital signal, $x(t)$, before it is sent through the transmission digital media. Then, the received signal, $x(t)$, is decoded to its original structure, $m(t)$. In the case of the analog data signaling structure, the transmitted bit stream, $b(t)$, is mapped to an analog signal, $h(t)$, before it is sent through the transmission analog media. The received $h(t)$ is demodulated to its original $m(t)$.

Serial data communications are based on the change of voltage levels using 1-bit time 1 or 0 (high or low). The structure of the serial data packet transmission consists of four successive parts: (1) a start bit; (2) 6-, 7-, or 8-bit-length data bits (0–0.8 V TTL), which are sent out one after another; (3) a parity bit used to indicate data transmission errors; and (4) at least one 1-, 1.5-, or 2-bit-length stop. A parallel port with several data buses can transmit several bits simultaneously, in contrast to serial port transmitting through only one data bus each time.

The network segment signaling rate or the rate of sending data bits is measured by the number of times the signal switches its voltage levels per second. This is measured by the baud rate, which is 1 over the shortest signal transmission time in bits per seconds. Hence, this rate can be used to assess the data packet size to be transferred per second.

6.3.3 OSI/ISO Standards

The Open System Interconnection (OSI) model proposed by the International Organization for Standardization (ISO) provides a widely accepted structure for network protocols. It also ensures the universal compliance of the data transferred between the network devices and automation components. This model consists of a seven-layer stack onto which the data packet is processed. Each layer performs some specific tasks while providing the service to the adjacent layer above. Also, each layer shares information only with its adjacent OSI layers. The OSI network model layers consist of the

- *Physical layer* providing the mechanical and electrical characteristics of the physical network segment through its transmission media constraints (e.g., allowable signal magnitude and frequency, and remote or active connection and disconnection)
- *Data link layer* providing synchronization of data packet transmission over the network segment, including the error detection and correction mechanism
- *Network layer* starting, transmitting, and terminating the data packet transmission process between network segments, as well as executing data routing and addressing functions
- *Transport layer* sequencing the data packet transmission over the network segment, including executing transmission confirmation functions
- *Session layer* synchronizing data packet transmission between any nodes over the network segment through activation of the start and stop exchange controls
- *Presentation layer* coding and decoding the data packet transmitted over the network segment
- *Application layer* determining the transmission protocol and processing the data packet to be transmitted

6.4 Constraint Specifications of an Industrial Network

The industrial automation networks have to avoid the disruption or deterioration of their performance due to an aggressive environmental impact from explosives, dust, the presence of water, vibration, hot or cold temperatures, electromagnetics, wetness, or a combination of these zones. Thus, depending on the zone, the transmission support media accordingly require specific protection. In front of the explosive zone with some inflammable products within the gas, steam, or dust zone that could cause a fire, a risk estimation is required, along with network redundancy. Then, a disaster recovery plan can be designed accordingly. In the dust zone with the presence of water and a high mechanical vibration, the protection equipment requirements (a cabinet and/or cable protection) are classified with the code IPxy IKz, where x, varying from 0 to 6, is the protection index against the solid wastes; y, varying from 0 to 8, is the protection index against the liquids; and z, varying from 0 to 10, is the protection index against the mechanical shock. Within the hot and cold temperature zones, the protective equipment is based on the temperature range, while in the case of unstable temperatures, some provisional tools (e.g., meteorological stations) are used to monitor and assess the degree of lightning and thunder exposure, for example. In the electromagnetism zone, the equipment is required to be in compliance with a lower immunity level of electromagnetism field exposure. The exposure can be from a media-based disturbance, usually at low frequency, or an air-based disturbance, usually at high frequency. The protective measure can either use some transmission cables to separate the power supply lines from the data transport lines, or set the hertz-based zones of the electromagnetism field emission far from each other and the power supply lines. The wet zones are classified based on the humidity rate of the ambient air. In the case of the humidity operational zones, a heating system can be used.

6.5 Communication Protocols for Industrial Automation Systems

Several communication protocols are selected and deployed to operate the industrial network, depending on their application requirements. Thus, in order to design, implement, and troubleshoot industrial data transmission networks, it is necessary to have a good understanding of their protocol-based technical specifications, such as the transmission characteristics, frame structure, and transmission speed, as presented in this section. When all nodes send data simultaneously, the data can be corrupted (i.e., data collision occurs), which could degrade the communication network performance.

There are various methods dealing with data collision and arbitration issues, including

- Collision Sense Multiple Access/Collision Detection (CSMA/CD), such that when a data collision is detected over the network segment, each node stops transmitting requests and answers for some time before starting again.
- Collision Sense Multiple Access/Bitwise Arbitration (CSMA/BA), when two nodes transmit requests and answers simultaneously, their node addresses are used to determine the order of processing.
- Master–slave communication method, when the communication can only be launched by the master device, and slave devices are only able to respond to its requests. This method can be limited in the case of multiple masters (nodes) on the same network.
- Token ring, which is passed sequentially over the network nodes in order to allow them each time to send or receive data over the network. This method is not able to manage a data packet transmission priority.

A brief technical description of these industrial protocols is presented in the following paragraphs.

6.5.1 4-20mA

This is a lower-speed data transmission protocol based on the current signal as carrier between a single transmitter and several receivers. It is commonly used for data packets between field automation devices (sensors) to control units.

6.5.2 HART

The Highway Addressable Remote Transducer (HART) is derived from the 4-20mA protocol, and it is suitable for slow data transmission rate. The 4-20mA current signal carrier uses the frequency shift keyed for data packet encoding between 1200 and 2200 Hz. Its transmission rate is 1200 baud, and it allows the transmission of transducer measurements for the monitoring and diagnostics of industrial applications. The HART protocol is based on point-to-point connections over a master–slave network transmission structure.

6.5.3 Interbus

Interbus is a ring-based network using a RS-485 optical fiber or an infrared-based medium over the peer-to-peer connections. It uses a full-duplex operation offering scability of this data network structure.

6.5.4 Modbus

Modbus communicates using a master–slave network transmission structure. Master devices can send the data packet to individual slaves or broadcast it to all slaves. The structure of request from the master consists of a slave to broadcast an address, an action requested, an application data packet, and an error detection control to verify the data packet contents. Accordingly, the slaves are expected to submit the requested data packet to the master node or execute the action requested. Slaves respond only in peer-to-peer communication toward the master node. A slave's typical response data frame is (1) its address and the confirmation of the field action taken, (2) any application data to be returned, and (3) an error checking field. When a data transmission error is detected or enables the performance of an action, the slave responds by sending an acknowledgment message. The ModBus protocol uses either RS-232 (up to about 20 m) or RS-485 (up to about 1200 m) as the physical layer, or a master–slave medium access with a transmission rate varying between 1200 bits and 115 Kbits/s. The Modbus protocol is restricted to addressing 247 devices on one data link that may be connected to a master station. To extend such a network, a Modbus master bridge is used to receive the Modbus queries on the serial interface and transmit them over the network.

The Modbus protocol frame structure contains four basic elements: (1) the device address, (2) the function code defining the message type, (3) the application of the data block, and (4) the error numeric check value. There are two Modbus serial transmission data frame structures: ASCII and RTU. In ASCII format, a data packet is encoded using hexadecimal values (0 … 9 and A … F). Hence, the Modbus/ASCII frame structure consists of the address, function code, data, checksum, and Carriage Return/LineFeed protocol. There is an elapsed time of less than 1 s between two consecutive data packets transmitted. Also, it uses the longitudinal redundancy check (LRC) while with Modbus/RTU data frame is done in a binary format. There is a minimum of 3.5 characters before each data packet can be transmitted. Hence, the Modbus/RTU frame structure consists of silence, address, function code, data, checksum, and silence. In addition, it uses the cyclic redundancy check (CRC). It should be specified that Modbus/ASCII uses two hexadecimal characters to encode a message. Commonly used addresses are from values 0 up to 247, all to be assigned to the automation field device, except 0, which is used as the broadcast address. More specifically, among Modbus codes they include read coil output status using code 01, read input status using code 02, read holding registers using code 03, read input registers using code 04, force single coil using code 05, preset single registers using code 06, and read exception status using code 07. Modbus protocols use two methods for the communication error checking field: the data character frame error through parity checking (even, odd, parity) and the message frame error through frame checking (CRC and LRC). Typical Modbus transmission speeds are 38,000 kbps for RS-485 and up to 20 kbps for RS-232.

6.5.5 Profibus

Profibus is a master–slave, serial, and linear bus communication protocol without the terminators commonly used for the control application of industrial systems. A master device initiates the interconnected slaves cyclically. Its query data frame contains the device output data or device setting (e.g., set-point temperature of fermentation tank), while the response data frame contains the input data (e.g., the sensed temperature value). The Profibus connection is a half duplex over a copper wire cable (a shielded, twisted pair). The data transmission rates vary up to 12 million baud, with a maximum data packet length of 244 bytes over a maximum distance of 100 m. Up to 126 nodes can be connected in five segments separated by repeaters. Each node has one master connected to the slave devices for less than 1.2 km without repeaters. Up to 32 stations can be supported for twisted shielded pair cable. Ring-based topology Profibus–Process Automation (PA) connecting control stations with automation field devices allows the execution of self-diagnostic functions, real-time control, and the assessment of process supervision suitable in explosive hazardous areas. This protocol allows us to simultaneously transmit data and the power for automation field devices through a coaxial pair cable with a transmission speed up to 31.25 Kbits/s. It is based on the Manchester Coded Bus Powered (MBP) transmission technique. With Profibus–Field Bus Message Specification (FMS), the transmission speed is 1.5 Mbits/s for a total distance of 400 m without repeaters. Profibus–Decentralized Periphery (DP) is designed for a high-speed data transmission up to 12 Mbits/s, and it is suitable for direct communication between the control unit and decentralized nodes. The copper-based Profibus-DP network has a net topology. The normalized Profibus-DP cable color code is violet, while the Profibus-PA cable is blue and the optical fiber is yellow. A typical data frame of the Profibus protocol occurs in the following order: start delimiter (1 byte), data packet length (1 byte), length repeated (1 byte), receiver address (1 byte), sender address (1 byte), function code (1 byte), destination service access point (1 byte), application data (up to 244 bytes), frame checking sequence (1 byte), and end delimiter (1 byte).

6.5.6 DeviceNet

DeviceNet is a high-speed, byte-level, serial bus, multidrop communication network used to interconnect automation devices, such as sensors and actuators. Communication for DeviceNet could be through either:

- *Polling*, when each network device sends or receives an update of its status upon request.
- *Strobing*, when a device broadcasts an updated status of requests to all devices. Each node responds sequentially in the following order: 1, 2, 3, 4, etc.

Start of data packet	11-byte arbitration method	Control field (packet address)	0–8 byte data packet	15-bit error CRC	ACK	End of data packet

FIGURE 6.2
DeviceNet message frame. (Data from DeviceNet.)

- *Cyclic*, when a device periodically updates its status toward other devices. This is combined with the change of state data transmission.
- *Change of state*, when a device sends messages only when there is a change in their status. This reduces the utilization time of the network.

Figure 6.2 presents the DeviceNet message frame structure.

A DeviceNet network may have up to 64 devices using 1 device on each node, with an address range from 0 to 63. There are three baud rates for DeviceNet networks: 125, 250, and 500 Kbits/s. This limits the length of the network for a maximum total distance of 500 m, with a maximum distance between nodes (up to 64) of 10 m within the bus topology.

6.5.7 ControlNet

ControlNet provides high-speed transmission of time-critical input/output (I/O) devices, especially interlocking data and messaging data. ControlNet is based on a data producer–consumer model that uses the object messages to exchange information. Here, an object message is a piece of information to be exchanged with one or more nodes on the network. The ControlNet network supports two types of connections: (1) peer-to-peer connections between one producer and one consumer and (2) multicast connections between one producer of data and more than one consumer. The repeaters are required to create topologies, such as the star or ring topology. The arbitration method consists of the Concurrent Time Domain Multiple Access (CTDMA) and Time Slice Multiple Access, while the message error checking is done through modified Consultative Committee on International Telegraph and Telephone (CCITT) standards with a 16-bit polynomial. Each node can send only one Media Access Control (MAC) frame containing a data packet of a size up to a maximum of 510 bytes during each transmission opportunity, for a maximum total network length of 1000 m. Also, all nodes on ControlNet can simultaneously access the same data from a single source through synchronization. ControlNet operates on multiple modes, such as (1) master–slave, where a single master polls multiple slaves without collisions; (2) multimaster; (3) peer to peer; and (4) CTDMA. Using either star, tree, or ring topologies, the ControlNet transmission medium is either the coaxial cable or optical

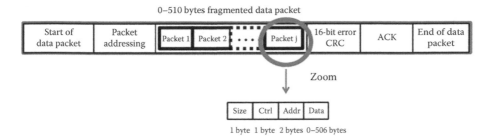

FIGURE 6.3
Structure of ControlNet data packet frame. (Data from ControlNet.)

fiber, and it can interoperate with Ethernet and DeviceNet protocols. The maximum number of nodes allowed is 99, with a distance up to 100 m. Figure 6.3 illustrates the typical structure of the ControlNet data packet frame.

6.5.8 Industrial Ethernet

Several variations of Industrial Ethernet include ProfiNet, EtherCAT, EtherNet/IP, ModBus, and Ethernet Powerlink, mainly due to their data latency. Using the bus or star network topology, the Industrial Ethernet maximum network length is 2500 m, with a maximum allowable distance between nodes of about 500 m over either the twisted-pair or optical fiber. The typical Ethernet data frame structure is summarized in Figure 6.4.

6.5.9 Serial Synchronous Interface

Here, the network node synchronizes with discrete clock signal (pulse train) data packet transmission between automation field devices (sensors and controllers) and controller units.

Hence, during each pulse, a data packet is transmitted. This is followed by 25 ms of elapsed time before the next transmission. Data are shifted out of the register when the sensor node receives clock pulses from the controller node, for example. When the clock is held at a high level and the minimum dwell time has elapsed, new data are available to read from the register.

0–8 bytes start of data packet	6 bytes receiver address	6 bytes sender address	2 bytes length address	0–1500 bytes data packet	0–4 bytes checksum

FIGURE 6.4
Structure of Ethernet data packet frame. (Data from Moyne, J.R., and Tilbury, D.M., *Proc. IEEE*, 95 (1), 29–47, 2007.)

6.5.10 WorldFIP

WorldFIP is a field bus communication network interface between the field-level devices of the automation systems (i.e., controller, sensor, and actuator). The maximum number of nodes on the WorldFIP network is 256, with a maximum of four repeaters. The physical layer transmits bits through the Manchester code. The WorldFIP frame consists of three parts: (1) the start of the frame sequence (PRE + FSD), (2) the control field and data (CAD) containing only the binary information 1 and 0, and (3) the end of the frame sequence (FED). There are two types of address assignment: (1) the variable addressing associated with the sender (a 16-bit coding for up to 65,536 variables) and (2) the message addressing where each transmitted message has sender and receiver addresses using 24 bits, as well as the network address segment and the station address within the segment. There are two types of nodes in the WorldFIP: the bus arbitrating node managing accessibility to the transmission medium (bus) and the producer or consumer node. Any node can be either a producer node or a consumer node, or even both types at the same time. WorldFIP's data link layer also provides acknowledged and unacknowledged message transfer services, which makes the point-to-point data exchanged more reliable up to 2.5 Mbps.

6.5.11 Actuator Sensor Interface Bus

AS-I is a two-wire interface connecting the controller with its I/O devices (e.g., sensors, encoders, analog I/O, and push buttons and actuators). Either star, ring, or bus network topologies can be used at a distance up to 100 m. A 24 V direct current (VDC) power supply is provided up to 8 A over the network. An AS-I network can only have a maximum of 32 nodes and a cable length up to 100 m without repeaters. The master polls some requests to the slaves and receives the slaves' specific addresses sequentially. The swap of slave devices is supported, while multiple master operations are not supported. The AS-I network error detection mechanism is operated automatically by the AS-I masters, which detect any invalid slave transmission and schedule repeats. This network handles only small volumes of data since it is implemented in the lowest level of the control architecture.

6.6 Audit of Industrial Network

Among requirements to the migration or updating of the industrial data network there are the technical network audit and modeling to assess some network performance indices, especially in the network transmission latency, the congestion rate, the throughput, the network utilization, and the data losses and efficiency, as illustrated in Figure 6.5.

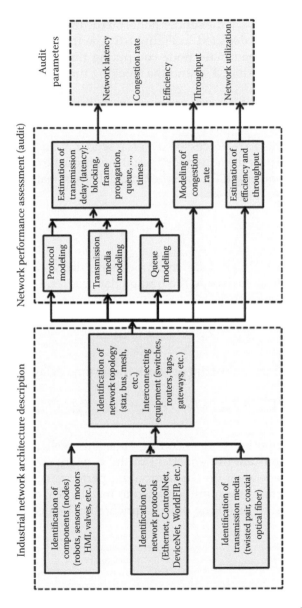

FIGURE 6.5
Network audit methodology schematic.

6.6.1 Estimation of Industrial Network Latency

Based on information from the network architecture and protocols, it is possible to evaluate the time delay of data traveling between the sender node and receiver node. Usually, this time delay can be split into three parts, representing the time delays at the sender node, on the transmission media, and at the receiver node, as depicted in Figure 6.6. The time delay at the sender node, T_{sender}, includes the sender node computing time, T_{scomp}; the encoding time, $Tscode$; the queue time, T_{q_p}; and the blocking time, T_b, such that

$$T_{sender} = T_{scomp} + T_{scode} + T_{q\_p} + T_b \tag{6.5}$$

Once the packet data are processed over the sender node, the delay related to the transmission over the network physical media can be accessed through the network segment length, T_p, and the time delay associated with framing the data packet size, T_f, such as

$$T_{net} = T_f + T_p \tag{6.6}$$

Similarly, at the receiver node the processing delay can be estimated through the time delay to decode the framed data packet, T_{dcode}, and the processing time, T_{dcomp}, such that

$$T_{rec} = T_{dcode} + T_{dcomp} \tag{6.7}$$

The estimation of the industrial network traffic time delay is given by

$$T_d = T_{scomp} + T_{scode} + T_b + T_f + T_p + T_{q\_p} + T_{dcode} + T_{dcomp} \tag{6.8}$$

Because the computing and encoding times, as well as the decoding times, are nearly constant (independent of the network size, as well as the data packet, but dependent on the computing performance), the modeling of data traffic time delays is mainly related to the physical and protocol network parameters, such as (1) the data fragmentation modeling, (2) the data

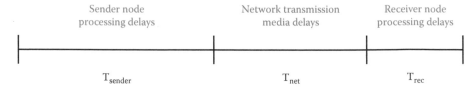

FIGURE 6.6
Timing diagram of data exchange for a generic protocol.

processing (blocking and queuing), and (3) the data transmission and even data transfer (propagation and routing). As such, it can be approximated as

$$T_d \approx T_b + T_f + T_p + T_{q\_p} \qquad (6.9)$$

6.6.1.1 Modeling of Transmission Media

The transmission media is characterized by the protocol and the network topology involved, and it is defined by the total time taken by a frame to pass through it. This time is divided into (1) the frame time, T_f, or the time required to transfer a frame into a transmission medium at a specific data rate, and (2) the propagation time, or the time required to transport the information through a cable with a specific length. As such, this transmission network elapsed time depends on the data packet size, rate, and physical network segment length. The model of the transmission media is given by the transmission time such that

$$T_f \approx \frac{useful\ frame\ length}{data\ transmission\ rate} = \frac{message\ frame\ size\ (bits)}{data\ rate\ (bits/s)} \qquad (6.10)$$

while the propagation time is given by

$$T_p \approx \frac{cable\ length\ (m)}{propagation\ speed} = \frac{length\ of\ the\ network\ cable\ (m)}{Typical\ transceiver\ speed\ (m/s)} \qquad (6.11)$$

Typical transmission speeds are 2×10^8 m/s and 3×10^8 m/s over the copper and coaxial media and the optical fiber, respectively.

6.6.1.2 Modeling of Data Fragmentation

Data fragmentation is usually done through a gateway device. Such a device is designed to guide the data traffic over at least two different networks. Its data processing time, which involves time delays for the encapsulation and de-encapsulation of the data packets, can be neglected. Therefore, a gateway can be modeled as a network commutating device capable of fragmenting data packets in order to meet the data transfer size requirements of at least two different networking protocols. Consider two data networks, i and k, with T_i and T_k their maximum transmission units (MTU), respectively, as depicted in Figure 6.7. It is desired to transfer data from network i to network k connected through a gateway R1.

If $T_i \leq T_k$, there is no fragmentation required.

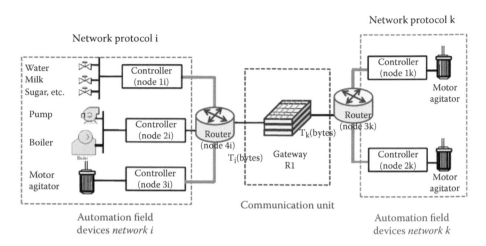

FIGURE 6.7
Gateway connecting two different networks.

If $T_i > T_k$, the resulting number of fragmented packets is computed as follows.

The size N of each fragment of data packet is given by

$$N = 8 \times \left\lceil \frac{T_k}{8} \right\rceil \tag{6.12}$$

where $\left\lceil \frac{A}{B} \right\rceil$ is any integer up to $\frac{A}{B}$. Consider the overhead size of the data packet (message) frame, N_{nd}, such that the resulting number of fragments, p, is given by

$$p = \left\lceil \frac{T_i - N_{nd}}{8 \times \left\lceil \frac{T_k}{8} \right\rceil - N_{nd}} \right\rceil + 1 \tag{6.13}$$

The data fragmentation process defines the transmission time such that it causes an increase by $N_{nd} \times (p - 1)$ extra bytes to be sent over the physical transmission media. Hence, it multiply by the number of packets to be processed on the network 2 is multiplied by p. This causes an overall increase in the transmission time.

6.6.1.3 Estimation of Routing Time

The time estimation of the routing of data packets within industrial networks can be done using the Bellman–Ford algorithm. The Bellman–Ford algorithm is commonly used by some network commutating devices (i.e., switch, router, etc.) to derive the shortest path with the lowest cost when routing a data packet from one network to another. This is done using the estimation of the network link cost at each node and selecting the best alternative (shortest path). Hence, the cost of a network link is given by

$$d = link\_cost = \frac{l}{v} + \frac{L}{C} \tag{6.14}$$

where l, v, L, and C are the cable length, transmission media propagation speed, length of the data packets (message) to be sent, and transmission rate, respectively. The Bellman–Ford algorithm can be formulated such as

Initialization
$V = \alpha$

$d_\alpha = 0, \ d_i = \infty \ \forall i \neq \alpha$

Iterations. While $V \neq \emptyset$
Pick a node i from V and remove it
For each link (i, j) originating from i, with given weight distance $a_{i,j}$
If $(d_j > d_i + a_{ij})$, then
$d_j \leftarrow d_i + a_{ij}$,
Insert j inside V

Once the data packet is sent toward the receiver node through the transmission media, its arrival time depends on the network segment length from the route selected, as well as the data packet size.

6.6.1.4 Estimation of Blocking Time

The blocking time, T_b, is the waiting time of a message from a sender node prior to a sender node passing through the network segment and after its processing is complete. If there is congestion or data collision, it depends on the arbitration method and waiting time. The blocking time is a key characteristic of the network performance. Among arbitration methods of network protocols there are (1) token passing (Profibus, Modbus, and ControlNet), (2) Carrier-Sonac Multiple Access (DeviceNet and Ethernet), and (3) bus arbitration table (WorldFIP).

With the *ControlNet protocol*, before sending the data packets, the node must await a receiving token from the logically previous node. Hence, the blocking time, T_b, can be derived from the transmission time and the token rotation time of previous nodes. If the time period of data packets of each node and the size of data packets are known, the blocking time, T_b, is given explicitly as

$$T_b = T_{rd} + T_{gd} + \sum_{j \in N_q} \min\left(T_{net}^{(j,n_j)}, T_{nd}\right) + \sum_{j \in N_{nq}} T_{htk}^{(j)} \qquad (6.15)$$

with T_{rd}, N_{nq}, N_q, T_{gd}, n_j, T_{nd}, T_{htk}, and T_{net} being, respectively, the remaining time left by the sender node to end data packet transmission; the number of nodes without a requested message; the number of nodes with a requested message in their queue; the elapsed time when all data packet transmission is stopped, except the synchronizing data packet; the number of data packets waiting in the queue of the jth node; the estimated token holding time of a node; the travel time of a token between two successive nodes; and the network transmission time (in the case where data packets are waiting at the queue of the jth node while simultaneously holding a token, $T_b = T_{net}^{(i,n_j)}$).

Example 6.1

Consider that a ControlNet-based network is made up of eight nodes $\{1, 2, 3, 4, 5, 6, 7, 8\}$ with a transmission delay of $T_{nd} = 827.2$ μs, such that (master) node 7 is expected to pass a token to nodes 4, 6, and 8 receiving those requests. Then, those nodes can be classified into two groups, and $N_{nq} = \{1, 2, 3, 5\}$ and $N_q = \{4, 6, 7, 8\}$.

When there are N_m master nodes requesting multiple data transmissions while using the maximum token holding time, the blocking time is given by

$$T_b = \sum_{i \in N_{nd} \backslash \{j\}} \min\left(T_{net}^{(i,n_i)}, T_{nd}\right) \qquad (6.16)$$

with the constraints such that even if it has more data packets to be transmitted, the node token holding time is bounded by

$$T_{nd} \text{ i.e. } T_{net}^{(i,n_j)} \leq T_{nd}$$

In addition, the time spent on the guard band period can be derived from

$$\begin{cases} T_{gd} = \dfrac{1}{3} \times T_{tr} \\ T_{tr} \leq N_m \times T_{nd} \end{cases}$$

where T_{tr} is the token rotation time. Therefore, it is bounded by

$$T_{gd} \leq \frac{1}{3} \times N_m \times T_{nd}$$

Also,

$$
\left\{
\begin{array}{l}
T_{rd} \leq T_{nd} \\[4pt]
T_{htk}^{j} \leq T_{nd} \\[4pt]
T_{net}^{(i,n_j)} \leq T_{nd}
\end{array}
\right.
$$

which is equivalent to

$$
T_{rd} + \sum_{j \in N_q} \min\left(T_{tx}^{(i,n_j)}, T_{nd}\right) + \sum_{j \in N_{nq}} T_{htk}^{j} \leq T_{nd} + \sum_{j \in N} T_{nd}
$$

Hence,

$$
T_b \leq (N+1) \times T_{nd} + \frac{1}{3} \times N_m \times T_{nd}
$$

After simplification, it is considered that

$$
T_b = \left(N + 1 + \frac{1}{3} \times N_m\right) \times T_{nd} \tag{6.17}
$$

Note that the Modbus and Profibus protocols use token passing as the arbitration method, making their blocking time similar to that of ControlNet protocol. A direct estimation of the blocking time delay for the Ethernet protocol is difficult due to the additional waiting time before the retransmission after the collisions between data packets. The binary exponential backoff (BEB) algorithm is used to detect a collision between two or more data packets (message) on the Ethernet protocol. Here, an emitting node stops sending data packets and waits a random length of time before the round-trip retransmission is chosen between 0 and $(2i - 1)$ slots times (with i for the ith node data collision). However, for more than 10 collisions, the time interval is set to be 1023 slots. Hence, the total blocking time can be estimated from the BEB algorithm using the residual time, T_{rd}, from the node i until the network is idle and $E\{T_k\}$ the expected time of the kth collision. The blocking time, T_b, in the DeviceNet protocol is given by

$$
T_b^{(k)} = T_{rd} + \sum_{\forall j \in N_{hp}} \left\lceil \frac{T_b^{(k-1)} + T_{bit}}{T_{pr}^{(j)}} \right\rceil T_{tx}^{(j)} \tag{6.18}
$$

If the data transmission frequency is constant, then the *j*th message has a period given by $T_{pr}^{(j)} = T_{pr}$. This blocking time could be bounded such that

$$T_b^k \leq T_{nd} + N_{hp} \times \frac{T_{nd}}{T_{pr}} + \frac{T_{nd}}{T_{pr}} \sum_{j \in N_{hp}} T_b^{(k-1)} \tag{6.19}$$

This is equivalent to

$$T_b^k \leq (k-1) \times T_{nd} \times \left(1 + \frac{N_{hp}}{T_{pr}} + \frac{N_{hp}^2 \times T_{nd}}{T_{pr}} \right) \tag{6.20}$$

where N_{hp} corresponds to the set of high-priority nodes. Such nodes can induce a higher blocking time when they are queued.

6.6.1.5 Estimation of Frame Time

The frame time, T_f, is the delay of the message during the transmission that directly depends on the size of the data, N_d; the nondata packet size, N_{nd}; the size of padding bits, N_p, to meet the minimum frame size requirements; and the bit time or time of transmission of a bit in the transmission medium, T_{bit}, related to the protocol chosen. Consider N_s as the data size suitable for the stuffing mechanism, the frame time can be derived as

$$T_f = [N_d + N_{nd} + N_p + N_s] \times 8 \times T_{bit} \tag{6.21}$$

The values of N_{nd}, N_p, N_d, N_s, T_{bit} are described for the Profibus, Modbus, ControlNet, and DeviceNet protocols, as presented in Figure 6.8.

Note that in the DeviceNet protocol, if the bit value 1 appears more than five times in succession, then a bit value of 0 is added, and vice versa, while in the Profibus, Modbus, and ControlNet protocols the Manchester biphase encoding is used without bit stuffing being required.

6.6.1.6 Estimation of Propagation Time

The propagation time, T_p, depends on the signal transmission speed and the distance between the sender and receiver nodes.

$$T_p = \frac{l}{v} \tag{6.22}$$

where l is the length of the network segment and v is the signal transmission speed. Table 6.2 and Figure 6.9 summarize some physical network parameters for various protocols.

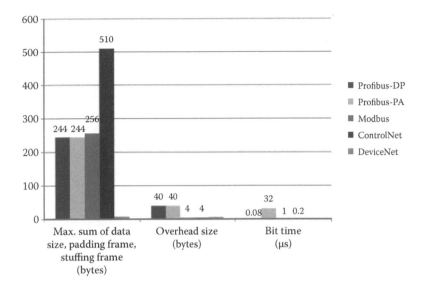

FIGURE 6.8
Parameter values for various network protocol framing times. (Data from F.-L. Lian et al., *IEEE Control Systems Magazine*, 21(1), 66–83, 2001.

TABLE 6.2

Physical Media for Profibus, Modbus, ControlNet, and DeviceNet

Physical Media	Protocols
Twisted pair	Profibus-DP/PA Modbus DeviceNet
Fiber	Profibus-DP/PA ControlNet
Coax	ControlNet

Source: Data from ControlNet, DeviceNet, and Profibus.

6.6.1.7 Estimation of Mean Queuing Time

Network data traffic can be either cyclic, when data packets are transmitted through periodically refreshed network commutating devices (e.g., gateways, routers, and switches), or acyclic, in the case where the data status is refreshed according to a node request or an event, such as for the configuration or diagnostics of process operations. The queuing time, T_q, is the time a data packet remains in the buffer at the sender node while the previous data packets in the queue are processed. It is related to the blocking time of the previous data packet in the queue, the transmission periodicity, and the processing time of the data packet. Figure 6.10 illustrates the model corresponding to the queue of the buffer memory at a network node.

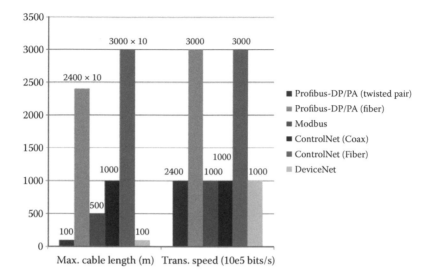

FIGURE 6.9
Signal transmission speed and maximum cable length allowed for various protocols. (Data from ControlNet, DeviceNet, and Profibus.)

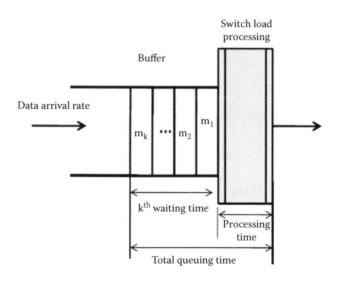

FIGURE 6.10
Waiting time of a buffer memory at a sender node.

Consider the following parameters such that λ is the frequency of data arrival in the buffer, and μ is the rate at which the data exits the buffer, indicating the data transmission capacity of the output port. Then, the system load, A (in E), is given by

$$A = \frac{\lambda}{\mu} \tag{6.23}$$

Define the average number of data frames in the waiting part of the buffer memory by

$$N = \frac{A}{1-A} \tag{6.24}$$

Hence, the queuing time at the sender node is given by

$$T_q = \frac{1}{\mu(1-A)} \tag{6.25}$$

The path of the data packet transmission network depends on the network protocol. Then, for a given transmission time, $T_{max} = 10$ s it is possible to calculate the mean of the node time delay as

$$T_{d\_avg} = \frac{1}{N} \sum_{i \in N_{node}} \left[\frac{\sum_{j=1}^{M^{(i)}} T_{net}^{(i,j)}}{M^{(i)}} \right] \tag{6.26}$$

Here $T_{net}^{(i,j)}$ denotes the transmission time between nodes i and j, $i \neq j$.
where N is the total number of network nodes, N_{node} is a subset of sender nodes, and $M^{(i)}$ is the data packet requested at node i. In the case where all messages are sent periodically such that $M^{(i)}$ is equal to the total running time divided by the period of messages, the *throughput* or transmission rate is given by

$$Throughput = \frac{length\ of\ useful\ frame\ information}{transmission\ time\ delay} \tag{6.27}$$

Table 6.3 summarizes some delay estimation techniques for commonly encountered protocols.

TABLE 6.3

Summary of Estimation of Data Network Transmission Delay for Different Protocols

Estimation Step	Protocol	Equation	Comments
Blocking time T_b	ControlNet, Modbus, Profibus	$T_b = \left(N + 1 + \dfrac{1}{3} \times N_{nd} \right) \times T_{nd}$	Processing time at sender node, as well as elapsed time associated with arbitration method used to resolve data collision
	DeviceNet	$T_b^k = (k-1) \times T_{nd} \times \left(1 + \dfrac{N_{hp}}{T_{pr}} + \dfrac{N_{hp}^2 \times T_{nd}}{T_{pr}} \right)$	
Frame time T_f	ControlNet, Modbus, Profibus, DeviceNet	$T_f = [N_d + N_{nd} + N_p + N_s] \times 8 \times T_{bit}$	Frame time related to the protocol and the physical network segment characteristics
Propagation time T_p	ControlNet, Modbus, Profibus, DeviceNet	$T_p = \dfrac{l}{v}$	Propagation time related to the length of the network link and data size
Queuing time T_{queue}	ControlNet, Modbus, Profibus, DeviceNet	$T_q = \dfrac{1}{\mu(1-A)}$ or $T_{q\_p} = \dfrac{\sigma}{D_o}$	Queuing time associated with the idle time of data before being sent to the network link
Data fragmentation p		$p = \left[\dfrac{L_1 - N_{nd}}{8 \times \left[\dfrac{L_2}{8} \right] - N_{nd}} \right] + 1$	To estimate the data packet fragmentation
Protocol benchmarks	Modbus, Profibus, DeviceNet	$T_b + T_f + T_p + T_q \leq 2 \; ms$	To access network processing and transmission performance
	ControlNet	$T_b + T_f + T_p + T_q \leq 0.5 \; ms$	

6.6.2 Modeling of Network Congestion Rate

Network congestion occurs when its maximum capacity is passed. This causes network data losses, which can decrease through regulation of the data packets sent. When the data size to be sent increases, the delay increases due to congestion in the network. The modeling of network congestion enables evaluation of the number of data packets lost at the level of the commutating devices (e.g., routers, switches, and hubs) in the network when its buffer memory is full at the time of data arrival.

Consider L_B (bits) to be the size of the buffer of a commutating device, L (bits) the minimum size of frames in the network, and $\displaystyle\sum \sigma_i$ (bits) the amount of data present and waiting to be processed by the commutating device. Thus, congestion occurs when

$$\sum \sigma_i > L_B \qquad (6.28)$$

The number of lost packets is estimated by

$$N_{cogt} = \frac{\sum \sigma_i}{L} \qquad (6.29)$$

6.6.3 Modeling of Network Efficiency

Network efficiency is the evaluation of the performance of a communication network through assessment of the rate of message contention or collision. As such, an efficiency of 100% means that there is no collision in the network ($T_q = T_b = 0$), such that all time delays from the network are considered to be optimal. The efficiency of a network, P_e, is defined as the ratio between the frame and propagation times to the total elapsed time used to send data packets (queuing, blocking time, and frame and propagation times), which is given by

$$P_e = \frac{\sum_{i \in N_{node}} \sum_{j=1}^{M^{(i)}} T_{net}^{(i,j)}}{T_{d\_sum}} = \frac{\sum_{i \in N_{node}} \sum_{j=1}^{M^{(i)}} T_{net}^{(i,j)}}{T_q + T_b + T_f + T_p} \qquad (6.30)$$

where

$$T_{d\_sum} = N \times T_{d\_avg} \qquad (6.31)$$

Hence, when P_e is tending to 100%, the network performance is near optimal.

6.6.4 Estimation of Industrial Network QoS

Transmission data losses usually occur over the commutating devices (switches, routers, etc.). As such, in order to access the data network QoS, it is convenient to quantify criteria related to how the network data are fragmented up to their flow over the network (traffic). The QoS of an industrial network consists of assessing (1) the data flow rate, (2) the transmission latency, (3) the network data loss estimation based on the network and traffic characteristics (protocol, transmission media, topology, and commutating

TABLE 6.4

Generic Methodology for QoS Assessment of Industrial Networks

QoS Assessment Step	Estimation Method or Mathematical Model
Data fragmentation to derive message N_L fragments from message volume	$N_L = interger\ of\left(\dfrac{a}{R_u}\right) + 1$ with a being the packet of data size to be sent over the network (bits), R_u the complete packet of data size (bits), and N_L the number of data packet fragments
Data traffic parameter definition	$\sigma = L_u + Nb_{address}$ with $Nb_{address}$ being the size of addresses (bits)
Traffic modeling of packet of data entry into the network	$b(t) = \sigma + \rho t$
Modeling of input data streams $(B_1,...,B_k)$ into network commutating device (router, switch, etc.)	$B_1 = \displaystyle\sum_{i=2}^{m}\left(S_i + \rho_i\left(u_1 + \dfrac{R_i}{D_i}\right)\right) + (D_1 - TD_o)u_1$ $u_1 = \dfrac{S_1}{D_1 - \rho_1},$ $B_k = \displaystyle\sum_{i=1,i\neq k}^{m}\left(S_i + \rho_i\left(u_k + \dfrac{R_i}{D_i}\right)\right) + (D_k - TD_o)u_k - \rho_1\dfrac{R_1}{D_1} + R_k u_k = \dfrac{S_k}{D_k - \rho_k} - \dfrac{R_k}{D_k}$
Modeling of the transit time of input data streams within multiplexing components from the commutating device	$TD_{mux} = \dfrac{1}{TD_o}min_k \overline{B_k}$
Modeling of the waiting time of the commutating device's FCFS buffer	$TD_{fcfs} = \dfrac{1}{D_o}\dfrac{D_i - TD_o}{D_i - \rho}S$
Modeling of the transit time in the demultiplexing component of the commutating device	$T_{dmax} = 0$
Modeling of the transit time in the commutating device without the service classifier	$T_{com}^1 = T_{fcfs} + T_{mux} + T_{dmux}$

(Continued)

TABLE 6.4 (CONTINUED)

Generic Methodology for QoS Assessment of Industrial Networks

QoS Assessment Step	Estimation Method or Mathematical Model
Estimation of the service classifier time from the commutating device	$T_{s\_class} = \overline{TD_i} = (T - \varepsilon_i) + \dfrac{S_i + \rho_i \varepsilon_i}{R}$ with $\varepsilon_i = \dfrac{S_i}{C_{in} - \rho_i}$ and $Q \le D_i$ $T = \dfrac{\sum_{j \ne i} \phi_j}{C}$ and $R = C\,\dfrac{\phi_i - R_{i,max}}{\sum_j \phi_j - R\,max}$
Modeling of the transit time into a commutating device with the service classifier	$T_{com} = T_{com}^1 + T_{s\_class}$
Estimation of the propagation time in the transmission media	$T_p = \dfrac{Cable\ lenght\ (m)}{Propagation\ speed \left(\dfrac{m}{s}\right)}$
Computing data packet losses	$\dfrac{R_B - \sum S_i}{\sigma}$
Computing the transmission rate (bits/s)	$Transmission\ rate = \dfrac{Frame\ length\,(bit)}{T_{net}\ (s)}$

device modeling), and (4) the modeling of network congestion over the commutating devices. Table 6.4 summarizes the QoS evaluation steps and variables used. From Table 6.4, S_i, ρ_i, R_i, D_i, D_o, T_{com}^1, ε_i, Φ_j, $L_{i,max}$, $\overline{TD_i}$ being, respectively, the maximum arrival of data packets from the ith node (bits), the data transmission rate from the ith node (bits/s), the data packet frame size from the ith node (bits), the maximum data arrival transmission rate (bits/s), the maximum data departure transmission rate (bits/s), the routing delay from the ith node within a commutator with first come, first served (FCFS) (s), the constant data rate variation from the ith node, the data rate weight from the ith node within the service classifier, the maximum data packet frame size from the ith node (bits), and the maximum routing delay from the ith node within the commutator with FCFS (s).

Based on the protocol commissioned, it is possible to capture the congestion quantitative description using equations developed in this section to derive key industrial parameters of the technical network audit. Table 6.5 summarizes a methodology for auditing the industrial network.

TABLE 6.5

Summary of Main Steps for Industrial Network Auditing

Steps	Formula/Method	Description
Queuing Time at Sender Node		
Estimation of the network latency based on the protocol model	$T_f = (N_d + N_{nd} + N_p + N_s) \times 8 \times T_{bit}$	Blocking time and frame time
Modeling of the queuing time at each node (switch, hubs, etc.)	$T_q = \dfrac{1}{\mu(1-A)}$	Queuing time of a data packet in the sender node buffer before being transmitted over the network
Transmission Medium		
Latency modeling on the transmission media (twisted pair, optical fiber, etc.)	$T_f = \dfrac{\text{size of useful data frame (bits)}}{\text{transmission rate}\left(\dfrac{\text{bits}}{\text{s}}\right)}$ $T_p = \dfrac{\text{cable length } (m)}{\text{propagation speed over the medium}}$ Hence, it is given by $T_f + T_p$	Sum of the transmission and propagation delays
Latency Estimation		
Estimation of the total transmission delay of a packet in the network	$T_d = T_q + T_b + T_f + T_p$	Transmission delay of a packet of data
Congestion Estimation		
Congestion in the modeling by estimation of the number of packet losses	$N_{cgt} = \dfrac{\sum \sigma_i - L_B}{L}$	Number of packets of data arriving at the commutating device when its buffer is full
Efficiency Estimation		
Modeling of the transmission efficiency	$P_e = \dfrac{\sum i \in N_{nd} \sum_{j=1}^{M^{(i)}} T_{tx}^{(i,j)}}{\sum T_d}$	All time delays associated with the transmission time over the total elapsed time required to send data packet

6.7 Network Performance Criteria, Benchmarks, and Selection Trade-Off

Typical industrial networks are a mixture of topologies, network devices (repeaters, hubs, switches, etc.), and protocols that require higher data availability between a fixed set of process nodes (e.g., sensors, actuators, and operation centers). As such, those data networks have their specific strengths and weaknesses inherent from their design, along with some conflicting design requirements. For example, excessive delays can negatively affect the system performance, while the number of nodes in the system could require high resources (e.g., switches and routers). The time delays and the overhead have to be kept as low as possible. Furthermore, depending on the transmission method used, the transmission process could display a higher noise-to-immunity ratio while providing high transmission speeds, such as with carrier band in contrast to baseband. Therefore, it is suitable to investigate the trade-off between some data network design requirements in order to derive a procedure to size it and develop guidelines to evaluate the conflicting design objectives. In addition to the network data rate and latency, among other criteria commonly used to compare different network designs there are the level of immunity to noise, the amount of bit error and field bus fault occurrences, the protection requirements (i.e., the power availability over the transmission wire consumption), the maximum network length and number of nodes allowed, and the optional tools, such as the diagnostic capabilities embedded.

6.7.1 Efficiency, Average Delay Time Standards, and Data Transmission Rate

Usually, data released by the sender node over the data network can be done through either the zero policy, random policy, or scheduled policy. Sender node data latencies refer to the amount of time required for a data packet to transit between a sender node and a receiver node. Commonly encountered industrial control operations can tolerate latencies of 10–50 ms. Figures 6.11

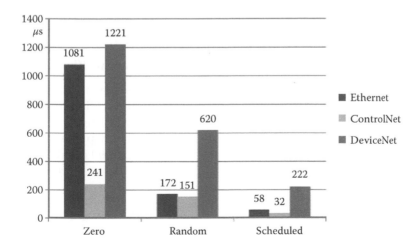

FIGURE 6.11
Average time delay for various data-releasing policies. (Data from F.-L. Lian et al., Network protocols for networked control systems, in Handbook of *Networked and Embedded Control Systems*, ed. D. Hristu-Varsakelis, W.S. Levine, Birkhauser, Basel, 2005.)

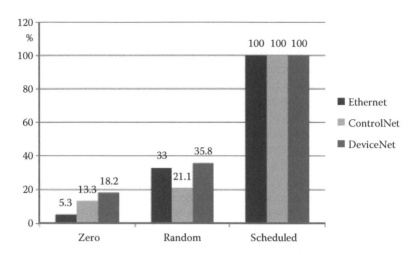

FIGURE 6.12
Network efficiency (%) for various data-releasing policies. (Data from F.-L Lian et al., Network protocols for networked control system, in *Handbook of Networked and Embedded Control Systems*, ed. D. Hristu-Varsakelis, W. S. Levine, Birkhauser, Basel, 2005.)

and 6.12 display standard values for the network efficiency and the average delay time for various networks.

The data rate is a measure of the amount of data transferred for a fixed duration. In the case of digital communication, the rate is given by the amount of bits that can be transferred. Thus, the data latency is characterized by the speed of the network. The data rate defines the data latency, as well as the transceiver propagation delay, the media propagation delay, and the protocol overhead.

6.7.2 Media-Based Consideration and Network Bandwidth

The data packet propagation depends on the transmission network physical characteristics. For example, the data packet propagation delay through the fiber-based network is approximately the speed of light. The protocol time delay is due to the protocol overhead required (e.g., parity, addressing, error check, handshaking bits, and data size). Figures 6.13 and 6.14 display the raw data rates and the message protocol latencies for several networks.

6.7.3 Signal-to-Noise Ratio, Error Rate, and Data Throughput

The industrial environment defines the communication signal-to-noise ratio (S/N) between the high-current field devices, such as high-voltage transformers, generators, turbines, welding equipment, and motors. Immunity to noise sources must maintain reliable network operation by setting the proper configuration for the grounding and shielding of transceiver devices.

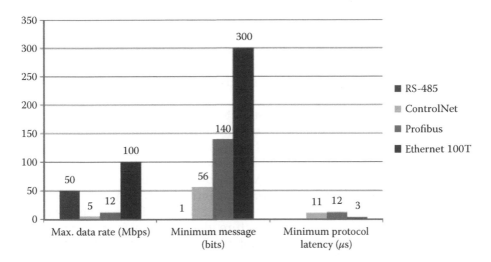

FIGURE 6.13
Benchmarks of data transmission and latency for data rate between 1 and 100 Mbps. (Data from C. Kinnaird, *Specifying Industrial Field Bus Networks for Automation*, Texas Instrument, Dallas, Texas, EDN Asia, 2007.)

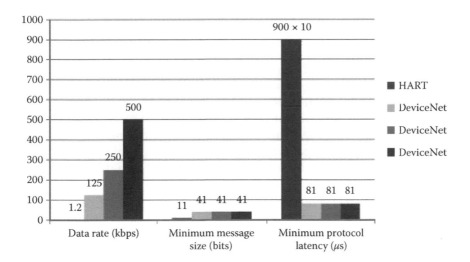

FIGURE 6.14
Benchmarks of data transmission and latency for a data rate of <0.5 Mbps. (Data from C. Kinnaird, *Specifying Industrial Field Bus Networks for Automation*, Texas Instrument, Dallas, Texas, EDN Asia, 2007.)

The maximum data transfer rate for a network segment is proportional and bounded by the bandwidth, the S/N, and the signal characteristics generated by each network device. For example, a decrease of the bit rate without hysteresis in the signal produces a decrease in the device receiver threshold, as well as causing high switching from noise during the signal transmission. Inversely, the high signal levels on the network are expected to cause an increase in the S/N. With the baseband transmission method, it is suitable to operate at low speeds to circumvent the noise disturbance signal, while the carrier band and broadband methods allow higher transmission speeds with reduced noise effects.

Similarly, the error rate can be estimated from the S/N, the noise, and the signal interference levels. Hence, the data transmission rate and the error rate depend on the type of application. In the case of a low-error-tolerance application, such as with motion control, it is suitable to ensure a higher reliability of the data transmission network. As transmission rates increase, error rates increase up to the case where the protocol retransmits data packets to avoid a high expected data rate over a low real data rate. As data packets are transmitted over the protocol frame structure, the data transfer rate containing useful information is less than the bit rate. Hence, the throughput or effective transmission rate can be derived through the ratio of transmission data containing useful information.

6.7.4 Response Time and Exceptional Range

The interval between the time at which the node sends requests and the time elapsed to the receiving node responding to them is called the response time. It depends on the network transmission rate and the transmission method (cyclic or acyclic polling). In addition, the industrial networks require some long interconnections, called network range. This allowable network length is limited by the signal losses in the transmission media and the electrical noise pickup during the transmission. The signal cable attenuation is defined as the signal loss measured at 10 MHz per 304 m of cable. Due to obstruction, during signal propagation, the signal is attenuated independently of the transmission media (copper, fiber, or wireless). For example, with ControlNet using 1786-RG-6 cable, which has an attenuation of 13.5 dB, the maximum segment length is

$$Max\ allowable\ (high-flex)\ cable\ length$$
$$= \frac{(20,29-number\ of\ taps\ in\ segment \times 32\ dB)}{Cable\ attenuation\ at\ 10\ MHz\ per\ 304m}$$

Twisted-pair copper wire has higher losses than other transmission media. It range from 1.5 to 5 dB per 100 m at 1 MHz, and the data can be transmitted at a distance up to 6.1 km on two wires or up to 12.2 km on four wire lines. It is possible to achieve 30 km on the unloaded lines (unlimited for loaded lines) or up to 1.5 km on the power lines. Optical fiber has losses of 0.3 dB/1 km. Hence, for the large network with repeaters, connectors, routers, and so forth, the fiber cable length is given by

Maximum cable length (km)

$$= \frac{(Power\ Budget\ -Total\ attenuation\ of\ connectors,\ bulkhead,\ etc.)}{Fiber\ cable\ attenuation\ in\ db/km}$$

For some network types, the signal attenuation over network length, as well as the maximum allowable nodes per network, is summarized in Figures 6.15 through 6.17.

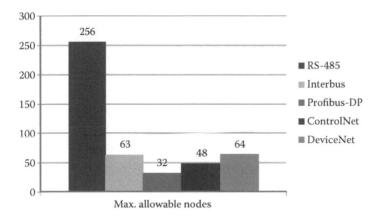

FIGURE 6.15
Allowable number of nodes for the standard networks. (Data from C. Kinnaird, *Specifying Industrial Field Bus Networks for Automation*, Texas Instrument, Dallas, Texas, EDN Asia, 2007.)

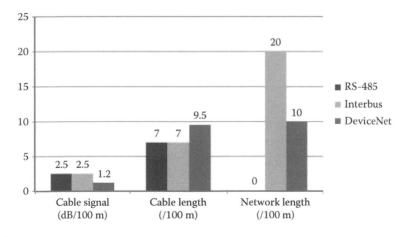

FIGURE 6.16
Network limits for a data rate of <0.5 Mbps. (Data from C. Kinnaird, *Specifying Industrial Field Bus Networks for Automation*, Texas Instrument, Dallas, Texas, EDN Asia, 2007.)

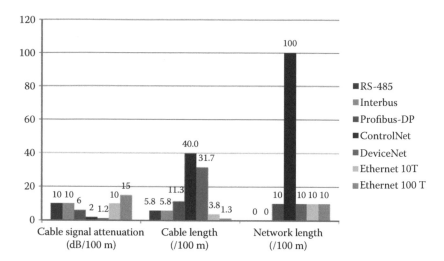

FIGURE 6.17
Network limits for a data rate between 5 and 100 Mbps. (Data from C. Kinnaird, *Specifying Industrial Field Bus Networks for Automation*, Texas Instrument, Dallas, Texas, EDN Asia, 2007.)

6.7.5 Industrial Network Sizing and Selection Methodology

Generally, some key factors to consider when choosing a best-fit network for a given process automation application are (1) the amount of data needed by the application, (2) the relevance of network redundancy, (3) the transmission speed of the network by differentiating between the raw speed and the actual data throughput, (4) the length or size of the messages and how it affects other communication properties, (5) the existing cabling type, and (6) the end-user needs. Recall that the usual challenges facing digital communications are the trade-offs of those factors during different possible network designs. Furthermore, in the case of the hybrid data network, the method of representing, encoding, and transmitting the data is called the protocol, which is owned and customized with respect to the specific field devices by each manufacturer. This raises the network design challenge to ensure interoperability or the ability to communicate intelligibly with other devices. Thus, the compliance check must be performed at each stage of the design to guarantee strict adherence to the protocol specifications.

A physical and media sizing procedure for the data network can be summarized as follows:

1. Identify the I/O process devices and the interface within the data network (i.e., sensors, actuators, etc.) and analyze the process operation's (i.e., expected time delays) environmental and geographical constraints.

2. Choose the network topology and the protocols according to the process constraints derived from the analysis above.

3. Select the transmission media (copper, optical fiber, etc.) based on the protocol functional constraints.

4. Derive the signal attenuation per network link.

5. Based on the constraints from the protocol and the signal attenuation, determine the repeater requirements and number of taps. Also, the number of terminating resistors should be derived, as well as chosen the connectors, hubs, switches, and so forth.

6. Choose the connection for programming devices.

7. Decide whether to use redundant media depending on environmental operating risk assessment.

8. Estimate the propagation delay, especially

   a. The maximum propagation delay through the data network

   b. The maximum propagation delay through the redundant network

9. If the resulting maximum propagation delays are higher than the delays expected from the process operation, choose another network topology and protocols.

In addition to the network sizing steps listed above, the subsequent selection procedure of an optimal network topology and communication protocol must take into account specific performance criteria, such as *technical design criteria*, which are (1) the network temporal criteria (e.g., the speed of data transmission, the response time, and the time to update all process variables), (2) the network topological criteria (e.g., the maximum length with repeaters, the maximum distance between equipment, and the network wiring structure with connectors), and (3) other technical design criteria, including priority management, the cooperation model (client/server, producer/consumer, etc.), the mode of transmission (point to point, multipoint, or broadcast), the protocol efficiency (data length and length field), the protocol security measures (acknowledgment and flow), and data recovery through partial network redundancy. Before the evaluation and commissioning of the network design, some *strategic design criteria* must be considered. Among those criteria there are network benchmarking in relation to the standards over the application fields, network component availability from suppliers, network scalability, compliance with other industrial networks, and maintenance operations. In addition, some other design requirements for the data network may include (1) the dependability or autonomous operations, (2) the time-tagged data, (3) the distributed information of the global time and the guaranteed delivery time, and (4) real-time versus non-real-time traffic.

Example 6.2

Consider the interconnection of devices involved in the motion control through a data network characterized by the transmission of numerous data from the process variables. This network has some

TABLE 6.6

Typical Network Requirements for Some Industrial Applications

Industrial Application Constraints	System Performance Requirements	Network Design Requirements
Motion control application with a throughput (e.g., 200 parts/h), allowing operating precision up to 10 μm for a workpiece dimension around 1 m.	The motion control should have a settle time of 200 ms for a velocity of 2 m/s with a position resolution of ± 1 μm.	The data signaling rate should be higher than 10 Mbps while having less than 18 nodes per network link and a total network distance of 10 m within the allowed noise immunity and low power requirements.
Process control applications should include operating constraints on temperature, pressure, refined oil flow, etc., to be maintained continuously within the comfort zone. In addition, the process should be scalable for workpieces up to 8 m in size.	The control system sampling rate of 100 measurements per second from the sensors, with a resolution of 16 = bit resolution per sensor. Control remote units are geographically dispersed and distributed in the design of the shop floor network.	The data signaling rate should be less than 1 Mbps, while more than 45 nodes per network is allowed for a total network length of 100 m with the ground offset tolerance and lower power requirements.

bounded transmission delays varying from 1 ms up to 1 s, and with the data rate between 50 Kbit/s and 5 Mbit/s for a distance interval ranging from 10 m up to 4 km. Among operating and environmental constraints there are (1) the environmental conditions related to the operating temperature, some vibrations, the electromagnetic disturbances, the water, and the dust; (2) the high data availability requirements; (3) the clock synchronization; and (4) safety for the field operator. Table 6.6 summarizes typical network requirements, relating process control design constraints with the network's performance and design requirements.

Example 6.3

Usually, network redundancy is achieved by connecting the network commutating devices with redundant ports and implementing two cabling system networks such that during the occurrence of an event of a cable failure or a degraded cabling system, the data will flow through the redundant network. Here, it is desired to estimate the maximum propagation delay between channels A and B of an optical-based data network with the following characteristics: (1) two nodes separated by three optical fiber–based repeaters in series on both channels A and B, and (2) channel A uses 3500 m segment on the fiber versus 3000 m segments on channel B, as illustrated in Figure 6.18.

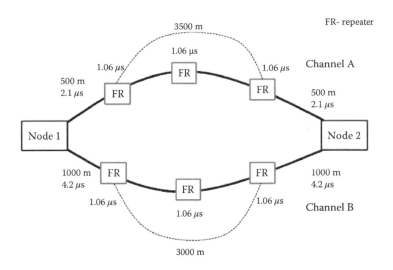

FIGURE 6.18
Two-channel network.

The total time delays for channel A can be derived from the following:

- Derive delay 1 given by 500 m × 4.17 ns/m = 2.1 μs.
- Derive the delay 2 fiber repeater: 901 ns + 153 ns = 1.1 μs.
- Derive delay 3 given by 3500 m × 5.01 ns/m = 17.5 μs.
- Derive the delay 4 fiber repeater: 901 ns + 153 ns = 1.1 μs.
- The total delay for channel A is 44.6 μs.

Similarly, the total time delays for channel B can be derived from the following:

- Derive delay 1 given by 1000 m × 4.17 ns/m = 4.2 μs.
- Derive the delay 2 fiber repeater: 901 ns + 153 ns = 1.1 μs.
- Derive delay 3 given by 3000 m × 5.01 ns/m = 15.0 μs.
- Derive the delay 4 fiber repeater: 901 ns + 153 ns = 1.1 μs.
- The total delay for channel B is 42 μs.

Skew between channels = (*delay through A*) − (*delay through B*) = 44.6 μs − 42 μs = 2.6 μs

$$Skew\ between\ channels = 2.6\,\mu s$$

Hence, the maximum cable length (*L*) is given by

$$L = (13.3 - 1.0\,\text{db})/1.0\,\text{db}/\text{km} = 13.3\,\text{km}$$

Assume that the shortest-distance fiber cable is preterminated such that the attenuation for an average-distance cable section must be less than 13.3 dB. Therefore, there is 3 dB for the two connectors and 10 dB for 1 km of optical fiber cable, as well as for the two terminating resistors (connectors). Usually, the network segment has several bus connectors and trunk cables.

Exercises and Problems

6.1. a. List four causes of data packet time delay.

b. List three functions of a data network system.

c. List the fundamental measures to assess a data network's system performance.

d. Is Ethernet a network protocol?

e. Derive the elapsed time required to transmit an ASCII data packet over a serial line that is 10 data bits long and has no parity. Redo it when it is 12 data bits long.

f. List two arbitration methods for resolving data collisions over a network topology.

g. How is it possible to transfer additional data with a signal that was coded according to HART?

h. Is it possible to have 0 mA instead of 4–20 mA in a protocol? Justify your response?

i. What does the Hamming distance describe?

j. Which coding does Ethernet use?

6.2. It is required to remotely monitor and control a motor speed using the ControlNet transmission protocol depicted in Figure 6.19 and the devices summarized in Table 6.7. Based on the configuration, the ControlNet is expected to meet the request packet interval (RPI) of 56 ms, also known as the scheduled I/O data transmission frequency between the variable speed drive panel view and the logic control unit. The configuration should be considered a network update time (NUT) of 244 ms. Note that the waiting time by a network for each node to respond (slot time) is 131 μs.

a. Making all necessary assumptions, how many nodes are there in this network (including connectivity, unscheduled nodes, and unscheduled spare nodes for future traffic)? How many taps are required? Considered that each node can send or receive all kinds of data (scheduled, maintenance, and even unscheduled)

b. After approximation, derive the maximum propagation delay of a ControlNet network with coax and fiber media and adequate numbers of repeaters and taps.

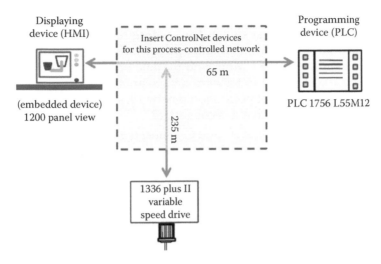

FIGURE 6.19
ControlNet-based industrial network.

TABLE 6.7

Processing Delay for Some ControlNet Devices

ControlNet Transmission Device	Processing Delay
1786-RPA, RPCD (repeater adaptor/copper)	815 ns
1756-CNBR (controller for redundant NW)	100 ns
1784-PCC (computer interface cards)	901 ns
1786-TPS/1786TPYR (T-tap, Y-tap)	94 ns
1786-RPFM (drives)	153 ns
1756 A10/A (controller)	753 ns
1336 plus II (variable speed drive)	1410 ns
1200 E (panel view)	402 ns
1756-CNB/E (bridge)	533 ns
1786-FSKIT (200/62.5-micron fiber cable)	5.01 ns/m
1786-CTX (RG-6 coax cable)	4.17 ns/m

  c. Sketch a high-level ControlNet-based network schematic ensuring that field device communicate properly (especially avoiding unrealistic data flow).

  d. How can the network be modified to ensure data flow redundancy?

Note that the number of repeaters is based on the distance the signal has to travel.

Recall that the maximum segment length is given by

$$max\ segment\ length = 1000\,\mathrm{m} - \left[16.3\,\mathrm{m}\ (number\ of\ taps - 2)\right]$$

6.3. Consider two network switches used to direct the data packet transmission. Assume that switch 1 is sending data packets of 1200 bits each to switch 2 with a 10-data-packet buffer size, while the network bandwidth is 1.5 Mbps and the data propagation time is 4 ms. Switch 1 is expected to forward data to another link when switch 2 is congested. At what point should network switch 2 send a signal to switch 1 indicating that it is congested?

6.4. Consider three nodes, A, B, and C, connected by a store-and-forward switch (switch with buffer), as shown in Figure 6.20. Here, the network link connecting each node and the switch with buffer has bandwidths and propagation delays as depicted in Figure 6.20.

  a. Derive the elapsed time in one direction of data transmission when sending a 1200-bit packet between nodes A and B to node C?

  b. Consider a 200-bit acknowledgment transmitted after node A sends a data packet to node C and, before transmitting the following packet. Estimate the network transmission delay after a cyclic transmission between nodes A and B to node C. What would the average network bandwidth be?

6.5. List all key features in the Modbus, Profibus-PA, HART, ControlNet, and WorldFIP protocol data header structures.

6.6. Two packet streams arrive at a switch node as illustrated in Figure 6.21.

  a. Derive the average time delay of a packet at the switch node from B.

  b. How many packets from sender node A are received at the switch node on average?

  c. What is the total number of packets at the switch node from two senders, A and B? What is the average packet delay at the switch node buffer?

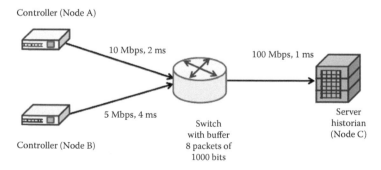

Controller (Node A)

10 Mbps, 2 ms   100 Mbps, 1 ms

5 Mbps, 4 ms   Switch
with buffer   Server
historian
(Node C)
Controller (Node B)   8 packets of
1000 bits

FIGURE 6.20
Industrial network consisting of three nodes and one switch.

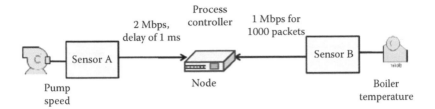

FIGURE 6.21
Two-sender node and one-switch node.

6.7 Consider a network with a link bandwidth (or bit rate) C; the expected number L of bits in a packet; the cable distance d, equal to 1.7 km; the propagation delay T, the transmission delay D; and a throughput given by $\frac{1}{(1+3.5B)}$, with B being the ratio between propagation delay and transmission delay.

  a. Derive the relationship between transmission delay, D, with L and C. What is the relationship between B with C, L, and T?

  b. Derive the propagation delay, T, if the signal is traveling over the fiber-optic medium at light speed.

  c. Given $L = 1000$ bits, derive B and the throughput. If the packet length increases by twofold, how does the throughput vary?

6.8. Consider five nodes, called A, B, C, D, and E, that have connections on a single 10 Mbps data network. Assume that node A requests 2 Mbps for its connection, node B requests 1 Mbps, node C requests 3 Mbps, node D requests 4 Mbps, and node E requests 10 Mbps. List at least three methods for fairly allocating the 10 Mbps bandwidth from the 20 Mbps requested bandwidth to the five nodes and analyze the trade-offs between these different fair methods.

6.9. Among the network devices below, identify those that could be used to connect a controller programming device to a data network:

  a. Tap on a segment

  b. ControlNet access cable connecting network access ports (NAPs)

  c. USB port of a 1756-CN2(R) series B communication module

  d. USB-to-ControlNet connecting cable replacing the 1784-PCC communication module.

  Consider a redundant network for a two-tank fermentation process, as depicted in Figure 6.22.

  e. Determine the number of taps, cable type, and repeaters using a ControlNet protocol to design a communication network such

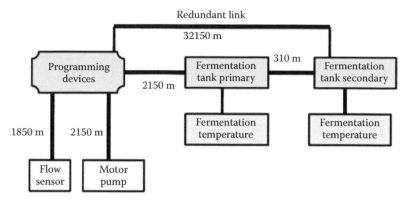

FIGURE 6.22
Layout of two-tank fermentation process with a redundant network.

that the minimum propagation delay is ensured for the fermentation process.

f. Using the network devices for the ControlNet protocol, draw the resulting communication network for this fermentation process.

6.10. Choosing the Profibus network protocol and using network devices (e.g., tap, repeater, terminator, bridge, cable media, router, and switch), design a communication network for the shop floor assembly process depicted in Figure 6.23. Recall that the distance between HMI and the controller unit is 355 m, and there is 75 m between the control unit and sensor or actuator.

6.11. As illustrated in Figure 6.24, consider a network connecting node A (host master) to node B (slave workstation) and transmitting a

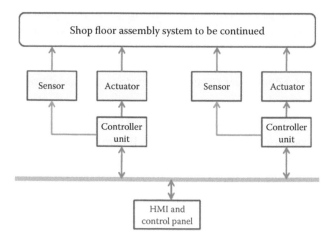

FIGURE 6.23
Layout of shop floor assembly process.

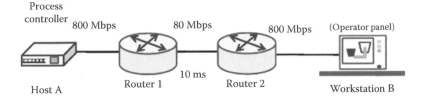

FIGURE 6.24
Two-router and two-host communication network.

message length of 2000 bytes with a fixed packet size of 1000 bytes (including protocol specific header) and a packet header of 40 bytes. Assume that the propagation time between network access routers 1 and 2 and host master A, as well as the router processing time, is negligible (near 0 s).

a. Using the network depicted in Figure 6.24, derive the percentage of bytes corresponding to the transferred message information.

b. Derive the transmission time of message M from host master node A to workstation node B if the datagram packet transfer switching method is chosen.

6.12. By choosing the DeviceNet or ControlNet protocol and using network components such as tap, repeater, terminator, bridge, or cable media, as well as data network connecting devices (router, switch, etc.). Design by sketching an industrial network for the shop floor assembly process depicted in Figure 6.25. Recall that the distance for a low-bandwidth network segment is less than 150 m, and for high-bandwidth segment, less than 975 m (hint: specific minimum devices and components, as well their location in the network).

6.13. Consider the optical fiber–based network depicted in Figure 6.26. Here it has the following characteristics: the cable length between the controller node and actuator node is 1705 m, and all field buses connecting each sensor to the controller node are 80 m. The data transmission rate is 12 Mbps with a corresponding data size interval of 1200–72 bytes (maximum to minimum).

a. Derive the average field bus time delay.

b. Analyze the effect of this field bus time delay on the closed-loop control system using discrete root locus analysis.

6.14. Consider the Profibus network depicted in Figure 6.27.

Derive the minimum bidirectional network latency with characteristics such that the link between all sensors to the programming device (PLC with embedded HMI) has a length of 1300 m and a data rate of 250 bytes every 15 ms, while the link between programming devices (PLC/HMI) and all actuators has a length of 1300 m and

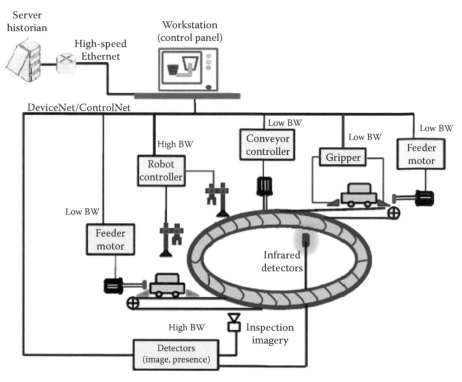

FIGURE 6.25
Typical car assembly shop floor layout.

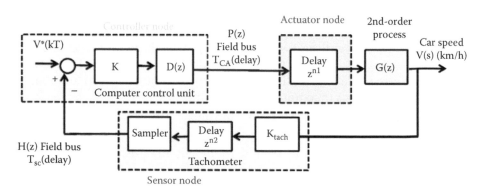

FIGURE 6.26
Closed-loop control network control system using a field bus network.

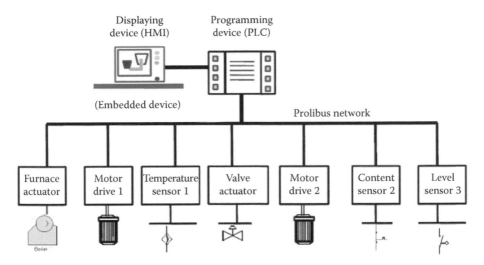

FIGURE 6.27
Typical architecture of a control system with the Profibus protocol.

data transferring flow of 100 bytes every 100 ms. Similarly, the network link between the transmitters and programming logic device (PLC/HMI) has a length of 1900 m and data rate of 250 bytes every 20 ms, while the link between the programming device PLC to the motor drive has the following characteristics: length of 1300 m and transfer flow of 150 bytes every 150 ms. Note that this data network is operating on a cyclic communication scheme. Specify the suitable field bus transmission media for each segment.

6.16. There are 19 nodes representing different car motion control subsystems, depicted as follows: 3 electronic engine controllers, 1 engine temperature sensor, 1 electronic brake controller, 1 wheel speed measuring system, 1 electronic transmission controller, 1 electronic fluid transmission system, 1 vehicle position sensor, 1 cruise control, 1 vehicle direction, 1 vehicle distance recording system, 1 vehicle fuel level recording system, 1 intelligent speed adaptation system, 2 additional nodes for the data acquisition system, 1 node for an in-vehicle interface, and 2 nodes for monitoring and displaying program units. The transmission rates per network device and equipment listed above are the engine temperature at a cyclic rate of 1000 ms, the electronic brake controller at a cyclic rate of 100 ms, the wheel speed information transmission at a cyclic rate of 100 ms, the fluid transmission at a cyclic rate of 1000 ms, the electronic transmission controller at a cyclic rate of 10 ms, the electronic transmission controller at a cyclic rate of 100 ms, the vehicle position at a cyclic rate of 5000 ms, and) the cruise control or vehicle speed at a cyclic rate of 100 ms.

a. Derive various network-induced transmission time delays considering that the controller processing delay and natural delay of the vehicle are negligible.

b. Derive the network latency for an AS-I network protocol.

c. Derive the maximum process bandwidth (allowable time delay or network sampling) when the process stability is maintained in spite of the overall performance degradation.

6.17. Consider an airport luggage management system interconnected by Modbus networks with a token passing sequence as specified in Figure 6.28.

It is required to assess the congestion at the bridge by estimating a data network blocking time during the full utilization of this Modbus

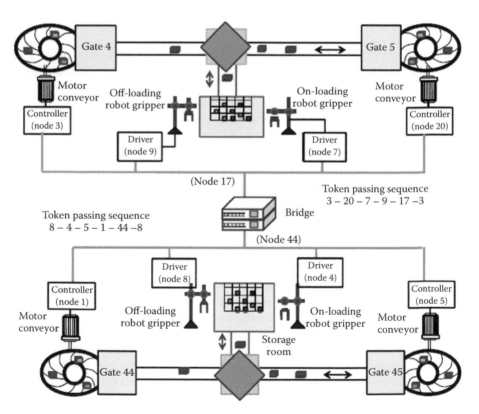

FIGURE 6.28
Token passing in Modbus networks for airport luggage management.

protocol, as depicted in Figure 6.20 (Hint: blocking time). Assume that all coaxial network segments have the following data transmission characteristics: 1020 m, 200 bytes, every 10 ns.

Bibliography

Alberto L., I. Widjaja, *Communications Networks: Fundamental Concepts and Key Architectures*, 2nd ed., McGraw Hill, New York, 2003.

Anurag K., D. Manjunath, J. Kuri. *Communication Networking: An Analytical Approach*, Morgan Kaufmann, Burlington, MA 2004.

ControlNet International (OVDA), ControlNet specifications, ControlNet International, Ann Arbor, Michigan, 2003.

Cruz R., A calculus for network delay, part 1: Network elements in isolation, *IEEE Transactions on Information Theory*, 37(1), 114–131, 1991.

Decotignie J.D., P. Pleinevaux, *A Survey on Industrial Communication Networks*, Springer, Berlin, 1993.

Djiev S., *Industrial Networks for Communication and Control*, Djiev Publications, Sofia, Bulgaria, 2011.

Lian F.-L., J.R. Moyne, D.M. Tilbury, Performance evaluation of control networks: Ethernet, ControlNet, and DeviceNet, *IEEE Control Systems Magazine*, 21(1), 66–83, 2001.

Lian F.-L. et al., Network protocols for networked control systems, in *Handbook of Networked and Embedded Control Systems*, ed. D. Hristu-Varsakelis, W.S. Levine, Birkhauser, Basel, 2005.

Kim S., H.H. Kazerooni, High speed ring-based distributed networked control system for real-time multivariable applications, in *ASME International Mechanical Engineering Congress and Exposition, Dynamic Systems and Control, Parts A and B*, pp. 1123–1130., Anaheim California, USA, 2004.

Kinnaird C., *Specifying Industrial Field Bus Networks for Automation*, Texas Instrument, Dallas Texas, EDN Asia, 2007.

Mansour I.E., Rondeau, T. Divoux, Industrial systems communications: Design and integration, *International Journal of Computer Integrated Manufacturing*, 14(6), 545–559, 2001.

Moyne J.R., D.M. Tilbury, Emergence of industrial control networks for manufacturing control, diagnostics and safety data, *Proceedings of the IEEE*, 95, 29–47, 2007.

Park J., S. Mackay, E. Wright, *Practical Data Acquisition for Instrumentation and Control*, Newnes (Copyrighted Elsevier), Boston, 2003.

Reynders D., S. Mackay, E. Wright, *Practical Industrial Data Communications: Best Practice Techniques*, Butterworth-Heinemann, London, 2005.

Sen S.K., *Profibus: Fieldbus and Networking in Process Automation*, CRC Press, Boca Raton, FL, 2014.

Stallings W., *Data and Computer Communications*, 9th ed., Prentice-Hall, Upper Saddle River, NJ, 2011.

Strauss C., *Data Communications: Practical Electrical Network Automation and Communication Systems*, Newnes (copyrighted Elsevier), Boston, 2003.Tanenbaum A.S., D.J. Wetherall, *Computer Networks*, 5th ed., Prentice Hall, Upper Saddle River, NJ, 2011.

Thomesse J.P., *A Review of the Fieldbuses*, Annual Reviews in Control, Elsevier, Amsterdam, 1998.

Thompson L.M., *Industrial Data Communications*, 4th ed., Research Triangle Park, North Carolina, ISA, 2007.

Tipsuwan Y., M.Y. Chow, *Control Methodologies in Networked Control Systems*, Elsevier, Amsterdam, 2003.

Vendors Association, DeviceNet specifications, Open DeviceNet, Vendors Association, OVDA, Ann Arbor MI, 2002.

Wilamowski B.M., J.D. Irwin, *Industrial Communication Systems*, 2nd ed., CRC Press, Boca Raton, FL, 2011.

Zurawski R., *The Industrial Communication Technology Handbook*, CRC Press, Boca Raton, FL, 2005.

Appendix A: Boolean Algebra, Bus Drivers, and Logic Gates

TABLE A.1

Compiled Boolean Theorems

$A + B = B + A$	$\overline{\overline{A}} = A$
$A \cdot B = B \cdot A$	$(A + B) + C = A + (B + C)$
$A + 0 = A$	$(A \cdot B) C = A (B \cdot C)$
$A \cdot 0 = 0$	$(A \cdot B) + (A \cdot C) = A (B + C)$
$A + 1 = 1$	$(A + B) \cdot (A + C) = A\,BC$
$A + A = A$	$A + AB = A$
$A \cdot A = A$	$A + \overline{A}B = A + B$
$A + \overline{A} = 1$	$A(A + B) = A$
$A \cdot 1 = A$	$\left(\overline{A \cdot B}\right) = \overline{A} + \overline{B}$
$A\overline{A} = 0$	$\overline{(A + B)} = \overline{A} \cdot \overline{B}$

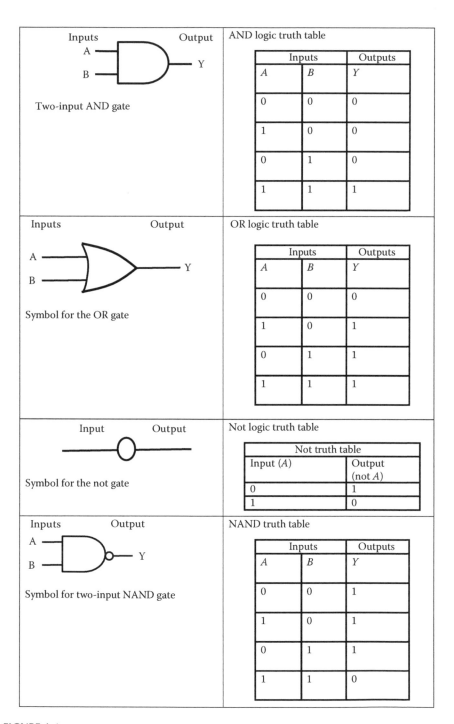

FIGURE A.1
Some logic gates, flip-flops, and drivers.   (*Continued*)

Inputs Output	NOR truth table
  Symbol for two-input NOR gate	

Inputs		Outputs
A	*B*	*Y*
0	0	1
1	0	0
0	1	0
1	1	0

Inputs Output	EOR truth table
  Symbol for two-input EOR gate	

EOR truth table

Inputs		Outputs
A	*B*	*Y*
0	0	0
1	0	1
0	1	1
1	1	1

FIGURE A.1 (CONTINUED)
Some logic gates, flip-flops, and drivers.

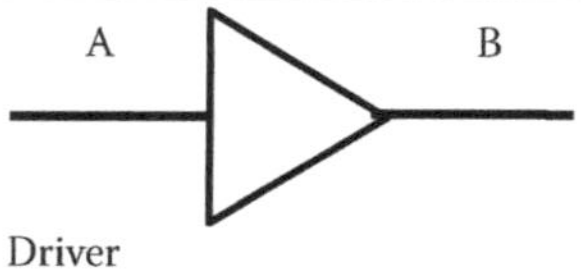

Driver logic

A	B
0	0
1	1

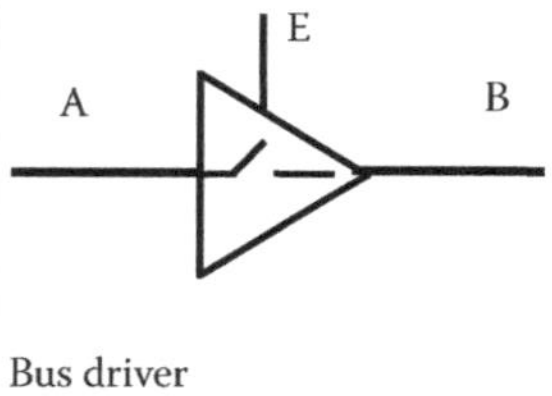

Bus driver logic

E	A	B
0	0	No connection
0	1	No connection
1	0	0
1	1	1

Truth table R–S latch with NOR gate

R	S	Q	$\overline{Q}$	Comment
0	0	Q	$\overline{Q}$	Hold state
0	1	0	1	Reset
1	0	1	0	Set
1	1	0	0	Avoid

Driver

Bus driver

RS-type latch

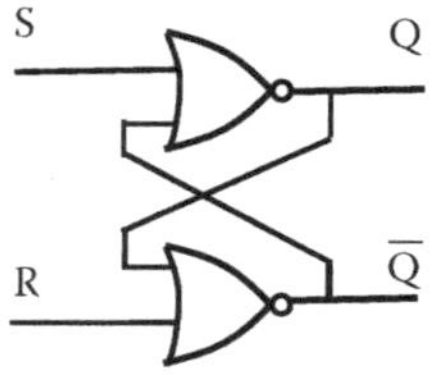

(a) $\overline{R}$ – $\overline{S}$ latch gate with NOR gates

FIGURE A.2

Other logic devices: drivers and bus drivers.

(Continued)

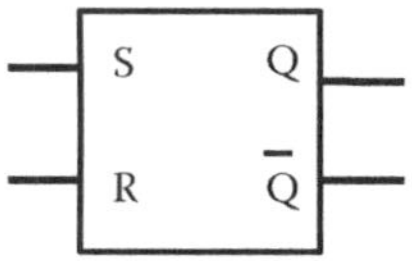

R – S latch stable table

Inputs		Current		Next	
R̄	S̄	Q	Q̄	Q	Q̄
0	0	0	1	1	0
0	0	1	0	1	0
0	1	0	1	1	1
0	1	1	0	0	1
1	0	0	1	0	1
1	0	1	0	1	0
1	1	1	0	1	0

(b) $\overline{R} - \overline{S}$ latch with NOR gates

(b) $\overline{R} - \overline{S}$ latch

(a) Truth table R – S latch

R̄	S̄	Q	Q̄	Comment
0	0	1	1	Avoid
0	1	0	1	Reset
1	0	1	0	Set
1	1	Q	Q̄	Hold state

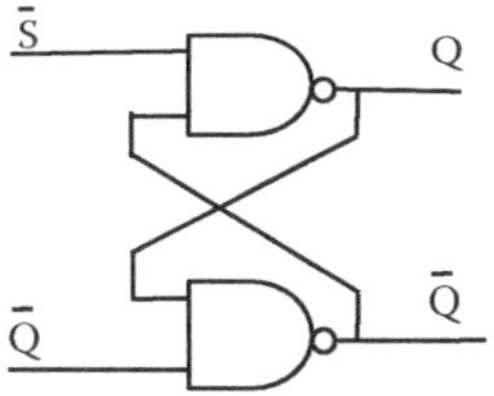

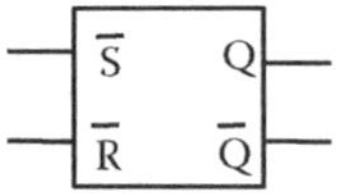

$\overline{R} - \overline{S}$ latch gate
with NAND gates

FIGURE A.2 (CONTINUED)
Other logic devices: drivers and bus drivers.

(*Continued*)

Gated $\overline{R} - \overline{S}$ latch

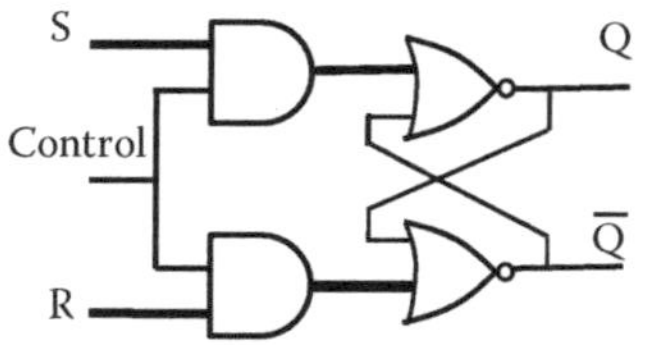

(c) $\overline{R} - \overline{S}$ flip flop gate with NAND gates

(b) Truth table $\overline{R} - \overline{S}$ flip flop

Control	S	R	$Q(t+\delta)$
0	0	0	$Q(t)$
0	0	1	$Q(t)$
0	1	0	$Q(t)$
0	1	1	$Q(t)$
1	0	0	$Q(t)$
1	0	1	0
1	1	0	1
1	1	1	0

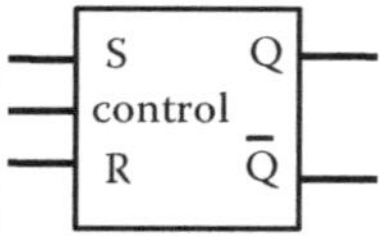

(d) $\overline{R} - \overline{S}$ flip flop gate with NAND gates

(c) State transition table of SR latch

$Q(t)$	$S(t)\ R(t)$		
	01	11	10
0	0	d	1
1	0	d	1
	$Q(t+1)$		

FIGURE A.2 (CONTINUED)
Other logic devices: drivers and bus drivers.

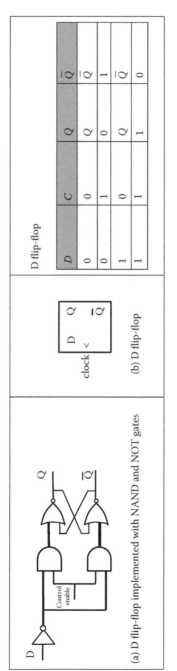

The D flip-flop figure consists of three parts:

(a) D flip-flop implemented with NAND and NOT gates

Inputs: D, Control enable; Outputs: Q, $\overline{Q}$

(b) D flip-flop

Logic symbol with inputs D and clock, outputs Q and $\overline{Q}$

D flip-flop truth table:

D	C	Q	$\overline{Q}$
0	0	Q	$\overline{Q}$
0	1	0	1
1	0	Q	$\overline{Q}$
1	1	1	0

Timing diagram of D flip-flop (signals: control, D, Q)

FIGURE A.3

D-Type (delay flip-flop).

(Continued)

(a) Truth table D flip-flop

Control	D	Present State Q(t)	Next State Q(t+1)
0	0	0	0
0	0	1	1
0	1	0	0
0	1	1	1
1	0	0	0
1	0	1	0
1	1	0	1
1	1	1	1

(b) Gated D flip-flop transition or state table

	control(t)		D(t)	
	00	01	11	10
Q(t)	0	0	1	0
	1	1	1	0

$$Q(t+1) = \overline{control} \cdot Q(t) + control \cdot D(t)$$

(c) Truth table JK flip-flop

J	K	Q(t)	Q(t+1)
0	X	0	0
1	X	0	1
X	0	1	1
X	1	1	0

Algebraic function

$$Q(t+1) = J\overline{Q}(t) + \overline{K}Q(t)$$

(a) JK flip-flop

S   Q
J
clock $<$
K
R $\overline{Q}$

(d) Truth table T flip-flop

T	Q(t)	Q(t+1)
0	0	0
0	1	1
1	0	1
1	1	0

Algebraic function

$$Q(t+1) = T\overline{Q}(t) + \overline{T}Q(t)$$

(b) T flip-flop

T   Q
clock $<$
   $\overline{Q}$

FIGURE A.3 (CONTINUED)
D-Type (delay flip-flop). (a) JK flip-flop (b) T flip-flop.

Adder

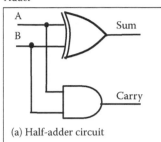

(a) Half-adder circuit

(e) Half-adder circuit logic

A	B	Sum	Carry
0	0	0	0
1	0	1	0
0	1	1	0
1	1	0	1

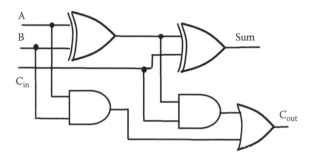

(b) Full-adder circiut

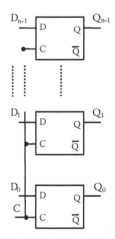

(c) An n-bit register or buffet constructed with D flip-flops

FIGURE A.4
D-Type (delay flip-flop). (a) Half-adder (b) Full-adder (c) D flip-flop based register.

Index

Page numbers followed by f and t indicate figures and tables, respectively.

T - #0332 - 071024 - C468 - 234/156/21 - PB - 9780367735029 - Gloss Lamination